The Acheulian Site of Gesher Benot Ya'aqov

Volume III

Vertebrate Paleobiology and Paleoanthropology Series

Edited by

Eric Delson
Vertebrate Paleontology, American Museum of Natural History,
New York, NY 10024, USA
delson@amnh.org

Eric J. Sargis
Anthropology, Yale University
New Haven, CT 06520, USA
eric.sargis@yale.edu

Focal topics for volumes in the series will include systematic paleontology of all vertebrates (from agnathans to humans), phylogeny reconstruction, functional morphology, Paleolithic archaeology, taphonomy, geochronology, historical biogeography, and biostratigraphy. Other fields (e.g., paleoclimatology, paleoecology, ancient DNA, total organismal community structure) may be considered if the volume theme emphasizes paleobiology (or archaeology). Fields such as modeling of physical processes, genetic methodology, nonvertebrates or neontology are out of our scope.

Volumes in the series may either be monographic treatments (including unpublished but fully revised dissertations) or edited collections, especially those focusing on problem-oriented issues, with multidisciplinary coverage where possible.

For other titles published in this series, go to
www.springer.com/series/6978

The Acheulian Site of Gesher Benot Ya'aqov

Volume III

Mammalian Taphonomy. The Assemblages of Layers V-5 and V-6

Rivka Rabinovich

Institute of Earth Sciences and National Natural History Collections, Institute of Archaeology, The Hebrew University of Jerusalem, Berman building, Edmond J. Safra campus, Givat Ram Jerusalem, 91904, Israel

Sabine Gaudzinski-Windheuser

Palaeolithic Research Unit, Römisch-Germanisches Zentralmuseum, Schloss Monrepos, D-56567 Neuwied

and

Johannes Gutenberg-University Mainz, Institute for Pre- and Protohistoric Archaeology

Lutz Kindler

Palaeolithic Research Unit, Römisch-Germanisches Zentralmuseum, Schloss Monrepos, D-56567 Neuwied

and

Johannes Gutenberg-University Mainz, Institute for Pre- and Protohistoric Archaeology

Naama Goren-Inbar

Institute of Archaeology, The Hebrew University of Jerusalem, Mt. Scopus Jerusalem, 91905, Israel

 Springer

Rivka Rabinovich
Institute of Earth Sciences and National
   Natural History Collections
Institute of Archaeology
The Hebrew University of Jerusalem
Berman building
Edmond J. Safra campus
Givat Ram Jerusalem
91904, Israel
rivka@vms.huji.ac.il

Lutz Kindler
Palaeolithic Research Unit
Römisch-Germanisches Zentralmuseum
Schloss Monrepos, D-56567 Neuwied

and

Johannes Gutenberg-University Mainz
Institute for Pre- and Protohistoric
   Archaeology
kindler@rgzm.de

Sabine Gaudzinski-Windheuser
Palaeolithic Research Unit
Römisch-Germanisches Zentralmuseum
Schloss Monrepos
D-56567 Neuwied

and

Johannes Gutenberg-University Mainz
Institute for Pre- and Protohistoric
   Archaeology
gaudzinski@rgzm.de

Naama Goren-Inbar
Institute of Archaeology
The Hebrew University of Jerusalem
Mt. Scopus Jerusalem
91905, Israel
goren@cc.huji.ac.il

ISSN 1877-9077   e-ISSN 1877-9085
ISBN 978-94-007-9994-3   ISBN 978-94-007-2159-3 (eBook)
DOI 10.1007/978-94-007-2159-3
Springer Dordrecht Heidelberg London New York

Cover illustration: Drawing of a Fallow Deer by Amir Balaban. Photographs of fossil bones from the Acheulian site
of Gesher Benot Ya'aqov by Gabi Laron

Cover design: Noah Lichtinger

Printed on acid-free paper

Springer is part of Springer Science+Business Media (www.springer.com)

A Volume in the Gesher Benot Ya'aqov Subseries

Coordinated by

Naama Goren-Inbar

Institute of Archaeology, The Hebrew University of Jerusalem

In memory of Dr. Eli Lotan, our student, colleague and mentor, whose scientific curiosity and vast scope of interests are engraved in the memory of all who knew him and mourn his loss.

Foreword

Each period and part of the globe has its landmark sites, the ones that seem to define what a particular period is all about or the nature of some important step or threshold in the cultural and biological evolution of our species. For the Plio-Pleistocene, all eyes of course are fixed on Africa, the continent where the human story began. And two sites in particular, FLK-Zinj in Olduvai and FxJj50 in Koobi Fora, are clearly the archaeological standards, the reference points for our understanding of this remote page in our evolutionary history, and the ones to which every other site of the same period is compared.

But once hominins began to leave Africa and spread into the rest of the Old World, a process that began about 1.8 Myr, the archaeological record of Israel moves center stage, and for almost every subsequent major development in the human career one or another site in Israel has become a standard by which such developments are characterized and evaluated. For example, for many years the earliest undisputable human habitation outside of Africa was the 1.4 Ma site of 'Ubeidiya in the Jordan Valley, only recently surpassed by the spectacular and somewhat earlier remains discovered at Dmanisi in Georgia and the redating of the famous paleontological localities in Java where in the nineteenth century the first *Homo erectus* fossils were discovered.

Similarly, in the intense and fascinating debates that surround the origins of anatomically and behaviorally modern humans, there is hardly a student of prehistory anywhere who hasn't heard of the Middle Paleolithic caves of Skhul and Qafzeh. These classic sites occupy a central position in our ongoing attempts to understand where people with modern anatomy and cognitive capacities came from, and what role they may have played in the demise of Eurasia's beetle-browed Neanderthals.

The 21 ka site of Ohalo II, exposed during an extended drought by the retreat of the Sea of Galilee, provides us with startlingly early evidence for the beginning stages of the harvesting, grinding, and baking of wild cereal grains, marking the first of a series of dramatic steps toward the "broad spectrum revolution" and the emergence of the world's first sedentary farming villages.

And now the nearly 800 ka Israeli site of Gesher Benot Ya'aqov (GBY), also located in the Jordan Valley and not all that far from 'Ubeidiya, is emerging as a unique and spectacular record of human lifeways during this remote period of the early Middle Pleistocene. GBY, the focus of this timely and important study, not only provides evidence for a second major wave of human expansion out of Africa but, thanks to the painstaking work of project-director Naama Goren-Inbar, together with Rivka Rabinovich, Sabine Gaudzinski-Windheuser, Lutz Kindler, and their many collaborators, GBY is also yielding a record of unparalleled detail about the lifeways and capabilities of these ancient and hitherto poorly known hominins. For example, systematic plots of the spatial distribution of literally thousands of tiny burned flint microchips at GBY revealed the presence of "phantom" hearths, thereby providing some of the most compelling evidence that hominins already had control of fire more than three-quarters of a million years ago. Thanks to its largely waterlogged condition, GBY also preserves an unparalleled wealth of organic remains, including thousands of fruits, seeds, and pieces of wood, some burned, as well as delicate fossil crabs, amphibians, fish, and molluscs. As the work on this marvelous organic record progresses, we are learning not only about the ancient lakeside environment in which these hominins lived, an invaluable framework in its own right, but we are gaining insights into the unexpectedly varied resource base available to these archaic human foragers. Through ongoing collaborative research with other archaeologists, paleontologists, paleobotanists, geologists, zoologists, isotope chemists, and a host of other specialists, the GBY team led by Naama Goren-Inbar is steadily piecing together a picture of this early period in human history that will serve as a standard for the Eurasian early Middle Pleistocene for many years to come.

The present volume, a detailed look at the bones of some 15 different taxa of medium- to large-sized mammals recovered during seven seasons of excavation at GBY, is an extremely important contribution to our knowledge about the lifeways of these Lower Paleolithic foragers. The site preserves a marvelous record of the animals that hominins procured and butchered on or near the shore of paleo-lake Hula nearly 800 ka. And there are some important insights and surprises here that readers will find of great interest. For example, while most scholars have long abandoned the idea that Middle Paleolithic humans (i.e., Neanderthals and their contemporaries elsewhere) were scavengers rather than hunters, opinion is much more divided about how their Lower Paleolithic predecessors obtained meat. Through the present study, GBY adds its voice to a steadily growing chorus of zooarchaeologists arguing that early Middle Pleistocene hominins, too, were capable hunters, at least by about a million years ago, if not before. At GBY this conclusion is drawn from several lines of evidence, most notably the presence of the full array of skeletal elements for a number of the more important taxa, suggesting that GBY's foragers had early access to whole carcasses, as well as the fact that many of the taxa are well represented by adults, even the elephants (*Palaeoloxodon antiquus*).

There is also a widespread view among paleoanthropologists that Lower Paleolithic sites tend to be heavily dominated by bones of megafauna and that regular use (hunting) of medium-sized ungulates like deer did not become commonplace until much later, perhaps as recently as 300 ka or 400 ka. While the remains of megafauna, both elephants and hippo, are clearly present at GBY, the smaller fallow deer (*Dama* sp.) was the principal mammalian resource exploited by the site's inhabitants, very likely hunted, not scavenged, and probably brought to the site intact, or nearly so. GBY's inhabitants were clearly familiar with the anatomy of their prey and, judging by the abundance of cutmarks and percussion marks, they thoroughly butchered and processed these animals for both meat and marrow.

As is necessary in any comprehensive, contemporary zooarchaeological study, the authors devote a lot of effort to taphonomic issues. This is absolutely essential for several reasons. Obviously, any study that wishes to contribute to our understanding of human behavior must first demonstrate that the bones preserved in an archaeological site of such great antiquity reflect the activities of humans and not the foraging proclivities of hyenas and other carnivores, or the selective transport and winnowing by the moving waters of the nearby lake. Moreover, while there are plenty of *bona fide* cutmarks and humanly induced impact fractures on the GBY bones, there are also lots of curious striations that are probably not a product of butchering or subsequent food processing. In order to figure out how these faunal assemblages came into being, and what produced the striations, the authors conducted an interesting series of tumbling, trampling, and burial experiments which are clearly described in the volume. The gist of their findings is that the GBY assemblages are largely the product of human activities. They find very little evidence that carnivores played more than a minor role in the formation of the assemblages and that density-mediated attrition of the more fragile bones has not seriously impacted the faunal remains. They also show quite convincingly that, despite GBY's proximity to an ancient lake, running water had little or no effect on the composition or spatial arrangement of the remains. As to the striations, they conclude that trampling of bones lying on or in the muddy matrix of the shoreline, by the site's human inhabitants and by animals coming to the lakeshore to drink, were the principal agents responsible for the damage.

This is an interesting and important volume, and an extremely valuable contribution to our growing understanding of the lifeways of Eurasian hominins in the more remote periods of the Paleolithic. Gesher Benot Ya'aqov adds to a steadily growing view that sees hunting of medium- to large-sized prey as an ancient human foraging strategy, emerging not in the Late Pleistocene or late Middle Pleistocene, but much earlier, perhaps as much as a million years ago, and possibly even earlier.

University of Michigan, Ann Arbor, Michigan   John D. Speth
February 2009

Preface

Human colonization of the Old World is generally viewed to have been feasible due to the emergence of larger-brained hominins characterized by more advanced abilities than those of their ancestors. *Homo erectus* is considered to be the first hominin to have left Africa, and hence responsible for the earliest sorties "Out of Africa." The presence of early hominins in Eurasia, documented by hominin skeletal material and, more frequently, by the remains of their material culture, is evidence of their mobility along dispersal routes, of which corridors have been the most widely investigated.

While the dispersal routes and the mechanisms that enabled hominin colonization are still a matter of intensive debate, the evidence emerging from the Levantine Corridor and from the Acheulian site of Gesher Benot Ya'aqov is of undisputable importance. Recent excavations at the site, among the earliest in Eurasia (ca. 780 ka), uncovered a stratigraphic archive that aids in the reconstruction of the paleohabitats of the early occupants of Eurasia, along with providing unique insight into their behavior.

The site of Gesher Benot Ya'aqov is a unique phenomenon because of its cultural similarity to the African Acheulian Technocomplex—the only one of its kind in the Levant—expressed by techno-typological markers, and because of the waterlogged nature of its deposits that preserved early organic remains such as wood, bark, fruits and seeds. These aspects and others are further complemented by the impressive preservation of mammal bones, which will be described in this volume.

Though at times meager, the site's mammal paleontological collection is of great importance as it contributes to the study of the diverse biogeography of the Pleistocene Levant, as well as to the paleoecology of the northern Jordan Valley and the Hula Valley and its vicinity (segment of the Great African Rift System). By utilizing the Early and Middle Pleistocene data retrieved from Gesher Benot Ya'aqov and its subsequent analyses, we are now better able to reconstruct the paleo-Lake Hula environment and its unique ecological niche, along with shedding new light on the processes that allowed for the excellent bone preservation at the site.

Modern human interference serves as the greatest risk to the site. Boat trips stop here daily, as the excavation area acts as a ramp for dragging the boats out of the water. Despite this and destructive, unnecessary drainage activities that extensively destroyed the landscape (and which are slated to continue), the two remarkable excavation layers (V-5 and V-6; see below) remain exposed on the river bank. Over the course of our excavations, they have yielded a wealth of bones and stone artifacts. Such rich assemblages are undoubtedly due to the still mainly undecipherable social modes of hominin behavior and activities.

The site of Gesher Benot Ya'aqov stretches for some 3.5 km along the Jordan River. Recent excavations of its eastern bank are the first to have uncovered an extensive depositional sequence featuring several Acheulian archaeological horizons. This volume is dedicated to analysis and interpretation of the faunal assemblages that originated in two of these horizons, Layers V-5 and V-6. Stratigraphically and conformably located one above the other, they yielded the richest and most abundant fossil bone assemblages at the site. More precisely, it is the older of the two, Layer V-6, that contains the exceptionally well-preserved and varied mammal assemblage, as it has been protected by the overlying layer (V-5), comprised of a multitude of shells (coquina) that had become thoroughly cemented by the river waters.

By the time excavation of the Layers V-5 and V-6 layers began, we had already accumulated substantial experience and moderate understanding of the nature of the site's Acheulian horizons. Despite this, what was revealed upon exposure of the two layers was unmatched by any other previous experience at the site nor by our own naïve and oversimplified predictions; here was an exceptionally high concentration of mammal and other animal bones, reflecting a rich biodiversity and a high degree of human-caused fragmentation and damaged-induced markings (cut marks, percussion marks, etc.).

Due to the different nature of the two layers' content in comparison to the rest of the excavated site, efforts were made to excavate them as extensively as possible, but when what was supposed to be the final season culminated in August 1997, it became clear that we were far from achieving our goal. As a result, we decided to add a previously unplanned field season in September 1997, that would become the seventh and final season, during which extensive effort were made to expose as much as possible of Layer V-6. While we never fully reached our objective of excavating the entire two layers, we succeeded in progressing further and gained a wealth of data.

The good bone preservation and the high number of damage marks seen on them, both natural and hominin-induced, call for the launching of a project aimed at their detailed study. It was only natural that we collaborate with Prof. Sabine Gaudzinski-Windheuser of the Römisch-Germanisches Zentralmuseum, who had served as the sole taphonomy analyst of the large mammal assemblage from the older 'Ubeidiya site. The Gesher Benot Ya'aqov team, composed of the authors of this volume, designed a project that ended up as both a zooarchaeological and an experimental taphonomic study. The aim was to gain insight into site-formation processes, and in particular to learn about the role of post-depositional processes. We do not claim to fully understand the extent of the social and subsistence drives that led to the assemblages' formation, but we do see this study as a thorough presentation of the data and its interpretation.

Acknowledgements

Many individuals and several foundations supported the Gesher Benot Ya'aqov project, and it is due to their contributions that we are able to present this volume.

Many participants took part in excavations, and the subsequent sieving and sorting of the bone-bearing sediments that originated in Layers V-5 and V-6. The fieldwork was carried out with the help of Idit Saragusti, Gonen Sharon and Nira Alperson-Afil, the all outstanding students of the Institute of Archaeology of the Hebrew University, who acquired vast archaeological experience in the course of the project and participated time and again over many years. The zooarchaeological study profited immensely from the dedication and knowledge of Rebecca Biton. Special thanks again to Nira for her invaluable analysis of the spatial organization of the artifacts and bones. We thank also Shoshana Ashkenazi, for contributing her ecological knowledge to the project, for her invaluable comments and suggestions, and for granting us permission to use her crab database. To Smadar Gabrieli, who undertook the conjoinable bone project. Thanks are due to Uzi Motro for his work on the statistical aspects of the study. Mona Ziegler contributed to the documentation, and Daniela Holst helped tirelessly in carrying out the experiments themselves. Thanks also to Anna Belfer-Cohen for her valuable comments. Nira Alperson-Afil produced the index, and Michal Haber edited this volume with outstanding dedication, insight, and expertise.

We thank Gabi Laron who photographed the archaeological material (Figs. 2.11, 3.1, 4.1, 4.2, 4.3, 4.4, 4.5, 4.7, 5.30, 5.31, 5.34, 5.35, 7.7, and 7.8), and Noah Lichtinger for her work on the digital illustrations. We thank Daniel Even-Tzur for supplying cement mixer for some of the experiments that took place in the Department of Evolution, Systematics and Ecology (ESE) of the Hebrew University.

We are particularly grateful to the following paleontologists who allowed us to use their innovative and as yet unpublished data, such as their taxonomic identifications and scientific records: Vera Eisenman (Equidae), Bienvenido Martínez-Navarro (Bovidae), Adrian Lister (Elephantidae and Cervidae), and Tal Simmons (Aves). Special thanks to Andy Current (Natural History Museum, London) past and present mentor to Rivka Rabinovich.

We extend our thanks also to the German-Israel Science Foundation and the Römisch-Germanisches Zentralmuseum, Germany, who made this entire study feasible; they granted us the means to conduct the study as well as providing Rivka Rabinovich and Naama Goren-Inbar a unique opportunity to collaborate with Sabine Gaudzinski-Windheuser.

Many thanks also to the Irene Levi Sala Care Archaeological Foundation, the Leakey Foundation, the Israel Science Foundation, the National Geographic Society, the Israel Science Foundation (Grant No. 300/06 to the Center of Excellence, Project Title: "The Effect of Climate Change on the Environment and Hominins of the Upper Jordan Valley between ca. 800 ka and 700 ka ago as a Basis for Prediction of Future Scenarios"), and the Hebrew University of Jerusalem, whose support and contributions to the excavations, analyses and research aided in the presentation of this study.

We wish to thank the administrative staff of the universities and research institutions, whose, work behind the scenes, greatly assisted us in completing the present study: Frida Lederman and Benny Sekay of the Institute of Archaeology of the Hebrew University, Sarit Levi of the Department of ESE, the administrative staff of the Authority for Research and Development of The Hebrew University and Herbert Auschrat of the Römisch-Germanisches Zentralmuseum.

Many thanks are due to Eric Delson and Eric Sargis, editors of the *Vertebrate Paleobiology and Paleoanthropology* book series, who generously accepted our study for publication, and to the Springer editorial staff, particularly Tamara Welschot and Judith Terpos.

We are most grateful to A.K. Behrensmeyer and Peter J. Andrews, as well as to two anonymous reviewers, who read earlier versions of this manuscript and provided invaluable comments and corrections that improved the manuscript enormously.

Finally, we wish to thank two of our beloved friends and colleagues who were directly involved with the research of Gesher Benot Ya'aqov and this particular project, and who passed away during the final phases of writing this volume. Prof. Hezy (Jeheskel) Shoshani of the Department of Biology, Addis Ababa University, a world-renowned zoologist and a specialist in all that concerns extinct and extant elephants, was murdered in Ethiopia on June 3, 2008. His commitment, interest, and unmatched enthusiasm will always be remembered. Spurred by his endless curiosity and never-ending search for additional information, Hezy arrived at the site looking for elephant remains. Indeed, his wish came true and, as a great specialist of elephant hyoid bones, he identified several such bone elements and subsequently made them academically known.

Dr. Eli Lotan became a student of archaeology following his retirement from a long and very successful career as a veterinarian in the Jordan Valley. He earned both his BA and MA in Prehistoric Archaeology from the Institute of Archaeology at the Hebrew University of Jerusalem, becoming friend and colleague to students and teachers alike. He participated in numerous archaeological projects and, in due course, joined us at Gesher Benot Ya'aqov. Though the oldest team member, he was young in both body and spirit. Eli was responsible for most of the excavation and registration of the Jordan Bank, and contributed immeasurably to our observations in all that concerns the identification of fossil mammal bones in the field. His extensive knowledge and scientific curiosity, coupled with a pleasant nature and vast experience, were a source of great inspiration to us all.

Contents

List of Abbreviations Used in the Text and the Tables

Anm	animal-induced damage
AST	astragalus
BC	breadth of caput tali (after Kroll 1991)
BD	greatest breadth of the distal end (after von den Driesch 1976)
BFD	greatest breadth of the Facies articularis distalis (after von den Driesch 1976)
BG	breadth of the glenoid cavity (after von den Driesch 1976)
BOS SP	*Bos* sp.
BOVINI	Bovini gen. et sp. indet. cf. *Bison* sp., Bovidae gen. et sp. indet.
BP	greatest breadth of proximal end (after von den Driesch 1976)
BPW	greatest depth of proximal end (after von den Driesch 1976)
BSG	body-size group (with 6 options, as below)
BSGA	weight range (>1,000 kg, e.g., elephant)
BSGB	weight range (approx. 1,000 kg, e.g., hippopotamus, rhinoceros)
BSGC	weight range (80–250 kg, e.g., giant deer, red deer, boar, bovine)
BSGD	weight range (40–80 kg, e.g., fallow deer, caprinae)
BSGE	weight range (15–40 kg, e.g., gazelle, roe deer)
BSGF	weight range (2–10 kg, e.g., hare, red fox)
BT	greatest breadth of the trochlea (after von den Driesch 1976)
CAPR	*Caprini* indet.
CARN	Carnivore und.
CER	Cervidae sp.
CERP	Centre Européen de Recherches Préhistorique de Tautavel, France
CH1	crown height of first lobe of tooth
CH2	crown height of second lobe of tooth
D1	greatest depth of the lateral half (after Davis 1985)
DAMA	*Dama* sp.
DD	distal depth (after Eisenmann 1992)
DW	distal width (after Eisenmann 1992)
ELEP	*Palaeoloxodon antiquus*
FPH	femur proximal shaft longitudinally broken
FSH	femur shaft longitudinally broken
GAZ	*Gazella* cf. *gazella*
GB	greatest breadth (after von den Driesch 1976)
GBA	acetabulum width (after von den Driesch 1976)
GBY	Gesher Benot Ya'aqov
GL	greatest length (after von den Driesch 1976)
GLP	greatest length of the *processus articularis* (after von den Driesch, 1976)
GUI	General Utility Index
H	height of distal humerus (after Davis 1985)
HIPO	*Hippopotamus amphibius*

HOM	hominin (and hominin induced damage)
HSH	humerus shaft longitudinally broken
HUJ	Hebrew University Collections, Jerusalem, Israel
IQW	Institut für Quartärpaläontologie Weimar (Forschungsinstitut Senckenberg), Germany
JB	Jordan Bank (the area along the left bank of the Jordan River, where Layers V-5 and V-6 lie partially exposed on the surface, but are mainly submerged underneath the river and hence required underwater excavation, but not in accordance with the strike and dip of each layer)
LA	length of the acetabulum (after von den Driesch 1976)
LAR	length of the acetabulum on the rim (after von den Driesch 1976)
LG	length of the glenoid cavity (after von den Driesch 1976)
LM	lower molar
MANF	mandible fragment
MAU	minimum number of animal units
MB	greatest depth of proximal end (after von den Driesch 1976)
MCHDW	width of distal condyle metacarpal (after Davis 1985)
MCLC	diameter or height of distal condyle (metacarpal) (after Davis 1985)
MCSC	width of distal trochlea (metacarpal) (after Davis 1985)
MCSH	metacarpal shaft longitudinally broken
MGPF	University of Florence, Museum of Geology and Paleontology, Florence, Italy
MM	Musée de Préhistoire Régionale de Menton, Menton, France
MNE	minimum number of skeletal elements
MNHN	Muséum national d'Histoire Naturelle, Paris
MNI	minimum number of individual animals
MT	metatarsal
MTHDH	width of distal condyle (metatarsal) (after Davis 1985)
MTLC	diameter or height of distal condyle (metatarsal) (after Davis 1985)
MTPH	metapodial shaft longitudinally broken
MTSC	width of distal trochlea (metatarsal) (after Davis 1985)
MTSH	metatarsal shaft longitudinally broken
NHM	Natural History Museum, London
NISP	number of identifiable specimens
PD	proximal depth (after Eisenmann 1992)
PEL ISH	pelvis ischium
Ph 1	phalanx 1
Ph 2	phalanx 2
PH1PH	phalanx 1 proximal longitudinally broken
PH2D	phalanx 2 distal
PW	proximal width (after Eisenmann 1992)
RDS	radius shaft
RDSH	radius shaft longitudinally broken
RIBP	rib proximal
RIBSH	rib shaft longitudinally broken
SCB	scapula blade
SCD	scapula distal
SCDH	scapula distal longitudinally broken
SD	smallest breadth of diaphysis (after von den Driesch 1976)
SH	smallest height of the ilium shaft (after von den Driesch 1976)
SKFH	skull fragment
SPL	splinter
Str.	striation
SUS	*Sus scrofa*

TAU	The Zoological Collections, Tel Aviv University
TBD	tibia distal
TBSH	tibia shaft longitudinally broken
TFH	teeth fragments
TUSKFH	tusk fragment
VATLP	vertebra atlas proximal
VEL	vertebra lumbar
Ver	vertebra
Ver. Ar	vertebral articular surface
VTRS	spine
UNM	unidentified mammal bones

List of Figures

List of Tables

Chapter 1

Introduction

Abstract This volume presents a study of site-formation processes at Gesher Benot Ya'aqov (GBY), focusing on the qualitative and quantitative analyses of the faunal remains in order to shed light on the processes involved in the genesis of the site. The objective of this study is to obtain a better understanding of the different processes that caused various damage types to the GBY bone assemblage. We hope to gain insight into a variety of taxonomic issues of the assemblages, and to draw conclusions based on a detailed comparison of the fossil bone assemblage together with the results of taphonomic experiments. Although the conclusions presented here originally pertained to GBY, they do much to contribute to our understanding of site-formation processes beyond the Jordan Valley and the Levant.

Africa, the cradle of humankind, is a continuous source of information on human evolution and behavior. The earliest sites displaying hominin activities are located here (Leakey 1971; Isaac 1983; Domínguez-Rodrigo et al. 2005), and research undertaken at some of the sites supplies the background for discussions concerning the behavioral abilities of early Plio-Pleistocene hominins.

Subsistence strategies play a paradigmatic role in our understanding of the evolution of hominin behavior (Washburn and Lancaster 1968; Roebroeks 2001), providing motivation for research into a wide assortment of related subjects. For example, hunting is not solely the act of killing animals, but is linked to a variety of issues involving the social organization and planning of subsistence activities, the paleogeographic and ecological location of a settlement, and the technology applied towards food procurement and processing (Gamble 1986).

Numerous sites ranging in age from 1.6 to 1.9 Ma are known at Olduvai Gorge, Tanzania (Leakey 1971; Potts 1988) and Koobi Fora, Kenya (Isaac 1983; Bunn 1997). Among these, the faunal assemblages from the FLK *Zinjanthropus* (Olduvai) and FxJj 50 (Koobi Fora) sites have received by far the most attention. In addition to bone taphonomy, the study of additional aspects of site-formation processes has been of crucial importance in explaining their archaeological histories (Isaac 1967; Behrensmeyer 1975a, b, 1978; Binford 1984; Bunn and Kroll 1986; Schick 1987; Potts 1988; Bunn 2001).

The integrative approach to studying site-formation processes (Schiffer 1972, 1976, 1983, 1987) does not consider sites solely as products of anthropogenic activity, but integrates the range of processes contributing to the genesis of an archaeological site, which are themselves the topics of various disciplines (e.g., experimental archaeology, ethnoarchaeology, geoarchaeology, taphonomy, sedimentology, etc.).

Based on the studies of these African Plio-Pleistocene sites, some scholars consider early hominins as opportunistic scavengers and food gatherers who favored riverine and patchy woodland/parkland associations where carnivores left numerous carcasses. The use of local lithic raw materials for the production of tools (Delagnes and Roche 2005; Goldman and Hovers 2009) enabled hominins to process animal tissue and, less commonly, vegetal elements (e.g., Binford 1981; Potts and Shipman 1981; Isaac 1983; Bunn and Kroll 1986; Blumenschine 1986, 1988; Potts 1988).

In contrast, when we first encounter regular occupation in northern Europe some 600 ka (Roebroeks and van Kolfschoten 1994; Dennell and Roebroeks 1996; for a contrary opinion, see: Bermudez de Castro et al. 1999; Parfitt et al. 2005), we find clear evidence of hunting activities by these earliest settlers. The site of Schöningen, Germany has yielded a magnificent series of wooden spears dating from about 400 ka. They were found amid the remains of approximately 20 horses, with almost no other species present (Thieme 1996, 1997). Other indirect evidence of Middle Pleistocene hunting originates from the site of Boxgrove, England (MIS 13), where a horse scapula with a circular perforation, interpreted as a hunting lesion, was found among other bones of the same animal. The bones were covered with cut marks and had been systematically fractured for marrow extraction (Parfitt and Roberts 1999; Roberts and Parfitt 1999). It has also become evident that specialized hunting of particular animal species took place contemporaneously in other areas of northern Europe (Gaudzinski 1995, 1996,

R. Rabinovich et al., *The Acheulian Site of Gesher Benot Ya'aqov Volume III: Mammalian Taphonomy.*
The Assemblages of Layers V-5 and V-6, Vertebrate Paleobiology and Paleoanthropology,
DOI 10.1007/978-94-007-2159-3_1, © Springer Science+Business Media B.V. 2012

2005). Here, hunting involving a system focused primarily on high-quality food yield (Gaudzinski and Roebroeks 2000) is emphasized by isotopic studies revealing that Neanderthals stood at the apex of the trophic pyramid (Bocherens et al. 1999). In addition, numerous recently excavated Middle Palaeolithic sites from the Levant indicate intentional hunting by Neanderthals (e.g., Speth and Tchernov 1998, 2001; Speth 2004; Speth and Clark 2006).

Despite evidence of hunting activities by the early settlers of northern Europe, it is unclear whether these hominins brought this ability with them or whether hunting was an essential subsistence strategy that flourished in Europe in order to cope with specific climatic and/or environmental conditions. This present study of a Levantine Early and Middle Pleistocene site located at the gateway to Europe aids in better understanding how hominin subsistence evolved.

A taphonomic analysis of 17 faunal assemblages associated with stone tools reveals that at the Early Pleistocene site of 'Ubeidiya, Israel (1.4 Ma) hominins focused on meat exploitation. There, the hominin interaction with the fauna is demonstrated by the presence, though rare, of cut marks, while traces of bone-marrow breakage by hominins are missing. This scarcity of cut marks could indicate that the assemblages primarily represent a pattern resulting from the natural mortality of the local fauna; i.e., animals that died of natural causes in the same location where hominin activity resulted in the formation of lithic assemblages (Gaudzinski-Windheuser 2005).

Although the presence of background faunal elements cannot generally be excluded, the assumption that the majority of the bones at 'Ubeidiya originated from this source seems unlikely. Bone assemblages with scarce cut-marks from various stratigraphic units (time-slices) within the 'Ubeidiya Formation associated with artifacts in primary depositional contexts have been analyzed. The suggestion that all the faunal assemblages reflect natural background remains would then lead to the unlikely conclusion that animal resources played hardly any role in the subsistence strategies of the earliest Levantine hominins. However, if we were to follow the Aiello and Wheeler hypothesis (1995), we would have to infer a strong evolutionary link between hominin brain expansion and an increased intake of animal protein.

Early hominins in Africa probably obtained much of their meat by scavenging small- to medium-sized carcasses at natural death sites and felid kills. But the latter strategy would only provide a significant amount of meat if scavenging was confrontational, which some consider too risky (Domínguez-Rodrigo 2002). Furthermore, only marrow of long bones and the brain are regularly available after felids have consumed their prey (Blumenschine 1986; Bunn and Ezzo 1993). In this context, it has even been assumed that within the carnivore guild, early African hominins occupied an ecological niche that was focused on bone-marrow transport and breakage (Brantingham 1998).

Viewed against this background, the apparent scarcity or absence of traces of bone-marrow breakage at 'Ubeidiya is of particular interest, as it forms a marked contrast to the scavenging-based, bone-marrow focused subsistence proposed by some scholars for the early African hominins (Blumenschine 1995; Capaldo 1997). At 'Ubeidiya, marrow-processing was either only rarely practiced or was not a component of the subsistence strategies used there.

However, the absence of evidence for bone-marrow processing does not necessarily imply that marrow procurement was absent from early Levantine hominin subsistence strategies. Instead, it could indicate occupation of 'Ubeidiya that was restricted to particular seasons (e.g., Speth 1987), transport of marrow-yielding bones to different locales in the landscape, or alternatively, that insects, small mammals, birds, eggs, etc., provided the fat and protein intake needed.

A more likely explanation for the lack of bone-marrow processing at 'Ubeidiya is that hominins actively hunted and regularly consumed the meat of horses and cervids, species on whose bones cut marks regularly occur. This hypothesis is based on the assumption that scavenging opportunities were comparable for both the African and the Levantine Plio-Pleistocene hominins (Gaudzinski 2004a; Gaudzinski-Windheuser 2005).

GBY is among the very few sites in which detailed taphonomic analyses can be carried out. Located in the Hula valley of northern Israel (Fig. 2.1), this Acheulian site (approximately 780 ka) provides archaeological evidence of hominin diffusion/migration out of Africa and into Eurasia and, together with 'Ubeidiya, represents a major step in the colonization of the Old World. The uniqueness of this site and its research potential exemplify some fundamental questions concerning hominin behavior in the land bridge between Europe and Africa.

This volume presents a study of site-formation processes at GBY, focusing on the qualitative and quantitative analyses of the faunal remains in order to shed light on the processes involved in the genesis of the site. The objective of this study is to obtain a better understanding of the different processes that caused various damage types to the GBY bone assemblage. We hope to gain insight into a variety of taxonomic issues of the assemblages, and to draw conclusions based on a detailed comparison of the fossil bone assemblage together with the results of taphonomic experiments. Although the conclusions presented here originally pertained to GBY, they do much to contribute to our understanding of site-formation processes beyond the Jordan Valley and the Levant.

Chapter 2

The Acheulian Site of Gesher Benot Ya'aqov

Abstract Gesher Benot Ya'aqov (GBY) is located in the southern Hula Valley, which, in turn, is located in the northernmost segment of the Dead Sea Rift, part of the Great African Rift System. This region is an integral part of the "Levantine Corridor," a land bridge connecting Africa and Europe, through which the diffusion and biotic exchange of many organisms took place in prehistoric times. The Hula Valley has preserved data of a phenomenon of great importance in human history: archaeological evidence recording hominin diffusion/migration out of Africa and into Eurasia. The unique sedimentological and hydrological conditions prevailing in the Hula, along with extensive and intensive tectonic activity, resulted in the complex and minimal exposure of Plio-Pleistocene geological formations. One of these, the Benot Ya'akov Formation, has revealed many unique hominin artifacts, fossil bones, and a multitude of organic remains. Its examination has significantly contributed to our understanding of the paleoecological conditions that prevailed in the region, as well as enabling a comparison between the paleoecological systems of the Early and Middle Pleistocene in Africa and the Levant, areas in which hominins were active already in very early prehistory.

The GBY site stretches some 3.5 km along the Jordan River (33° 00' 30" N, 35° 37' 30" E), outcropping on both banks and submerged below the present channel. It is located 4 km south of the Hula Valley at an elevation of approximately 61 m above m.s.l. (Fig. 2.1).

The site was first excavated in the early 1930s by Garrod and Stekelis (Stekelis et al. 1937, 1938), their respective pits dug north of the bridges on the right bank of the river (for a detailed location, see Goren-Inbar et al. 1992a). Between 1935 and 1954, Stekelis conducted surveys as well as small pit excavations (Stekelis et al. 1937, 1938; Stekelis 1960). Small-scale excavations in the same area were also carried out by Gilead in 1967–1968 (Gilead 1968).

The area's geology was studied by Picard, Schulman, and Horowitz (Picard 1952, 1963, 1965; Schulman 1967, 1978; Horowitz 1973, 1979, 2001), and the lithic artifacts and

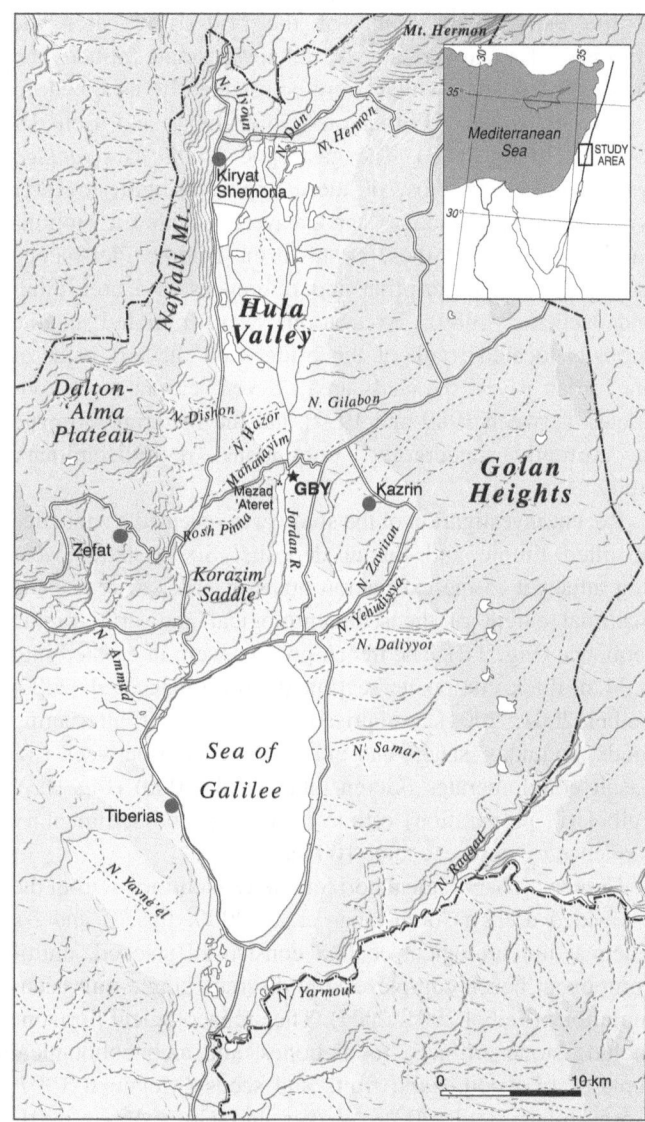

Fig. 2.1 Location map of Gesher Benot Ya'aqov

fossil bones they discovered generated great interest and led to some of the earliest Israeli Pleistocene multidisciplinary studies. Hooijer (1959, 1960) and Tchernov (1973, Tchernov

R. Rabinovich et al., *The Acheulian Site of Gesher Benot Ya'aqov Volume III: Mammalian Taphonomy.*
The Assemblages of Layers V-5 and V-6, Vertebrate Paleobiology and Paleoanthropology,
DOI 10.1007/978-94-007-2159-3_2, © Springer Science+Business Media B.V. 2012

et al. 1994) discussed the various paleontological aspects of the region, while Geraads and Tchernov (1983) described the hominin skeletal remains and compiled a revised list of mammals. Additional and detailed descriptions of these aspects of the GBY research have been previously published (Goren-Inbar and Belitzky 1989 and references therein).

2.1 The Renewed Excavations

Geological and archaeological surveys conducted at GBY in the early 1980s south of the bridges revealed the presence of many previously unknown small faulted and inclined deposits, which were subsequently assigned to the Benot Ya'akov Formation (Goren-Inbar and Belitzky 1989; Belitzky 2002). The gastropod *Viviparus apamaea* was identified in most of these exposures and, together with Acheulian handaxes and cleavers, enabled a preliminary assignment of these strata to the Middle Pleistocene (see a detailed description and references in Goren-Inbar and Belitzky 1989). Excavations were initiated in the deposits located south of the bridges on the left bank of the Jordan River (the study area). Seven seasons were conducted between 1989 and 1997, and the material retrieved is currently undergoing continuous multidisciplinary study.

Recent investigation of the study area revealed a sequence of tilted limnic and fluvial deposits. Six trenches were dug adjacent to the excavation areas in order to reveal the maximal extent of the sedimentological and stratigraphic sequence (Fig. 2.2). The trench profiles were integrated into a 34 m-thick composite section (Goren-Inbar et al. 2000; Feibel 2001, 2004), comprised of a series of alternating muds, coquinas, sands, and gravels bracketed between two basalt conglomerates (Goren-Inbar et al. 2000) (Fig. 2.3). Feibel (in preparation) estimates the exposed sedimentary sequence to represent some 100 ka.

Fifteen archaeological horizons were identified within the excavated areas (Goren-Inbar et al. 2000: figs. 2 and 3). Their sedimentological context consists of stacked, multi-component beach complexes and discrete, single-unit accumulations (Feibel 2001, 2004). The archaeological horizons include stone artifacts, fossil bones, and archaeobotanical remains of wood, bark, fruits and seeds (Melamed 1997; Goren-Inbar et al. 2001; Werker and Goren-Inbar 2001; Goren-Inbar et al. 2002).

The excavations were carried out in three different Areas: A, B and C, of which the study of the latter is the focus of the present study (Fig. 2.2). The typological, technological, and morphological analyses of the components of the archaeological horizons led us to assign them all to the Acheulian

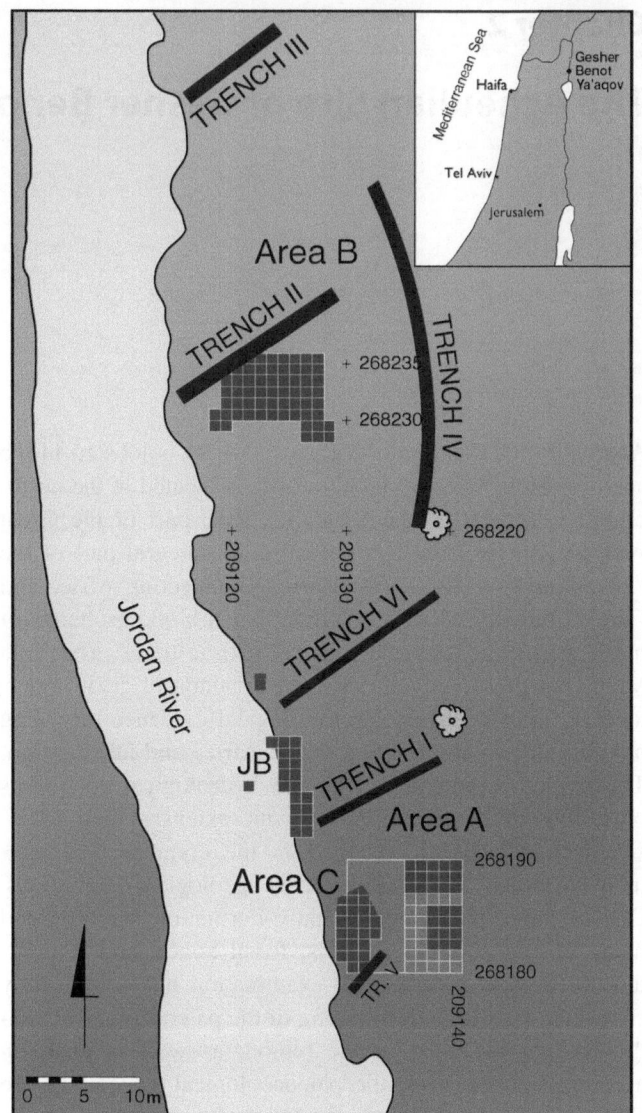

Fig. 2.2 The excavation areas and the geological trenches (the *shaded squares* in the grid have been excavated)

Industrial Complex, as well as to the phase characterized by bifaces made on large flakes (Sharon 2007). Among the other typological components that further supported the assumed African origin of the knapping tradition was the identification of cleavers, which are present throughout the cultural-stratigraphic sequence at the site (Goren-Inbar and Saragusti 1996; Goren-Inbar et al. 2000; Goren-Inbar and Sharon 2006). The detailed lithic analyses further illustrated that for a period of at least 50 ka, the *chaîne opératoire* (reduction process) of basalt bifaces had been an extremely conservative one with regards to typology, technology, size, symmetry, and regularization of the end-result (Sharon et al. 2011). This cultural conservatism has no negative implication on the cognitive level of the hominins, since it has also been demonstrated that the biface *chaîne opératoire* was

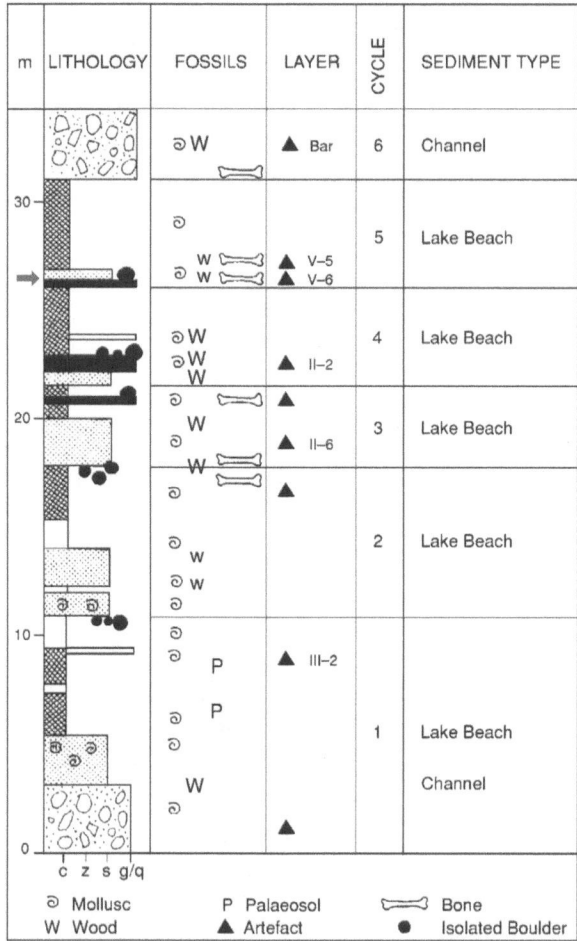

Fig. 2.3 Schematic composite stratigraphic section of the 34 m thick sedimentary succession of the Benot Ya'akov Formation excavated at GBY (Feibel 2001). Note the diversity of the sediment types and their occurrence in six sedimentary cycles. Note also the layers that contain wood, artifacts, molluscan shells and bones. On the basis of the Brunhes Matuyama Magnetic Chron Boundary (marked B/M) of 0.78 Ma these are correlated with the indicated Oxygen Isotope Stages. Wood—*w*, mollusks—*spiral*, artifacts—*full triangle*, palaeosol—*p*, clay—*c*, silt— *z*, conglomerate—*q*. (after Goren-Inbar et al. 2002, p. 23, fig. 9)

characterized by technological flexibility, which stemmed from know-how and decision-making employed in order to achieve the best results from a wide array of technological situations (Sharon et al. 2011).

This *chaîne opératoire* further contributed to the understanding of hominin behavior by illustrating the extent of hominin knowledge of the paleo-landscape and environment. The exploitation of basalt slabs for the production of bifaces necessitated their extraction from quarries located beyond the lake margins, and their transportation, along with giant cores, to the site (Goren-Inbar et al. 2011; Goren-Inbar 2011a). Coupled with other studies focused on the role of other raw-material exploitation at the site (flint and limestone), which were also not available at the lake margins,

it is evident that raw-materials transportation took place from exposures and rivers of higher elevations than those of the site. Other aspects of hominin mobility originate in a comparative study of lithic assemblages associated with the production of bifaces and experimental knapping aimed to mimic the GBY biface *chaîne opératoire*. The results indicate that fully modified bifaces were introduced to the site and were taken out from it. Others were brought in as partially finished tools (preforms), and in other cases the entire reduction sequence took place at the site (Sharon and Goren-Inbar 1999; Madsen and Goren-Inbar 2004; Goren-Inbar and Sharon 2006; Goren-Inbar 2011a).

The analysis of the lithic assemblages yielded insight into the level of technological evolution at the site, as expressed by the evidence of fire. Burned flint macroartifacts and microartifacts were found throughout the entire cultural sequence, where their presence was clearly not a result of natural fire but of a controlled one (Alperson-Afil and Goren-Inbar 2010). These finding contribute extensively to the reconstruction of hominin behavior as also relates to their subsistence and diet. The evidence of fire revealed in each of the archaeological horizons is based mainly on the high frequency of microartifacts. Their spatial analyses led to the identification of concentrations that are commonly referred to as "phantom hearths" (Alperson-Afil and Goren-Inbar 2010). Recently, in-depth spatial analyses of all the fifteen archaeological horizons yielded important results concerning the hearths' spatial organization. In addition, they provided significant information regarding the spatial distribution of other finds classes. These include fish bones, crabs, pitted stones, bifaces and their products, different raw material types, and botanical remains including burned wood fragments. These finds illustrate the ability of the GBY hominins to organize their space into what appears similar to modern (ethnographic) cognitive behavior of a mixed-gender community that practices division of labor and food sharing (Alperson-Afil et al. 2009; Goren-Inbar 2011a).

A striking pattern that emerges from our analyses is the variability of the finds retrieved from the different archaeological horizons. This is reflected by the differing quantities of medium and large bones, wood fragments, nuts of different species, specific tool types (e.g., handaxes and cleavers, pitted stones, massive scrapers), waste products (e.g., giant cores), and exploited raw materials. These results indicate that the occupations along the lake margins were not task-specific, but featured a variety of behavioral modes that left their mark on the landscape of the paleo-Lake Hula.

Among other reasons, the waterlogged nature of the site was due to the water regime of the paleo-Lake Hula. The water level fluctuated in a similar manner to that of Lake Hula in recent times, causing the deposit to change from relatively deep-water sediments to very shallow, littoral facies (Feibel 2001, 2004). Furthermore, the lignite horizons encountered in the section indicated that swamps, extremely rich in organic material, were present at the margins of the ancient lake, similar to those which existed around the now-drained lake (Dimentman et al. 1992).

Although a Middle Pleistocene age was originally assigned to the Benot Ya'akov Formation, newer research demonstrates that the formation is, in fact, older. The Brunhes Matuyama Boundary (780 ka) is located within the lower third of the sequence (Verosub et al. 1998; Goren-Inbar et al. 2000), placing it in the Early Pleistocene. The duration of the entire depositional sequence described above is estimated to be some 100 kyr (Feibel et al. 1998; Feibel 2004), based on the assignment of the observed sequence of six depositional cycles to MIS 18–20 (Feibel 2001, 2004).

The fossil fauna of some of the layers at GBY is rich and comprises many mammalian taxa of Eurasian and African origin, in addition to mollusks (Ashkenazi et al. 2010; Mienis and Ashkenazi 2011), birds (Simmons 2004), fish (Zohar and Biton 2011), crabs (Ashkenazi et al. 2005), turtles (Hartman 2004), micromammals (Goren-Inbar et al. 2000; Rabinovich and Biton 2011) and herpetofauna (Rabinovich and Biton 2011).

An unusually rich, diversified and dense bone assemblage was recovered from the coquina of Layer V-5 and the underlying clay layer (V-6) in Area C. This faunal collection includes what may be the largest *Dama* assemblage known from any Levantine Early-Middle Pleistocene site (Rabinovich et al. 2008a). The presence of extremely well preserved, identifiable bones, together with a rich lithic assemblage bedded within a diversified sedimentological context, has made for a highly promising taphonomic study designed to investigate issues concerning subsistence.

2.1.1 Area C and the Jordan River Bank

The fossil bone and lithic assemblages described in this study originated in two layers, V-5 and V-6, located in the southern part of the study area: Area C and the Jordan River bank in the vicinity of Area C (Fig. 2.2). The following section provides the stratigraphic, sedimentological, and excavation background for the assemblages derived from these two layers.

Because Recent flood plain deposits of the Jordan River covered the Pleistocene deposits in Area C, during the 1996 season a trench (Trench V) was dug perpendicularly to the strike of the layers exposed only on the river bank in order to better view the stratigraphy (Fig. 2.4). The trench's profile (northern face) was drawn, and the sequence of layers was identified, described, and each layer individually named (Fig. 2.5). The name of each layer is comprised of the trench number (in Roman numerals) and an additional number (in

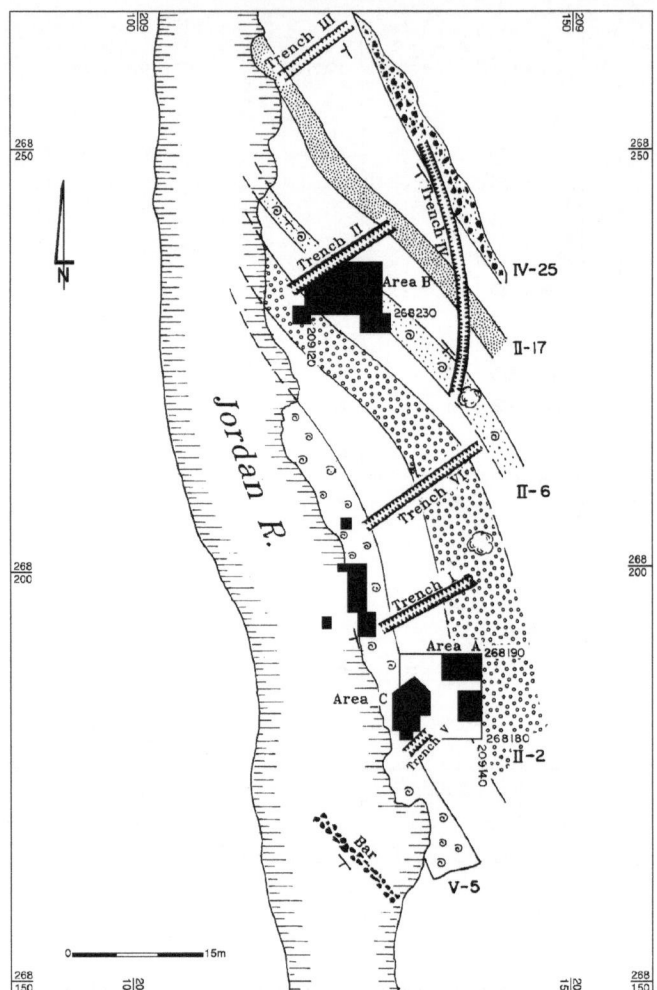

Fig. 2.4 Geological map of the study area (after Goren-Inbar et al. 2002, p. 22, fig. 7). (Selected layers mentioned in the text are marked by different symbols and by name reference: Layers II-2 and V-5 are coquinas; Layer IV-25 and the Bar are conglomerates.)

Arabic numerals) referring to the location of the layer within the sequence. Arabic numbers were given consecutively, starting from the westernmost (younger) exposed layer proceeding eastwards. Thus, Layer V-5 is younger than Layer V-6 and directly overlies it. The two layers are conformably bedded (Figs. 2.5, 2.6, and 2.7), and were exposed by erosion along the river bank and by excavations, in Area C, on land.

Following the excavation of Trench V, the adjacent Area C was excavated for three seasons (1996–1997), with work focused mainly on the two above-mentioned layers, V-5 and V-6. In addition, excavation of the layers' exposures in the river bank (JB) was also carried out (Figs. 2.8 and 2.9). Upon the discovery of large bones that had been exposed because of the very low water level, highly isolated, small-scale excavations of the layers were also carried out in the river itself (Fig. 2.2).

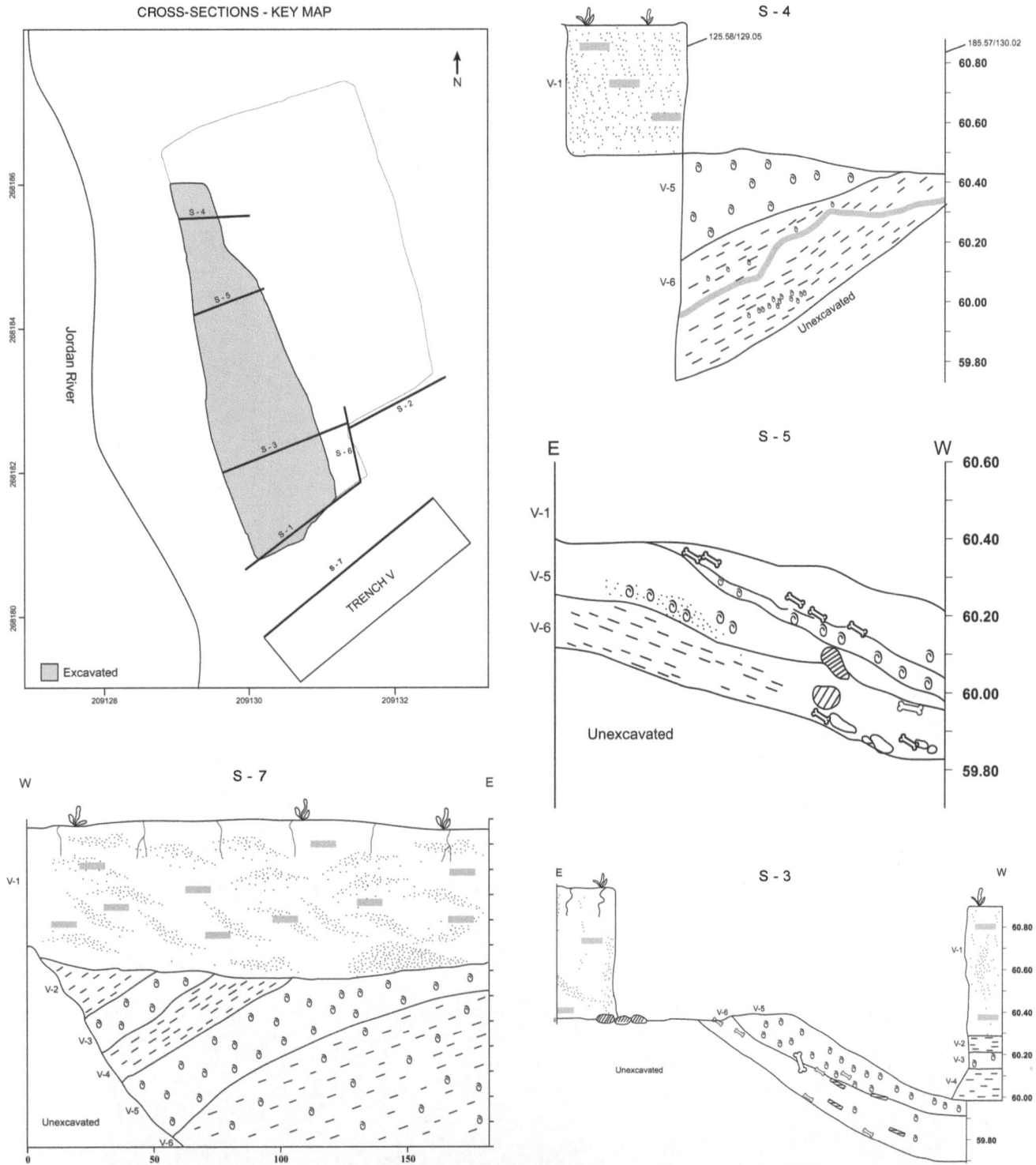

Fig. 2.5 *Upper left*: map of Area C with lines and names marking the location of each section (the key map reference is in Israel Grid coordinates). Sections underneath the same heading appear below the map and to its right (*Sections 3, 4, 7*; the horizontal scale in all sections is in 50 cm units)

Sedimentologically, Layer V-5 is coquina (a molluscan packstone), comprised primarily of the gastropod *Viviparus apamaea galileae* and *Bellamya* sp. (Mienis, pers. comm.; Spiro et al. 2009) with minimal contribution of small grain

matrix. It measures 30 cm thick, is partially exposed on its surface, and forms the cemented bank of the Jordan River in the study area in question (Fig. 2.10). It is the most extensively exposed layer in the study area, observed in the

Fig. 2.6 A view of the northernmost cross-section

Fig. 2.7 A view of the southernmost cross-section

Fig. 2.8 Excavation of the Jordan Bank (JB)

north in the bank lying west of Area B and discontinuously south of Area C (Figs. 2.4 and 2.11).

Layer V-6 is soft, calcareous dark mud (Feibel 2001, 2004), eroding quickly when exposed along the river bank. Its malacological assemblage is different from that of the overlying layer and includes *Melanopsis*, *Theodoxus*, and a variety of other species (Ashkenazi and Mienis 2005). It measures 1 m thick, but the excavations in Area C exposed only its uppermost 0.35 cm (Fig. 2.5).

The excavated volume and surface exposure of Layers V-5 and V-6 in Area C are 1.59 m^3 and 1.97 m^3 and 6.39 m^2 and 7.04 m^2, respectively. Each layer yielded extremely diverse, excellently preserved faunal assemblages. These assemblages were found together with high amounts of flint and, to a lesser extent, basalt artifacts (Sharon and Goren-Inbar 1999; Goren-Inbar et al. 2004; Goren-Inbar and Sharon 2006). Paleobotanical remains of wood, bark (Goren-Inbar et al. 2002), fruits and seeds were also recovered in significant quantities (Goren-Inbar et al. 2001).

Feibel (2004) described the sedimentary setting of GBY in a first-order cycle that consists of six second-order cycles that, in turn, reflect changes in the lacustrine environment.

The regular cycle of the paleo-Lake Hula suggests that no major tectonic events disturbed the sedimentary regime of the site. Based on the layers' location in the study area's stratigraphic sequence (Area C), they were assigned to the fifth second-order cycle (Cycle 5) of the site's cyclostratigraphic record and are considered to represent a sedimentation phase that took place during MIS 18.3 (Feibel 2004).

Indications for climatic change were observed in the six cycles of the GBY sedimentary sequence. These are based on stable isotopes, mollusks and plant paleocology that corroborate the cyclicity suggested earlier by Fiebel (Spiro et al. 2009, 2011). They suggest that the desiccation that created several micro-habitats during the drying and exposure of shallow water and shore habitats was what also enabled the honimin activity of Cycle 3 (i.e., Area B) and Cycle 5 (Area C). In Cycle 5, gentle fluctuations occured in the lake water level and are reflected in the appearance of the different species characteristic of both shallow and deeper water (e.g., Viviparidae) (Spiro et al. 2009, fig. 6). Layer V-6 contains rich vegetation of diverse aquatic, swamp and shoreline origins (Melamed 2003).

Fig. 2.9 Excavation of the Jordan Bank (JB)

The sympatric coexistence of both viviparid species in Layers V-5 and V-6—*Viviparus* that originated in the Orontes River systems, while *Bellamya* originated either in Asia or Africa, which require a low-fluctuating lake with a relatively mild and stable climate, indicate that the climatic conditions were mild and sub-tropical (Ashkenazi et al. 2010).

2.2 Excavation Methodology

Excavations in Area C exposed each layer along its dip and strike (Figs. 2.12 and 2.13), an excavation method not possible along the river bank, as the river water prevented any kind of exposure by flooding the excavated surface. This difficulty dictated an underwater excavation and necessitated grouping together the material from both layers, termed the Jordan Bank (JB) (Figs. 2.8 and 2.9).

During the pre-excavation surveys of the Benot Ya'akov Formation, it became evident that its exposures in the study area and vicinity are extremely small and situated structurally

in different positions due to deformations caused by tectonic activity (Goren-Inbar and Belitzky 1989; Goren-Inbar et al. 2000; Belitzky 2002). This complex geological structure necessitated an application of a standard grid to cover the entire area, which disregards the particular orientation and tilt of each exposed block. The Israel Grid was applied to the study area to serve as a reference system for all spatial locations (coordinates) concerning finds originating in the excavations. A suspending grid system was build over Area C but did not include the JB.

Excavations in Area C began with the documented removal (grid and datum) of the Recent and the Pleistocene re-deposited material (the Jordan River flood plain; "Unconformity") in order to expose the Middle Pleistocene bedding, and were carried out by conventional archaeological methods—from the surface downward. Upon exposure of the Middle Pleistocene deposits, the stratigraphy that became visible on the horizontal surface (a cross-section) was mapped. After this step, excavation of the Pleistocene deposits began by a exposure of the layers along their strike and dip (Figs. 2.6, 2.7, 2.12, 2.13, and 2.14). With the completion of the lateral exposure of the stone artifacts, organic material, and bones, the finds were mapped, photographed, and recorded before removal.

The excavated area, below the 1 m^2 hanging grid, was the grid's projection on the tilted surface, as illustrated by Goren-Inbar et al. (2002: 16, figs. 4 and 5). Therefore, the surface (and hence the volume) of the standard excavated unit is larger than 1 m^2, and its size is dictated by two angles—the strike and the dip of the excavated horizon. A rough calculation, based on the similar strike and dip of Layer II-6 in Area B (Goren-Inbar et al. in preparation), indicated that its surface is approximately 1.2 m^2. The excavated volume of each of the layers is based on a calculation produced by a GIS application (the entire volume removed from each of the excavated layers).

The standard unit of excavation was one fourth (a quadrant) of the projected surface of 1 m^2 on the tilted horizon of the excavated surface, to a depth of 5 cm. Initial and final elevations were taken in the highest and lowest corners of each 1 m^2 projection. Detailed distribution maps were drafted and three coordinates registered for all items measuring 2 cm and more, which included bone, stone, wood and bark. Each item was assigned a field catalogue number that appear both on the plans and on the registration forms (and later in the databases), and was classified as the item's "map number." In addition, each item was assigned an additional number that served as its catalogue number. Items smaller than 2 cm were placed in a "general bag" of an assigned quadrant, with accurate elevations documenting the initial and final heights of the excavated unit. The sediments from each excavated unit were water-sieved in the Jordan River with 2-mm mesh

Fig. 2.10 The exposed coquina of Layer V-5 in the JB

Fig. 2.11   View from the western bank of the Jordan River of the exposure of Layer V-5

sieves. Elevations (z coordinate, above m.s.l.) were measured from the tilted exposed surface of each excavation unit at the beginning and end of each excavation session. Two elevation readings were also taken for each find larger than 2 cm (stone artifacts, including unworked pieces, bone and organic material)—top and bottom, with both readings marked on the distribution maps.

The exposed horizon containing stone artifacts, bones, wood and bark was mapped in detail (1:5). The items were drawn as if viewed perpendicularly to the exposed horizon and therefore, this view is not related to the grid system (Fig. 2.14). Correlations between the grid and each map were marked where each point of the suspended grid system touched the tilted surface.

Two types of cross-sections (profiles) were documented, each at a scale of 1:10. The first type was drawn perpendicular to the layers and the second is cross-sections that were drawn along the grid lines. The first type was used in order to identify the layer's actual thickness and the relationship between the various superimposed layers (e.g., Fig. 2.7). This is a "phantom" cross-section that was re-drawn and added to in accordance with the excavation progress, creating a profile that, in essence, did

not exist in the field. The cross-section drawn along the grid lines was required in order to reconstruct the stratigraphic configuration of the entire excavated volume of Area C and produce a three-dimensional view of the different layers. It included the borders of the entire excavated area as well as particular locations within it.

2.2.1 Sediment Sorting and Its Analyses

Students of the Institute of Archaeology of the Hebrew University of Jerusalem carried out basic sediment classification of the sieved sediments. Stored in plastic bags, they were sorted according to their different components, with the fossil mammal bones (from micromammal to elephant) sent to the National Natural History Collections of the Hebrew University for further analysis. The lithic assemblages underwent detailed cataloguing, and later were techno-morphologically and typologically analyzed by detailed attribute analysis. This data has been stored in a compatible database along with that of the mammal fossil bones (see Chapter 4).

Fig. 2.12 Excavation of Area C

Fig. 2.13 Excavation of Area C, a view of the southernmost cross-section

Fig. 2.14 A view of the exposed Layer V-6

Chapter 3

Materials and Methodology

Abstract Analysis of the animal bones from Area C and the JB entails taxonomic identification followed by morphometric, taphonomic, and surface-modification analyses. Emphasis was also placed on a series of experiments, whose methodology is described below.

3.1 Systematic Description

Taxonomic identification of the bone assemblages was carried out by R. Rabinovich. In addition, several faunal specialists contributed to the identification of particular mammalian taxa. Taxonomic identification of the Cervidae was undertaken by R. Rabinovich together with A. Lister of the Natural History Museum, London (Rabinovich and Lister forthcoming); B. Martínez-Navarro of the Universitat Rovira i Virgili in Tarragona assisted in defining the Bovini (Martínez-Navarro and Rabinovich 2011; Martínez-Navarro 2004); and V. Eisenmann of the Muséum national d'Histoire naturelle, Paris identified the Equidae (MR 8569 et ESA 8045 du CNRS; Eisenmann n.d.).

Bone assemblages from prehistoric sites and recent mammalian collections from various institutions were used as a comparative data for our study. These collections are as follows: the Hebrew University Collections, Jerusalem (HUJ); The Zoological Collections, Tel Aviv University (TAU); the Natural History Museum, London (NHM; Paleontology, Zoology); Muséum national d'Histoire Naturelle, Paris (MNHN; AC); The Museum of Natural History, University of Florence, Museum of Geology and Paleontology, Florence (MGPF); Centre Européen de Recherches Préhistorique de Tautavel (CERP); the Musée de Préhistoire Régionale de Menton, Menton, (MM), and the Institut für Quartärpaläontologie Weimar (Forschungsinstitut Senckenberg) (IQW).

The GBY bone specimens were measured using calipers, with measurements given to the nearest 0.1 mm. Length, width, and thickness were measured in each bone. Unless otherwise stated, identifiable elements were measured according to the von den Driesch scheme (1976) (for further

details on each species, see Chapter 4). Length and width of the teeth were measured from the base of the crown, and crown heights were measured on tooth lobes from the buccal surface in mandibular teeth and from the lingual aspect in maxillary teeth. Information about the age and the sex were obtained from bones and teeth using data on tooth eruption and wear, as well as from stages of epiphyseal fusion (see details on each species in Chapter 4). Each bar of the black and white scales without legend in Chapters: 1–5, 7–8 is of 1 cm length.

3.1.1 Body-Size Groups (BSG)

Due to bone fragmentation, it was not always possible to assign a bone to a particular species. The unassigned bones were classified according to skeletal element and combined into body-size groups (BSG) (Table 3.1). The subdivision into the latter is based on Rabinovich (1998: table 2), and allows further analysis of the skeletal elements (Outram 2001) (see Chapter 5).

3.1.2 Skeletal Elements

Skeletal elements were quantified according to the following criteria: the number of identified specimens (NISP), the minimum number of individual animals (MNI), the minimum number of skeletal elements (MNE), and the minimum number of animal units (MAU) (Binford 1981; Lyman 1994). When describing taxonomic frequency, all the elements that were assigned to a particular taxon were counted (NISP values), including epiphyses and long bone shaft fragments. When calculating the MNI, the most common anatomical part of a particular taxon served as the basis for estimation, after having taken into account body side, age, and sex. The MNE was obtained by counting the epiphyses of a bone per body side, along with long bone shaft fragments whose anatomical position could be exactly determined

Table 3.1 Body-size groups (kg) of different taxa from GBY (after Rabinovich 1998)

Body-size group	Main species	Weight range
BSGA	Elephant	>1,000
BSGB	Hippopotamus, rhinoceros	approx. 1,000
BSGC	Giant deer, red deer, boar, bovine	80–250
BSGD	Fallow deer, Caprinae	40–80
BSGE	Gazelle, roe deer	15–40
BSGF	Hare, red fox	2–10

(Lam et al. 1998). When calculating the MAU, the number of documented skeletal elements was divided by the frequency in which the particular bone occurs in a skeleton (Binford 1984).

Because of the fragmented nature of the bone assemblages and the high occurrence of long bone shaft fragments, we considered the NISP data as best representing the GBY faunal assemblage, and thus most suitable for the analysis of skeletal remains (Rabinovich et al. 2008a). Taphonomic biases that could possibly influence these calculations and the assumptions inherent in using them were fully acknowledged (Lyman 1994).

Since complete bones are rare, each skeletal element, such as shaft fragments of the *Dama* humeri, was divided into sections based on its landmarks in order to record the fragments' anatomical location (adapted after Morlan 1994). The various degrees of the element's completeness is recorded: a complete bone equals one, less than half a bone equals four, half equals five, and more than half equals six (Campana and Crabtree 1987).

Many of the faunal elements were so fragmented that they could only be assigned to general categories such as teeth, flat bones (scapula, pelvis, ribs) and long bone shaft (fore and hind limbs) fragments. The long bone shaft fragments form most of the body-size groups. Nevertheless, their contribution to the archaeozoological record is a usual procedure in the analysis of Pleistocene bone assemblage, and has been adopted in this study (see Chapter 7) (Rogers 2000; Outram 2001; for a different opinion, see Faith and Gordon 2007).

Information on age and sex were also taken into account, with epiphyseal fusion, tooth eruption and wear, distinctive morphological traits and size difference serving as the various criteria. Unfortunately, due to the high degree of bone fragmentation, only a small sample of bones was included in the age and sex analysis. For each animal species, the relevant references are mentioned (see Chapter 4).

3.2 Bone Density and Economic Utility

As differential bone preservation due to post-depositional processes has been shown to be related to mineral density (g/cm^3), the skeletal-element representation for different species was analyzed with the use of bone mineral density data (Lyman 1994). Density can be measured with the application of photon densitometry, in which a photon beam of known strength passes through a defined part of the bone (scan site) (Kreutzer 1992; Lyman 1994). Bones with low mineral density are more easily affected by density-mediated attrition than high-density bones.

Density values have been ascertained for animal species representing different size groups. Analysis has shown that with regard to density values, only low inter-taxonomical variability exists; thus, density values obtained for a particular artiodactyl can be used for the analysis of the density-mediated survival of bones from a different artiodactyl within the same size group (Lam et al. 1999; Lam and Pearson 2005). Despite the fact that numerous methods are available for density estimation, complex processes are involved in the formation of animal bone assemblages and thus caution should be exercised when interpreting data and a variety of analyses should be used (see summary in Lam and Pearson 2005).

One factor that may be responsible for some density-mediated bone preservation is the transport or sorting of bones by fluvial mechanisms. Fluvial transport can be recognized on the basis of skeletal-element representation, since the process leads to a loss of more easily transportable bones (Voorhies 1969). The transport potential of different bones (density, size and shape) is known from experimental studies in flume channels (Voorhies 1969; Behrensmeyer 1975a), and the bones have subsequently been grouped according to their susceptibility to fluvial transport (Groups I–III) (Voorhies 1969; Behrensmeyer 1975a).

Recent studies have shown that differences in transportation occur between fragmented and complete bones, as well as between different taxa (Fernández-Jalvo and Andrews 2003). Multiple agents are responsible for the formation of bone assemblages: "Furthermore a fossil site is the result of sedimentation, but what has been most investigated is the initial transport, when bones start transport, rather than their final deposition" (Fernández-Jalvo and Andrews 2003: 146).

Skeletal-elements abundance is frequently used in order to evaluate economic utility at a site. The skeletal elements of the GBY mammalian fauna were correlated with various "Utility Indices," such as the Modified General Utility Index (MGUI), which combines the meat-, marrow-, and grease-utility indices calculated for each skeletal element based on the meat-, marrow-, and grease content of sheep and caribou found at a site (Binford 1978).

The quantitative composition of the GBY assemblages was calculated with bone density values obtained through photon densitometry (Lyman 1994: table 7.6), and with nutritional values derived according to the MGUI, unless otherwise specified (Lyman 1994: table 7.1; Outram and Rowley-Conwy 1998). We employed Spearman' rho correlations (Sachs 1984) to investigate these associations.

3.3 Bone Damage and Surface Modifications

For observing bone damage, we used a stereo light microscope with a magnification of 10×40 and an image analyzer system (with Media Cybernetics Image-Pro® Plus version 4.1 software for Windows®; photo editing with Adobe® Photoshop 5.0 and 7.0). Three-dimensional surface scans (with GOM ATOS 5.5 and Raindrop® Geomagic™ Studio 6 software) were also used for additional analysis (Römisch-Germanisches Zentralmuseum Mainz). Bone samples were analyzed with an analytical Quanta 200 Environmental Scanning Electron Microscope (ESEM; FEI Company) at the ESEM Center for Nanoscience and Nanotechnology at the Hebrew University of Jerusalem. Photographs for the figures were taken with all the above means (image analyzer, ESEM) and presented in different degrees of magnification in order to enhance certain details.

Bone assemblages from prehistoric sites and those derived from various comparative studies with known taphonomic histories served as the comparative collections for this study. Bone surfaces from the following sources were examined and served as the comparative material for the GBY taphonomic study: butchery experiments; rodent, porcupine and carnivore experimental collections; porcupine lairs and hyena dens, all part of the National Natural History Collections of the Hebrew University of Jerusalem.

The light microscope registered modifications such as striations, tooth scratches, cut marks, tooth marks, gnaw marks, and burned elements. Modifications were register per bone (striated bones, cut bones, etc.), with their exact anatomical position recorded in detail. Rabinovich and Gaudzinski-Windheuser each examined the same *Dama* bone assemblage, afterwards recording surface modification characteristics and their anatomical positions independently of each other ("blind test") in order to examine the validity of their identifications.

3.3.1 State of Preservation

Traces of pitting, erosion, exfoliation, smoothness, shininess, root marks, and discoloration on the bone surfaces indicate individual pre- and post-burial taphonomic histories (Voorhies 1969, 1970; Behrensmeyer 1978; Behrensmeyer and Boaz 1980; Gifford-Gonzales et al. 1985; Behrensmeyer 1988). Since the archaeological layers are waterlogged, the

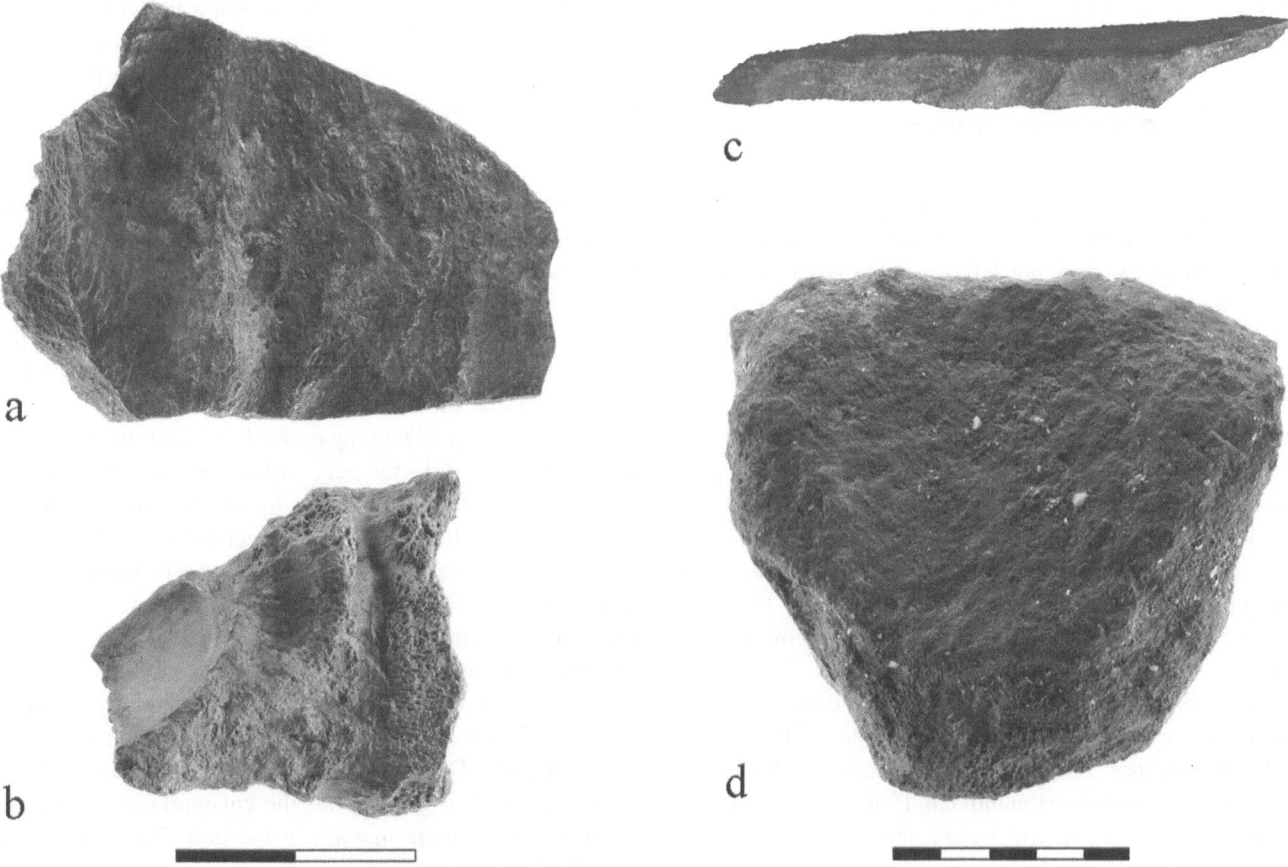

Fig. 3.1   Examples of bones displaying different states of preservation: **a** Type 1, **b** Type 3, **c** Type 4, **d** Type 5

bone surfaces are generally smooth and dark. In many cases, sediment containing complete and crushed mollusks was attached to the bones. Moreover, the depositional conditions often obliterated evidence of early stages of bone-burning and were probably also responsible for smoothing the fragment's general shape.

To a certain extent, Behrensmeyer's (1978) classification and interpretation of bone-weathering stages in East Africa may be applied to different environments (Andrews and Cook 1985; Tappen 1994), but does not suit the waterlogged nature of the GBY deposit that contained the bone assemblages. It appears that the waterlogged nature of the GBY deposit had a major effect on bone preservation. A set of categories was created in order to better describe the bones' surface preservation (Fig. 3.1):

Type 1: Shiny mottled brownish-black surface.
Type 2: Slightly weathered with a distinct blackish color; at times the matrix with mollusk fragments adheres to the bones.
Type 3: Slightly weathered with yellowish surface.
Type 4: Slightly weathered with a dull surface and a metallic bluish-gray color.
Type 5: Signs of weathering and exfoliation.
Type 6: Very weathered to the extent of causing morphological changes that modified the original surface of the bone.

3.3.2 Animal-Induced Damage

Bone fragments featuring jagged edges, tooth scratches, gnaw marks, puncture marks and digested bones indicate damage caused by carnivore activity. Other types of damage morphologies most closely resemble modification by rodents and small- to medium-sized carnivores. Our identification of carnivore teeth marks is based on criteria defined by Binford (1981), Haynes (1983), Blumenschine and Selvaggio (1988, 1991), and Blumenschine (1995). Each animal-modification damage mark was recorded by its skeletal anatomical position to allow further understanding of the role of the carnivore in the GBY faunal assemblage (Faith 2007). We used tooth marks to describe a puncture created by a tooth, tooth scratches to describe an elongated sign created by teeth, and gnawing marks to describe continuous scoring of the bone surfaces.

The basis for identifying animal-induced damage was the comparative assemblages derived from rodents (Rabinovich 1990), porcupines (Rabinovich and Horwitz 1994), and carnivores in captivity (Rabinovich 1990), as well as collected material from recent hyena dens (Kerbis-Peterhans and Horwitz 1992) and porcupine lairs (Rabinovich and Horwitz 1994).

Based on these data and other published sources, we were able to differentiate between the damage caused by rodents and small and large carnivores. However, when the damage pattern did not satisfactorily resemble any of the above, we added to the "unknown" category.

3.3.3 Hominin-Induced Damage

Cut-, hack-, and percussion marks were recorded, with first two defined according to the criteria of Binford (1981), Shipman (1981, 1986), and Shipman and Rose (1983). According to these scholars, cut marks made by stone tools are v- or u-shaped in cross-section and at times display shoulder effects and barbs. Comparisons were made with assemblages from known taphonomic origins, including butchery experiments.

Semi-circular notches, sometimes accompanied by intentional or unintentional retouches, were defined as impact marks. Their position was documented along with their location on either a single face or both bone faces. We have also observed bone flakes in which both ventral and dorsal faces lacked the outer cortical bone surface and/or displayed an impact cone. These flakes result from bone breakage. Also documented were bone fragments that exhibited signs of breakage, but not conically induced impacts (see Chapter 5).

We used the criteria defined by Blumenschine and Selvaggio (1988) in order to identify hominin percussion marks. These marks are sometimes confused with signs of carnivore modification, but one difference is that carnivore modification is rarely associated with microstriations (Blumenschine and Selvaggio 1988, 1991). According to Blumenschine (1995), another difference is that carnivore tooth damage produces an internal crushing in the impact depression, which is not found in anthropogenic-induced marks. However, it has recently been demonstrated that unmodified hammerstones do not necessarily produce striations, and that modified hammerstones frequently produce internal crushing (Domínguez-Rodrigo and Barba 2006). Therefore, we did not rely solely on the presence of internal crushing or striations to identify the percussion marks, but also took their anatomical location into account (Blumenschine and Selvaggio 1988, 1991; Blumenschine and Marean 1993). In order to avoid confusion, we used the term "percussion mark" throughout the text to describe signs of hominin-induced breakage.

Because of the variability in the quantitative way in which cut marks are recorded and its consequence on their interpretations (Domínguez-Rodrigo and Yarverda 2009), we have described in great detail the cut marks per bone element, including their anatomical location. Data can thus be examined through various interpretative ways, but can also be observed in its raw form.

3.3.4 Striations

In this study very fine cortical striae that cannot unambiguously be attributed to hominin- and carnivore-induced damage are defined as striations. In our samples these are very fine, irregular marks of various morphology and cross-section which distinctly differ from hominin-induced cut marks (a detailed description of these marks is given in Chapter 5: Section 5.2). These surface modifications also differ from carnivore damage as has been observed at GBY (a description of these marks is given in Chapter 5: Section 5.4). They randomly occur on bone surfaces, sometimes clustered in isolated patches and sometimes covering the entire bone surface.

Multiple reasons have generally been suggested for the formation of these striation types. They are mostly considered to have resulted from bone interaction with sediment. As their formation is usually regarded as associated with a particular sediment type, only limited research has been carried out that can contribute to their understanding (Shipman and Rose 1983, 1984; Andrews and Cook 1985; Behrensmeyer et al. 1986; Olsen and Shipman 1988; Fiorillo 1989; Oliver 1989; Nielson 1991; Blasco et al. 2008; Domínguez-Rodrigo et al. 2009; see also Gaudzinski-Windheuser et al. 2010).

Different experiments mimicking the sedimentary conditions at GBY were performed in order to gain better insights into the formation and origin of these striations.

3.4 Experiments

A variety of studies illustrate the contribution and importance of experiments in better understanding the fossil record through bone survival and surface modifications. Experimental taphonomy has shown that deposition environment, water presence, soil pH level, animal size and animal condition prior to burial (Denys 2002) are crucial factors in bone survival, as well as the lack of earlier collagen microbial degradation (Trueman and Martill 2002).

Water not only has the power to move and remove bones in and from assemblages, but to smoothen and polish them (Denys 2002). Fossil bones become more rounded than weathered, a trait that develops most rapidly in gravel environments. In contrast, when buried in clay and silts, dry or fresh bones remain almost unmodified (Fernández-Jalvo and Andrews 2003).

Various surface modifications were observed: striations (mainly on shafts) which are: "numerous, generally superficial, closely spaced, intersecting and of variable curvature, length and breadth" (Andrews and Cook 1985: 681); and striations linked to trampling (Behrensmeyer et al. 1986; Olsen and Shipman 1988).

Experiments in an artificial climatic and sedimentological surrounding, closely replicating the prehistoric sites, are a crucial tool for reconstructing an assemblage's taphonomic history, in which surface modifications, survival, and weathering are among the key assessment points (Fernández-Jalvo et al. 2002).

3.4.1 Experiment Methodology

Since a variety of striations occur repeatedly on the GBY bones, we designed a number of experiments that would test their possible taphonomic causes. As stated above, these experiments necessitated replicating the site's depositional conditions. The environmental and sedimentological aspects of the fossil- and artifact-bearing horizons of Layers V-5 and V-6 have been studied. As discussed in Chapter 2, the sediment of Layer V-5 is a coquina comprised of the gastropod *Viviparus apamaea*. The sediment consists mainly of complete and fragmented mollusk shells embedded in a clay matrix, with coarser sediment components absent (Feibel 2001). The reconstructed depositional history indicates that Layer V-5 sediments were deposited in the paleo-Lake Hula margin. Layer V-6 is characterized by "fine-grain offshore muds" (Feibel 2001: 137) of a different malacological assemblage (see details in Chapter 2).

A series of experiments involving trampling and sediment movement were designed in order to replicate the low-energy conditions typical of the paleo-Lake Hula. Sediment samples from the site that had undergone wet sieving were enriched with clay and used in the experiments. The state of fragmentation of particular sediment particles (i.e., mollusks) was documented before and after the experiments.

Dama sp. dominates the bone assemblages of Area C at GBY; the modern fallow deer (*Dama mesopotamica*) is a protected species in Israel, and therefore, our experiments made use of domestic sheep bones instead. These bones are similar in size to those of fallow deer and are available in large quantities in standard Israeli butcher shops. In addition to complete sheep bones, some long bone fragments of domestic cattle were also used. All bones were de-fleshed using metal knives and surgical scalpels, as well as stone tools, before the beginning of the experiments. In addition, the bones were not subjected to any additional cleaning (e.g., boiling) to remove grease or soft-tissue that could not be removed with scalpels. The process of de-fleshing and its resulting bone-surface modifications were meticulously documented in order to account for all modifications produced in the experiments' early stages, prior to burial, tumbling, and trampling.

We conducted three sets of controlled experiments:

1. Random conditions: scratching and burying measured the effects of both trampling and long-term burial under random conditions.
2. Tumbling: designed to replicate tumbling effects in a shoreline environment analoguous to the environmental conditions reconstructed for GBY.
3. Trampling: designed to replicate bone-trampling effects in a shoreline environment analoguous to the environmental conditions reconstructed for GBY.

The first experiment is aimed at identifying sediment particles responsible for striation formation observed on the GBY bones. This experiment was intended to achieve a general impression of bone modification following one year of burial.

Based on paleoecological and sedimentological data obtained at GBY, the tumbling experiments were meant to simulate sediment movement in a calm, lacustrine shoreline environment, whereas the trampling experiments were intended to mimic the role of animal/ or hominin activities in dry, muddy, or wet environments. For the two experiments, the original sediment—rich in both complete and fragmented mollusks from Layers V-5 and V-6 ("coquina"; Feibel 2001)—was used.

For the tumbling experiments, bones mixed with various amounts of water and sediment were placed in a concrete mixer rotating 25 times per minute. The bones were then tumbled up to 16 hours. For the trampling experiments, bones were embedded in a plastic box filled with "coquina" and trampled by an individual during different stages of water saturation of the sediment up to two hours.

Detailed descriptions of the experiments, together with detailed protocols for each experiment, are given in Chapter 6.

Three variables—i.e., bone tissue/outer bone surface, bone morphology, and bone-surface modifications—were analyzed before and after each experiment, since it was assumed that their relationship would shed new light on the operational sequence of the biostratonomic process.

Bone surfaces were photographed before and after the experiments using a stereo-light microscope with a magnification of 10×40, an image analyzer system (with Media Cybernetics Image-Pro® Plus version 4.1 software), and three-dimensional surface scans (with GOM ATOS 5.5 software and Raindrop® Geomagic ™ Studio 6 software). The resulting photographs were digitally edited using Photoshop 5.0 and 7.0. In the course of this study, we established a database containing over 2,000 high-quality photographs. Bone assemblages from other prehistoric sites and experimental studies, stored at the National Natural History Collections at the Hebrew University, served as comparative material for this taphonomic study.

Chapter 4

Systematic Paleontology

Abstract This chapter is dedicated to a detailed paleontological description of the medium- to large-sized mammal fauna discovered at Gesher Benot Ya'aqov (GBY), taking into account previous paleontological studies undertaken in this sector of the Dead Sea Rift.

The fossil fauna of Area C and the JB is diverse, and consists of mammalian taxa of Eurasian and African origin: *Ursus* sp., Canid, *Vulpes* sp., Carnivora indet., *Palaeoloxodon antiquus*, *Sus scrofa*, *Hippopotamus amphibius*, Megalocerini sp., *Cervus* cf. *elaphus*, *Dama* sp., Bovini gen. et sp. indet. (cf. *Bison* sp.), *Bos* sp., *Gazella* cf. *gazella*, Bovidae gen. et sp. indet, Caprini indet., *Equus* cf. *africanus*, and *Equus* sp. (Eisenmann n.d.; Martínez-Navarro 2004; Martínez-Navarro and Rabinovich 2011; Rabinovich and Lister in preparation) (Table 4.1). The fauna also includes other vertebrates: micro-mammalian species (Goren-Inbar et al. 2000), birds (Simmons 2004), turtles (Hartman 2004), reptiles and amphibians (Rabinovich and Biton 2011), and fish (Zohar and Biton 2011). The rich invertebrate assemblage includes crustaceans (freshwater crabs) (Ashkenazi et al. 2005) and mollosks (Ashkenazi et al. 2010; Mienis and Ashkenazi 2011).

4.1 Previous Faunal Studies

Published faunal assemblages from GBY (Table 4.2) (Hooijer 1959, 1960; Geraads and Tchernov 1983; Geraads 1986; Guérin and Faure 1988; Lister in Goren-Inbar et al. 1994) also include a small collection of bones from the "BAR" (Goren-Inbar et al. 1992b). Faunal specimens without provenance labeled "Gesher Benot Ya'aqov" (in various spellings) are found in the HUJ Collections, and some of the above-mentioned scholars based their analysis on these collections. Our study refers to these analyses, although caution is exercised given the fact that the specimens' exact provenance is uncertain. Recent destruction along the river bank has exposed assemblages from later periods, such as the Mousterian and Epipaleolithic (Sharon et al. 2002), suggesting that not every item collected during surveys of the river bank was necessarily part of the Early-Middle Pleistocene sequence of GBY.

4.2 Faunal Composition of the Present Study

A total of 3,953 bones were identified during our excavations of Area C and the JB. Of these, 1,504 were identified at species, genus or family level. In addition, 1,519 bones lacked diagnostic traits enabling further identification and were thus classified according to BSG. The remainder of the assemblage (N = 930) consisted of unidentifiable mammal bones (UNM) primarily comprised of bone splinters with an average length of 25 mm, width of 15 mm, and thickness of 6 mm (Table 4.3). Moreover, 844 highly fragmented teeth have been classified as "unidentified mammal."

Table 4.1 Frequency of animal species from Area C and the JB based on NISP

Taxon	V-5 NISP	V-6 NISP	JB NISP
Ursus sp.	1		1
Vulpes sp.	1		1
Carnivore indet.		3	2
Sus scrofa	3	5	5
Hippopotamus amphibius	4	13	5
Megaloceros sp.		2	1
Cervidae sp.	19	53	11
Cervus cf. *elaphus*	2	1	2
Dama sp.	285	545	355
Bos sp.	2	7	16
Bovini gen. et sp. indet. cf. *Bison* sp.		3	4
Gazella cf. *gazella*	6	8	12
Bovidae gen. et sp. indet.	2	9	5
Caprini indet.		1	1
Equus sp.	12	14	20
Equus cf. *africanus*	4	6	11
Palaeoloxodon antiquus	12	21	8
Total	353	691	460

R. Rabinovich et al., *The Acheulian Site of Gesher Benot Ya'aqov Volume III: Mammalian Taphonomy.*
The Assemblages of Layers V-5 and V-6, Vertebrate Paleobiology and Paleoanthropology,
DOI 10.1007/978-94-007-2159-3_4, © Springer Science+Business Media B.V. 2012

Table 4.2 Taxonomic identification of the GBY fauna, past and present. The current identification of the GBY fauna comes solely from Area C and JB and is based on the following studies: Martínez-Navarro et al. (2000), Martínez-Navarro (2004), Martínez-Navarro and Rabinovich (2011), Eisenmann (n.d.), and Rabinovich and Lister (in preparation)

| Taxon | Previous studies | | | | |
	Hooijer (1959, 1960)	Geraads and Tchernov (1983)	Geraads (1986)	Guérin and Faure (1988)	Lister in Goren-Inbar et al. (1994)
Ursus sp.					
Canid					
Vulpes sp.					
Carnivore indet.					
Sus scrofa	*Sus* cf. *scrofa*	+		+	
Hippopotamus amphibius	+			*H. behemoth/ amphibius?*	
Megaloceros sp.					
Cervidae sp.					
Cervus cf. *elaphus*	+			+	
Dama sp.	*Dama* cf. *mesopotamica*	?*Cervus* sp. indet.		?*Cervus* sp. indet.	
Bovini gen. et sp. indet. (cf. *Bison*)					
Bos sp.	*Bison priscus*	*Bos* sp.		*Bos* sp.	
Bovidae gen. sp. indet.					
Gazella cf. *gazella*		*Gazella* cf. *gazella*	*G. gazella*		
Caprini indet.					
Equus sp.	*E.* cf. *caballus*	*Equus* sp.		*Equus* sp.	
Equus cf. *africanus*					
Palaeoloxodon antiquus	*Elephas trogontherii*	*E.* sp. *antiquus/namadicus*[a]			*Palaeoloxodon antiquus*
	Stegodon mediterraneus	+			+
	Dicerorhinus merckii	*Dicerorhinus hemitoechus*		*Dicerorhinus hemitoechus*	

\+ indicates presence
[a]Beden, pers. comm.

The good state of bone preservation contrasts with their high degree of fragmentation, resulting in a relatively small sample size of bones that may be attributed to a particular species, even when taking into account the relatively small excavation volume of Area C (Table 4.4). It is highly likely that the majority of the bones that lack diagnostic features necessary for taxonomic determination (see Chapter 3) originated from the most abundant species within a particular BSG in the GBY assemblage.

Table 4.4 Density of bones by excavated layer (number of bones/excavated volume)

	Excavated volume (m^3)	Total number of bones	Density of bones (bones/m^3)
Layer V-5	1.39	974	700
Layer V-6	2.25	1,868	830

Table 4.3 Composition of the faunal assemblages from Area C and the JB

	Total number of recorded bones	Bones identified to taxon	Bones attributed to a body-size group (BSG)	Unidentified mammal bones (UNM)
V-5	974	353	401	220
V-6	1,868	691	709	468
JB	1,111	460	409	242

4.3 Systematic Description

Presented in this section is the raw faunal data of the most abundant species recovered at the site, compared to other assemblages from southern Lower Paleolithic Levantine sites (Early to Middle Pleistocene) (Table 4.5).

Table 4.5 Eurasian sites mentioned in the text, their abbreviations, countries, location of collection, source of measurement, reference and sources of recent specimens of *Dama mesopotamica*. The abbreviation of the collections are as follows: the Hebrew University Collections, Jerusalem (HUJ); The Zoological Collections, Tel Aviv University (TAU); the Natural History Museum, London (NHM; Paleontology, Zoology); Muséum national d'Histoire Naturelle, Paris (MNHN; AC); The Museum of Natural History, University of Florence, Museum of Geology and Paleontology, Florence (MGPF); Centre Européen de Recherches Préhistorique de Tautavel (CERP); the Musée de Préhistoire Régionale de Menton, Menton, (MM), and the Institut für Quartärpaläontologie Weimar (Forschungsinstitut Senckenberg) (IQW)

Dating	Site name	Site abbreviations	Country	Collection abbreviations	Measurement reference	Site reference
30,000	Ksar Akil	KSAR	Lebanon	NHM	Original data	Kersten (1989)
	Hayonim D	HD	Israel	HUJ	Original data	Rabinovich (1998)
	Qafzeh UP	QAF UP	Israel	HUJ	Original data	Rabinovich and Tchernov (1995)
	Kebara C	KEB C	Israel	NHM	Original data	Bate (1937)
	Kebara D	KED	Israel	NHM	Original data	Bate (1937)
	el Wad F-D	EWF, EWE, EWD	Israel	NHM	Original data	Bate (1937)
90,000	Qafzeh MP	QAF MP	Israel	NHM	Original data	Rabinovich and Tchernov (1995)
	Tabun B	TB	Israel	NHM	Original data	Bate (1937), Garrard (1980)
	Tabun C	TC	Israel	NHM	Original data	Bate (1937), Garrard (1980)
	Tabun D	TD	Israel	NHM	Original data	Bate (1937), Garrard (1980)
	Tabun Ea	TEA	Israel	NHM	Original data	Bate (1937)
	Tabun Eb	TEB	Israel	NHM	Original data	Bate (1937), Garrard (1980)
	Tabun Ec	TEC	Israel	NHM	Original data	Bate (1937), Garrard (1980)
	Tabun Ed	TED	Israel	NHM	Original data	Bate (1937), Garrard (1980)
	Tabun E/F	TE	Israel	NHM	Original data	Bate (1937), Garrard (1980)
	Zuttiyeh	ZUT	Israel	NHM	Original data	Bate (1927)
	Qum Qatafa	OUM	Israel	HUJ	Original data	Vaufrey (1951)
	Bear's Cave	ALMA	Israel	HUJ	Original data	Tchernov and Tsoukala (1997)
300,000	Holon	HOL	Israel	HUJ	Lister (2007)	Horwitz and Monchot (2007)
	Atapuerca TG, TN	ATP	Spain		Made (1998, 1999, 2001)	
	Arago	ARG	France	CERP	Original data	Moigne and Barsky (1999)
	Swanscombe	SWA	Britain	NHM	Original data	Lister (1996b)
	Isernia	ISER	Italy		Abbazzi and Masini (1996)	
	Boxgrove	BOXG	Britain	NHM	Original data	Pitts and Roberts (1999)

Table 4.5 (continued)

Dating	Site name	Site abbreviations	Country	Collection abbreviations	Measurement reference	Site reference
500,000	West Runton	WR	Britain	NHM	Original data	Lister (1996b)
	Akhalkalaki	AKH	Georgia		Vekua (1962, 1986)	
	Atapuerca TD6	TD6	Spain		Made (1998, 1999, 2001)	
	Atapuerca TD8	TD8	Spain		Made (1998, 1999, 2001)	
1 Ma	Untermasfeld	UNT	Germany		Kahlke, H.D. (1997)	
	Vallonnet	VAL	France	MM	Original data	Moullé (1990, 1997–1998)
	Selvella	SEL	Italy		De Giuli (1986)	
1.4 Ma	'Ubeidiya	UBE	Israel	HUJ	Original data	Geraads (1986), Gaudzinski-Windheuser (2005)
	Dmanisi	DMA	Georgia		Vekua (1995: table 36)	Bukhsianidze and Vekua (2006)
Recent comparison						
	Dama mesopotamica (Israel)	Recent		TAU	Original data	
	Dama mesopotamica (Iran)	NHM		NHM	Original data	

We refer also to a recently published species list from 'Ubeidiya, which summarizes the various studies on the site (Gaudzinski-Windheuser 2005: table 2 and see references herein; Belmaker 2009; Martínez-Navarro et al. 2009). Although Table 4.5 also lists the Eurasian sites mentioned in our study, a broader taxonomic discussion of Early and Middle Pleistocene species is beyond the scope of the current study.

Carnivores

Few carnivore remains have been discovered at the site, with the majority assigned taxonomically to a species level. In addition, three unidentified carnivore fragments from Layer V-6 and two unidentified fragments from the JB were also found. Carnivores have not been identified in previous faunal studies of GBY.

Ursus sp.

Ursus is represented by two incomplete lower canines discovered in the JB (no. 2,481, l) and Layer V-5 (no. 2,769). The tooth crown of one of the JB specimen is broken on the labial face. The maximum length at the tooth base enamel line is 17.97 mm and the width is approximately 13 mm.

The wear observed on the labial surface of the GBY bear canine is typical of older age, as has been noted in paleo- and modern bear populations (Miles and Grigson 1990; Stiner et al. 1998). Bear remains from 'Ubeidiya are identified as *Ursus etruscus* (Ballesio 1986; Martínez-Navarro et al. 2009). Two other species have been defined from later Levantine sites, *U. deningeri* (e.g., the "Bear's Cave") and *U. arctos* (see summary in Tchernov and Tsoukala, 1997). *Ursus arctos* continued into the Holocene (Kurten 1965; Tchernov and Tsoukala 1997).

Due to the incomplete nature of the GBY teeth and sexual dimorphism of the genus, taxonomic identification to species level cannot be determined.

Vulpes sp.

Two tooth fragments from Layer V-5 (no. 7,581, lower M) and the JB (no. 14,688, l) have been assigned to *Vulpes*. *Vulpes* sp. was also found at 'Ubeidiya (Gaudzinski-Windheuser 2005), and was recently defined as *Vulpes* cf. *V. praeglacialis* (Martínez-Navarro et al. 2009). The taxon

survived throughout the Pleistocene, and is found in the present-day Levant (Mendelssohn and Yom-Tov 1999).

Palaeoloxodon antiquus

Elephant remains are very rare in all three assemblages (Table 4.6). Their identification as *Palaeoloxodon antiquus* is based on teeth, and was carried out in collaboration with Lister (see also: Goren-Inbar et al. 1994; Lister pers. comm. 2005). Tusk and tooth fragments comprise the bulk of the finds; Tables 4.7 and 4.8 list the fragments' measurements.

The 12 remains from Layer V-5 comprise 4 tooth fragments, 6 tusk fragments, a posterior ramus of a hyoid bone (no. 1,652; Fig. 4.1) (Shoshani et al. 2004; Shoshani et al. 2007), and a carpal (no. 2,072). One of the tusk fragments (no. 4,902) exhibits striations, possibly resulting from damage sustained during the animal's lifetime (e.g., Haynes 1991) (Table 4.7).

The 21 elephant bones found in Layer V-6 comprise 7 tooth fragments (Table 4.7), 11 tusk fragments, a cervical vertebra (no. 2,020), a highly exfoliated and eroded carpal (no. 9,279), and an astragalus (no. 9,274, r; Fig. 4.2). The astragalus is similar to a female size known from European assemblages (Kroll 1991). Only eight elephant fragments were found in the JB: an M_2 (no. 753, r; Fig. 4.3), two tooth fragments (Table 4.7), and three tusk fragments, a pisiform carpal (no. 1,199), and an intermedium carpal (no. 1,201, r) (Table 4.8).

The M_2 (r) was measured according to Maglio (1973). It consists of 11 visible, worn plates. Two additional plates have been lost by anterior wear; therefore, the estimated plates count is 13 (Fig. 4.3). The maximum length of the M_2 is 273.94 mm, the maximum width of Plate 6 is 81.22 mm, the height of the last plate is 135.79 mm, the enamel thickness is 2.5 mm, and the lamellar frequency is 5. The tooth indicates that the age of the elephant's death was at least 26 years. The tooth's morphology reveals great similarities to a tooth from GBY described by Hooijer (1959, GBY C 2), although he assigned it to *E. trogontherii*.

Table 4.6 *Palaeoloxodon antiquus* from Area C and the JB

Skeletal element	Layer V-5	Layer V-6	JB
Skull fragment	1		
Tusk fragment	6	11	3
Tooth fragment	4	7	3[a]
Vertebra		1	
Carpal	1	1	2
Tarsal		1	
Total	12	21	8

[a]Complete tooth

Table 4.7 Frequency and measurements (mm; maximum length, width and thickness) of *Palaeoloxodon antiquus* tusk (TUSKFH) and teeth (TFH) fragments from Area C and the JB

Layer	Skeletal element	Length	Width	Thickness
V-5	TFH (no. 5,520)	13.00	6.71	3.48
V-5	TFH (no. 7,881)	20.02	8.48	7.75
V-5	TFH (no. 14,109)	11.62	7.25	3.56
V-5	TUSKFH (no. 4,902)	41.45	20.18	4.90
V-5	TUSKFH (no. 4,903)	20.40	9.65	2.40
V-5	TUSKFH (no. 4,905)	13.81	4.27	3.20
V-5	TUSKFH (no. 9,272)	33.65	14.43	2.79
V-5	TUSKFH (no. 12,261)	16.09	12.63	2.66
V-5	TUSKFH (no. 12,267)	17.16	10.19	2.71
V-6	TFH (no. 13,795)	21.23	11.99	6.68
V-6	TFH (no. 1,205)	24.65	7.45	3.94
V-6	TFH (no. 12,289)	33.80	15.07	3.35
V-6	TFH (no. 2,965)	9.00	4.64	2.80
V-6	TFH (no. 14,003)	20.77	19.68	5.94
V-6	TFH (no. 14,017)	17.44	6.05	2.47
V-6	TFH (no. 2,071)	53.65	28.61	4.17
V-6	TUSKFH (no. 5,408)	24.38	12.78	8.98
V-6	TUSKFH (no. 9,343)	32.46	12.60	3.46
V-6	TUSKFH (no. 4,033)	18.62	11.52	4.18
V-6	TUSKFH (no. 4,901)	25.39	19.43	5.34
V-6	TUSKFH (no. 8,139)	22.04	81.39	3.44
V-6	TUSKFH (no. 5,179)	9.83	10.16	3.35
V-6	TUSKFH (no. 5,180)	16.46	7.49	1.78
V-6	TUSKFH (no. 5,181)	25.44	9.89	6.87
V-6	TUSKFH (no. 5,183)	13.01	4.34	2.16
V-6	TUSKFH (no. 5,184)	13.72	9.07	2.17
V-6	TUSKFH (no. 6,468)	21.23	9.67	3.44
JB	TFH (no. 4,267)	16.54	13.46	7.89
JB	TFH (no. 12,844)	21.88	10.43	4.83
JB	TUSKFH (no. 1,083)	26.77	23.88	2.81
JB	TUSKFH (no. 765)	69.58	39.09	3.63
JB	TUSKFH (no. 1,112)	20.93	9.89	3.04

Table 4.8 Measurements (mm) of *Palaeoloxodon antiquus* postcranial elements from Layer V-6 and the JB

Layer	Skeletal element	GL[a]	GB[a]	BT[b]	BC[b]
V-6	Astragalus (no. 9,274, r)	144	135	120.42	139.73
JB	Pisiform (no. 1,199)	40.39	35.62		
	Carpal (no. 1,201)	99.14	101.76		

[a]Measurements after von den Driesch (1976): GL—greatest length, GB—greatest breadth
[b]Measurements after Kroll (1991): BT—breadth of Trochlea tali, BC—breadth of Caput tali

Fig. 4.1 Elephant hyoid from Layer V-5 (no. 1,652)

Fig. 4.2 Elephant astragalus from Layer V-6 (no. 9,274, r)

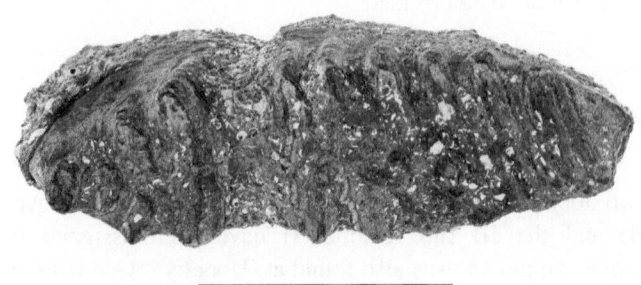

Fig. 4.3 Elephant M_2 from the JB (no. 735, r)

The site of GBY has the earliest occurrence of *Palaeoloxodon antiquus* in the southern Levant. *Palaeolodoxon antiquus* was also identified at Revadim Quarry (Marder et al. 1999; Rabinovich et al. 2005) and Holon (Davies and Lister 2007). In the southern Levant, proboscidean species are known from the Early and early-Middle Pleistocene (ca. 1.5 Ma–300 ka). This situation is very different from Eurasia where proboscideans continued throughout the Late Pleistocene (MIS 3 and perhaps later).

In addition to the finds from Area C and the JB, Area B also yielded elephant remains: a skull and other cranial elements attributed to *Palaeoloxodon antiquus* (Goren-Inbar et al. 1992b, 1994).

Hooijer identified two species at GBY: *Elephas trogontherii* and *Stegodon* (Hooijer 1959, 1960). While Tchernov and Shoshani (1996) and Lister (in Goren-Inbar et al. 1994; Lister 2004) redefined *E. trogontherii* as *P. antiquus*, identification of *Stegodon* was confirmed by

Tchernov et al. (1994). We thus conclude that the fossil fauna of GBY indicates the contemporaneous presence of two very large mammals.

The following proboscideans from the Levant have been taxonomically identified:

Mammuthus meridionalis from 'Ubeidiya (Beden 1986); *M. trogontherii* from Latamne in the Orontes Valley, Syria (Hooijer 1961; Guérin et al. 1993), where it was redefined as *M. trogontherii* or *M. trogontherii*-like (Lister 2004); *Palaeoloxodon antiquus* from Revadim Quarry (Marder et al. 1999) and Holon (Davies and Lister 2007); *Elephas* sp. from Evron Quarry (Tchernov et al. 1994), Oumm Zinat (Evron) (Horwitz and Tchernov 1989), and Tabun E/F (Bate 1937); *Stegodon* cf. *trigonocephalus* from Latamne (Guérin et al. 1993); and *Stegodon* sp. from Evron Quarry (Tchernov et al. 1994). *Stegodon* species are contemporaneous with *Mammuthus*, *Elephas* and *Palaeoloxodon*. *Stegodon*, *Mammuthus* and *Elephas* are represented in the above-mentioned sites only by teeth.

Hippopotamus amphibius

As early as 1959 Hooijer had defined *H. amphibius* at GBY (Hooijer 1959). The hippopotamus material in this study is composed only of tooth fragments: 4 from Layer V-5, 13 from Layer V-6, and 5 from the JB. The teeth from Layer V-5 include a canine tip (no. 2,574), a fragmented tip of a deciduous incisor (no. 3,132), an incisor fragment (no. 6,907), and an unidentified tooth fragment (no. 6,257). The teeth from Layer V-6 include part of a lower canine (no. 13,071, r; L = 121.68 mm, W = 33.83 mm) and an upper first deciduous premolar (no. 2,013, l) (Table 4.9). The rest of the sample consists of small eroded pieces of canines and molars (nos. 13,073, 13,106, 13,107, 13,071, 4,047, 13,074, 5,197, 9,271, 12,358, 6,752, 14,053).

From the JB, the tip of a lower canine (no. 759; L = 28.48 mm, W = 23.00 mm, maximum L = 140.24 mm, maximum W = 37.56 mm) and a molar fragment (no. 1,032) were found. The length and width of the former were

measured from the tip to the base and from the upper part of the tip, respectively. These two teeth belonged to young animals. Eroded teeth fragments form the rest of the sample (nos. 12,974, 775, 13,201).

It should be noted that the presence of deciduous teeth indicates young animals in all the layers. We were able to compare the upper first deciduous premolar (no. 2,013; Layer V-6) to modern specimens of *H. amphibius*. The size of each tooth falls within the range of variation of extant species (Table 4.9).

'Ubeidiya is the only site in the southern Levant where specimens of this genus were very abundant, and where two species have been identified: *Hippopotamus behemoth* and *Hippopotamus gorgops* (see summary in Gaudzinski-Windheuser 2005). *H. behemoth* has also been identified at Latamne (Guérin et al. 1993). Hippopotamus occurs only sporadically at Holon (Horwitz and Monchot 2007), Evron Quarry (Tchernov et al. 1994), and Oumm Zinat (Horwitz and Tchernov 1989).

Sus scrofa

Two phalanges (nos. 1,417, 1,750, third phalanx: DLS = 39.97 mm, Ld = 38.15 mm, GB = 24.24 mm) and an unidentified tooth fragment (no. 7,857) were discovered in Layer V-5. Three phalange fragments (no. 1,690; 86, r; no. 7,980, unfused second central phalanx), part of a femur (no. 1,653), and part of a proximal radius stem (no. 1,587, l) were found in Layer V-6. The JB yielded five skeletal elements, including an upper incisor (no. 13,108, I^3, l) whose measurements were taken near the base of the crown (L = 7.55 mm, W = 4.26 mm), two tooth fragments (nos. 2,550, 7,387), a carpal (no. 1,298, l), and a femoral shaft (no. 1,049).

Pleistocene specimens from the northern Jordan Valley reflect the large body size of the *Sus scrofa* in this region (comparable to the Natufian *Sus scrofa* of Mallaha (Eynan); see Bouchud 1987, table XLII). This observation is in accordance with GBY, as the large size of the third phalanx (no. 1,750) from Layer V-5 also suggests a large body size. The unfused lateral phalanx indicates an animal approximately 12 months old (Silver 1969).

Southern Levantine prehistoric fauna is not rich in pigs, but the species is nowadays extant (Mendelssohn and Yom-Tov 1999). Hooijer (1959) identified remains of *Sus* cf. *scrofa* in an unstratified deposit at Jisr Banat Yaqub, later confirmed by Geraads and Tchernov (1983). African pigs have been found at two sites, *Kolpochoerus olduvaiensis* at 'Ubeidiya (Geraads et al. 1986) and *K. evronensis* at Evron Quarry (Haas 1970; Geraads et al. 1986). At 'Ubeidiya, *Sus strozzi* (Geraads et al. 1986) was also found, recently assigned to *Sus* sp. (Gaudzinski-Windheuser 2005).

Table 4.9 Measurements (mm) of deciduous teeth of *Hippopotamus* from GBY and of extant *H. amphibius* (NHM, London)

Layer	Skeletal element	Base length	Base width	Height
V-6	dP¹ (no. 2,013, r)	15.69	11.15	21.59
NHM 1851.12.23.4	dP¹ (l)	13.35	9.35	15.05
NHM 84.524	dP¹ (l)	15.88	12.32	17.80
	dP¹ (r)	15.39	14.39	18.76

Sus scrofa was found at Revadim Quarry (Marder et al. 1999; Rabinovich unpublished) and *Sus* cf. *scrofa* was found at Holon (Horwitz and Monchot 2007).

Equus cf. africanus, Equus sp.

The equid material from GBY is composed of both teeth and postcranial elements. Species identification was undertaken by Eisenmann (n.d. 2006), who has concluded that part of the remains belong to a primitive form of ass, *Equus* cf. *africanus*. This species reached the size of the extant *E. africanus*. Its teeth were unlike those of a modern ass or hemione (*E. hydruntinus*), but resemble those of a modern zebra. Eisenmann also observes that the premolars have a very developed *pli caballin*, a feature rarely found in asses and hemiones (Eisenmann n.d.). Altogether, 16 specimens were found in Layer V-5, 20 in Layer V-6, and 31 in the JB.

Twelve specimens from Layer V-5 belong to *Equus* sp., comprising a molar fragment (no. 5,834), two incisors (nos. 2,022, 2,034), a distal longitudinally broken scapula (no. 1,620, l), part of a pubis (no. 1,626, l), an ilium shaft (no. 1,457, r), a residual metacarpal (no. 1,624. r), three residual metatarsals (nos. 2,035; 2,036; 2,004, r), a sesamoid (no. 1,673), and a tibia shaft (no. 5,887). Four additional specimens were assigned to *E.* cf. *africanus*: P^3 or P^4 (no. 2,002), a magnum (no. 2,037), a cuboid (no. 2,014), and a second phalanx (no. 2,011) (Table 4.10).

The *Equus* sp. remains from Layer V-6 (N = 14) are comprised of one mandible (no. 2,033, r), three lower teeth (nos. 2,007; 2,012, l; 2,003, l), and two tooth fragments (nos. 8,402, 10,150). The postcranial elements consist of a scapula fragment (no. 1,526), a radius (no. 1,610, l), two pelvis fragments (nos. 1,470; 1,469, l), a metacarpal (no. 1,561, l), a metatarsal (no. 1,539, l), a residual metatarsal (no. 2,010), and a sesamoid (no. 2,006). Six elements belong to *Equus*

cf. *africanus*: upper and lower teeth (nos. 2,008, l; 2,018, r; 2,019, r; 2,020, l), a humerus (no. 2,023, r), and a metacarpal (no. 2,017, r) (Table 4.10; Fig. 4.4).

Thirty-one specimens from the JB are assigned to equids. *Equus* cf. *africanus* is represented by 11 elements of the postcranial skeleton and teeth: 3 upper teeth (nos. 1,177, P^3, l; 1,281, P^3, l; 761, M^1, l), 3 lower teeth (nos. 11,785, M_1, r; 2,031, $M_{1/2}$, r; 918, $P_{3/4}$, r), a first phalanx (no. 1,202, l) (Table 4.10), a pisiform (no. 2,026, r), a semi-lunar (no. 1,282, r), a scaphoid (no. 1,280, l), and a proximal ulna (no. 13,093). The rest of the equid remains comprise teeth (nos. 2,027, P^1, r; 834; 1,216; 2,038, lower incisor, r; 2,098, dP_3, r; 2,030, dP_4, r; 2,029, $M_{1/2}$, l; 2,000, P_2, r; 2,001, lower unidentified M, l; 846, a tooth fragment), a mandibular symphysis (no. 2,032, l), a radius shaft (no. 1,267), 2 distal femur shafts (nos. 1,318; 1,321, l), a proximal femur shaft (no. 226, r), a proximal metatarsal (no. 1,319, r), a pelvis fragment (no. 905, l) (Table 4.10), phalange fragments (nos. 1,764, 12,973), and a sesamoid (no. 1,073).

Several remains from Layer V-6 belonged to animals less than one year old. In fact, these bones, including a mandible (no. 2,033, dP_2–dP_4, r), a deciduous incisor (no. 2,007), a

a

b

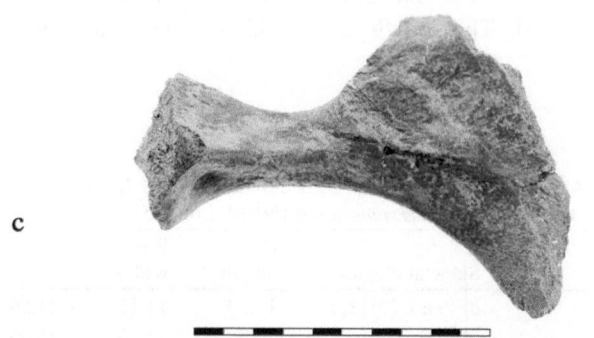

c

Fig. 4.4 a Lateral view of *Equus* sp. (foal) mandible from Layer V-5 (no. 2,033, r); **b** occlusal view of the same specimen; **c** pelvis fragment from Layer V-6 (no. 1,470)

Table 4.10 Measurements (mm) of *Equus* cf. *africanus* and *Equus* sp. postcranial elements from Area C and the JB

Layer	Skeletal element	GL[a]	PW[b]	PD[b]	DW[b]	DD[b]
V-5	Ph 2 (no. 2,011, l)	42.96	42.62	30.59	37.59	27.32
	Magnum (no. 2,037, r)	35.16	37.89			
	Cuboid (no. 2,014, l)	34.84	26.78			
V-6	HumD (no. 2,023, r)				67	70.7
JB	Ph 1 (no. 1,202, l)	78.01	45	34	36	
		LA[a]	LAR[a]	SH[a]		
	Pelvis (no. 905)	63.63	57.09	34.28		

[a]Measurements after von den Driesch (1976): GL—greatest length, LA—length of the acetabulum, LAR—length of the acetabulum on the rim, SH—smallest height of the ilium shaft
[b]Measurements after Eisenmann (n.d.): PW—proximal width, PD—proximal depth, DW—distal width, DD—distal depth

Table 4.11 Measurements (mm) of GBY *Equus* sp. and extant *Equus burchelli antiquarum* (1914.8.20.1, 1928.9.11.421, NHM, London)

Skeletal element[a]	1914.8.20.1		1928.9.11.421		GBY 2033
Mandible					
	L	R	L	R	R
Length (1)	247	245	240	240	
Length of diastema (15)	39	35	41	40	39
Symphysis width (22c)	25	25	24	25	32
Height at dP_4 (22a)	45	44	42	41	48
Premolar row (dP_2–dP_4) (6a)	99	99	90	92	93
Radius					GBY 1610
	L	R	L	R	L
Greatest length without epiphysis (GL)	190	188			220
Greatest breadth of proximal end (BP)	57	57			55
Greatest depth of proximal end (BPW)	29	29			33
Smallest breadth of diaphysis (SD)	24	25			25
Pelvis					GBY 1470
	L	R	L	R	L
Acetabulum length (LA), unfused	43	45			41
Acetabulum width (GBA), unfused	42	44			44
Ilium length	95	113			135
Ilium width	95	100			100.9
Smallest height of ilium shaft (SH)	23	24			25

[a]Measurements after von den Driesch (1976) displayed in parenthesis by their respective code or number according to body part

radius (no. 1,610, l), and a pelvis (no. 1,470), were probably part of the same animal. The bones' morphology and size were compared to modern newborn zebras (*Equus burchelli antiquarum*) (Table 4.11), suggesting the possible attribution of these specimens to an *Equus* cf. *africanus* foal.

From the JB, two lower premolars (nos. 2,098, 2,030) and a mandibular symphysis were assigned to a foal. This is based on a comparison with no. 2,033 from Layer V-6, and the fact that the fragmented mandibular symphysis is small and porous.

Hooijer (1959) identified a few teeth from Jisr Banat Yaqub (GBY) as *Equus* cf. *caballus*. Equid remains are quite scarce in the southern Levant. At 'Ubeidiya, *E.* cf. *altidens* and *E.* cf. *caballus* were identified (Eisenmann 1986). At Latamne, Hooijer (1961) identified a few teeth as *E.* sp., while in a more recent study of the site's fauna, *E.* cf. *altidens* was also uncovered (Guérin et al. 1993). *E. hydruntinus* was identified at the late Acheulian site of Oumm Zinat (Horwitz and Tchernov 1989). Thus, according to Eisenmann (1992), *E. caballus*, *E.* cf. *tabeti*, and *E. hydruntinus* (small ass) prevailed in the Middle Pleistocene Levant.

Bovidae

Several bovid species have been found in Area C and the JB. Among them are *Bos* sp. and *Gazella* cf. *gazella*. Elements that could not be assigned to a particular species were included in the Bovini gen. et sp. indet. (cf. *Bison* sp.) and Bovidae gen. et sp. indet. group. The latter is comprised mainly of poorly preserved and highly fragmented horn cores. However, the horn core fragments indicate the existence of a rich bovid community in the site's vicinity.

Species identification was undertaken in collaboration with Martínez-Navarro (Martínez-Navarro et al. 2000; Martínez-Navarro 2004; Martínez-Navarro and Rabinovich 2011) and in addition a small bovid assigned to Caprini sp. was defined recently.

Bos sp.

Specimens attributed to *Bos* sp. were found at the site. Their taxonomic determination is based on the enamel thickness of their premolars and molars (Martínez-Navarro and Rabinovich 2011). Two postcranial elements were discovered in Layer V-5: a metatarsal (no. 1,836, l) and a navicular cuboid (no. 1,829, l). Layer V-6 yielded the following seven remains: horn core pneumatic region (no. 13,366), teeth (nos. 10,160, P^2, l; 2,162, dP^4 l; 9,729, P_2, r; 13,019 and 13,020, lower M fragments), and a tarsal bone (no. 2,005, cuneiform, r). From the JB, 16 bones are attributed to this species, including a horn core fragment (no. 545), an upper tooth (no. 2,160, M^3, l), 3 lower molars (nos. 1,283, M_1, r; 2,090, M_3, l; 2,161, M_2, r), a lower second incisor (no. 897, l), and a mandibular fragment (no. 2,157, l). Also found were several postcranial elements: a vertebra fragment (no. 13,024), a proximal ulna (no. 225, l), a proximal metacarpal (no. 919), a tibia shaft (no. 910, l), two metatarsal fragments (nos. 2,080, l; 544, r), a calcaneus (no. 896, r); a tarsal (no. 2,016, l), and a second phalanx (no. 884) (Table 4.12).

Initial species identification by Hooijer (1959) was based on the identification of a horn core and teeth from GBY

Table 4.12 Measurements (mm) of (**a**) *Bos* sp. teeth from Layer V-6 and the JB (length and width measured from base of crown, crown height measured from buccal/lingual mid-point)

Layer	Skeletal element	Base length	Base width	CH1[a]	CH2[a]
V-6	dP^4 (no. 2,162, l)	22.50	21.50	22.04	21.42
	P^2 (no. 10,160, l)	18.40	14.67	10.11	
JB	I_2 (no. 897, l)	16.07	11.00	23.46	
	M_2 (no. 2,161, r)	26.35	18.81	48.49	45.43
	M_1 (no. 1,283, r)	26.64	17.46	41.47	42.78
	M_3 (no. 2,090, l)	44.34	19.57	40.55	33.52

[a]CH1—crown height of first lobe; CH2—crown height of second lobe

Table 4.12 (**b**) Measurements (mm) of *Bos* sp. postcranial elements from Layer V-6 and the JB

Layer	Skeletal element	GL[a]	BP[a]	BPW[a]	BD[a]	BFD[a]
V-6	Cuneiform (no. 2,005, r)	52.95	32.46			
JB	Ph 2 (no. 884)	50.48	36.00	36.90	31.41	25.73
	Calcaneum dist. (no. 896, r)				71.38	
	Maleolus (no. 2,016, l)	46.07	32.10			
	Metatars. dist (no. 2,080)				30.06	38.65

[a]Measurements after von den Driesch (1976): GL—greatest length, BP—greatest breadth of the proximal end, BPW—greatest depth of the proximal end, BD—greatest breadth of the distal end, BFD—greatest breadth of the Facies articularis distalis

as cf. *Bison priscus*. The specimens were later redefined as *Bos* sp. (Geraads and Tchernov 1983) (Table 4.2). Martínez-Navarro identified the specimens as *Pelorovis* (2004), but later redefined them as *Bos* (Martínez-Navarro et al. 2007; Martínez-Navarro and Rabinovich 2011) on the basis of anatomical landmarks on the skull as well as biogeographic considerations.

Numerous "large Bovid species"—*Pelorovis oldowayensis* and *Pelorovis* sp. (cf. *P. turkanensis*)—have been found at 'Ubeidiya (summarized in Gaudzinski-Windheuser 2005). At Latamne, Hooijer (1961) initially identified *Bison* cf. *priscus*; later, Guérin and others (1993) added the presence of *Bos primigenius*. *Bos primigenius* was discovered in several other Lower Paleolithic sites, such as Evron Quarry (Tchernov et al. 1994), Oumm Zinat (Horwitz and Tchernov 1989), and Holon (Horwitz and Monchot 2007). A large *Bos* specimen was found at Revadim Quarry, identified mainly by the thick enamel of the tooth fragments (Rabinovich, pers. com.).

Gazella cf. gazella

Gazelles are the most abundant species in the Late Pleistocene (Upper Epipaleolithic) record of Israel (Tchernov 1988; Rabinovich 1998), where nowadays three subspecies exist (Mendelssohn and Yom-Tov 1999). Thus, their presence in the early context of GBY is very important for our understanding of their existence in the local fauna even today.

Six incomplete gazelle bones were discovered in Layer V-5: a tooth fragment (no. 14,368), a distal radius (no. 1,659), a distal ulna (no. 4,764), a proximal part of a metacarpal (no. 6,374, r), a sesamoid bone (no. 1,430), and a second phalanx (no. 14,715). Eight fragments were found in Layer V-6: a horn core fragment of a young gazelle (no. 4,366), a hyoid

fragment (no. 6,492), mandible fragments (nos. 15,036, l; 12,514, r), a tooth fragment (no. 847), a scapula blade fragment (no. 13,086, l), a femoral shaft (no. 2,095, r), and a fragmented astragalus (no. 2,101).

Twelve gazelle bones were found in the JB: three tooth fragments (nos. 1,842, M^2, l; 899, M_2, l; 1,118, M_2, r), a mandibular fragment (no. 13,084, l), two atlas fragments (nos. 1,016, 1,033) probably from the same specimen, a proximal radius (no. 12,869, l), a pelvis fragment (no. 4679), a proximal metatarsal (no. 13,152), a distally unfused metapodial (no. 11,904), a complete first phalanx (no. 4,207; GL = 41.64 mm, BP = 14.14 mm, BP = 11.36 mm, BD = 9.24 mm, BFD = 6.87 mm), and a proximal portion of a second phalanx (no. 915; BP = 10.68 mm, BW = 13.87 mm).

Hooijer did not identify *Gazella* at GBY; this species was first identified in 1983 by Geraads and Tchernov, and was further described by Geraads (1986). In the recent study of the GBY "BAR" assemblage, remains of a skull with a horn core and an astragalus were assigned to *Gazella gazella* (Goren-Inbar et al. 1992b). Furthermore, though rare, *Gazella* was found in Early to Middle Pleistocene faunal assemblages at 'Ubeidiya (*Gazella* cf. *gazella*; Geraads 1986; later attributed to *Gazella* sp. by Gaudzinski-Windheuser 2005), Evron Quarry (*Gazella* cf. *gazella*; Tchernov et al. 1994), Revadim (Marder et al. 1999), and *G. gazella* at Holon (Horwitz and Monchot 2007). At Latamne, Hooijer (1965) attributed a tooth to *G. soemmeringi* after noting its large size.

Bovini gen. et sp. indet. (cf. Bison)

A few skeletal elements that were too fragmented in order to be attributed to a specific genus, but comprise the typical anatomy of Bovini (Martínez-Navarro and Rabinovich 2011), have been assigned to this group. These include a horn core fragment (no. 12, 637), a tooth fragment (no. 12,977), and a proximal metatarsal (no. 1,815, l), all from Layer V-6. A horn core fragment (no. 786), a P_2 (no. 11,819), a metatarsal fragment (no. 1164, l) and third phalanx (no. 844) were identified in the JB.

At Latamne, Hooijer (1961) identified *Bison* cf. *priscus* and at GBY he has identified a cf. *Bison priscus* horn-core (Hooijer 1959).

Bovidae gen. et sp. indet.

Numerous bovid horn core fragments were found in Area C (Layer V-5: nos. 2,869, 12,949; Layer V-6: nos. 1,710, 2,189,

4,668, 7,038, 10,098, 12,394, 12,730, 12,960, 12,982, the JB: nos. 774, 1,107, 2,203, 4,648, 13,841). These fragments lack any identifiable mark that would allow for species definition, but their size and thickness have enabled us to assign them to medium- and large-sized bovids.

Caprini indet.

Only scarce evidence exists for the presence of Caprini indet. at GBY, in the form of a distal scapula from Layer V-6 (no. 1,627, l; GLP = 40.81 mm, LG = 30.35 mm, BG = 25.81 mm), and a proximal ulna (no. 15,072, r) from the JB. In both shape and size, the distal scapula resembles wild goat known from Late Pleistocene sites (Rabinovich 1998). Unfortunately, the GBY fragments have not retained sufficient morphological traits to allow for exact species identification (Crégut-Bonnoure 2006). In previous faunal studies undertaken at the site, *Capra* sp. was reported from the Acheulian assemblage of the "BAR" (Goren-Inbar et al. 1992b). The presence of *Capra* sp. is also known from Tabun E/F (Bate 1937).

Early occurrence of *Capra* has been identified in Early Pleistocene levels at Dmanisi (*Capra dalii* nov. sp.; Bukhsianidze and Vekua, 2006) and in the latest Early Pleistocene levels at Akhalkalaki (*Capra* sp.; Vekua, 1986). Thus, *Capra* is present in Eurasia already during the Early Pleistocene, and in light of this evidence, the remains from GBY may represent a later distribution of this genus.

Cervidae

Three cervid species were identified at the site: *Megaloceros* sp., *Cervus* cf. *elaphus*, and *Dama* sp. The latter is the most prevalent of the three, thus providing a better base for detailed systematic and taphonomic research. This group is being studied in collaboration with Lister in order to better understand the origin of Cervidae in general and of *Dama* in particular in the region (Rabinovich and Lister in preparation).

The earliest identification of cervids at GBY was carried out by Hooijer (1959, 1961), who studied material from early excavations and surface collections of the site. The remains from the site comprised several postcranial elements defined as *Dama* cf. *mesopotamica* and *Cervus* cf. *elaphus*. Later, Geraads and Tchernov (1983) accepted Hooijer's identification of *Cervus* cf. *elaphus* but revised the identification of the Mesopotamian fallow deer to *Dama* sp. Geraads and Tchernov noted the complex characteristics and the unresolved origin of the medium-sized Pleistocene cervids.

In 1966 at 'Ubeidiya, Haas identified the following four species: *Cervus* cf. *ramosus*, *Dama* spec. and/or *C.* cf. *philisii*?, *C.* (*Euctenoceros*) cf. *senezensis*, and *Megaceros* sp. Geraads (1986) identified there only two species of cervid, the large *Praemegaceros verticornis* and the smaller Cervidae gen. et sp. indet. Within the latter, Pfeiffer (1999) identified the *Axis* sp. Pirro-Nord? and *Dama nestii*?.

The Lower Paleolithic site of Evron Quarry yielded a few cervid remains that were recently identified as cf. *Cervus*? *elaphus* and cf. *Capreolus* sp. (Tchernov et al., 1994), as well as the above-mentioned *Dama* sp. (Tchernov, 1988). Recently *Dama dama* cf. *mesopotamica* was identified by Lister (2007) at Holon.

Even though the faunal assemblages of Latamne (Guérin et al. 1993) and Evron Quarry exhibit similarities, the large cervid *Praemegaceros verticornis* (Geraads, 1986; Guérin et al., 1993) (referred to by Hooijer (1961) as *Orthogonoceros verticornis*) is known only from Latamne, where it is the only cervid present in the assemblage.

Megaloceros sp.

Three bones from the site have been assigned to the large deer *Megaloceros* sp. A metacarpal shaft (no. 1,490) and a proximal part of a metararsal (no. 1,669) were found in Layer V-6, and a metatarsal was found in the JB (no. 2,228, r). Although these bones are too fragmentary to be measured, more complete specimens collected from GBY's surface area point to this animal's small size, comparable to *Praemegaceros verticornis* from 'Ubeidiya (Geraads 1986).

The absence of a Megalocerine from GBY has been used in the past to illustrate differences between the older faunas from 'Ubeidiya and Latamne and the more recent GBY fauna. The taxonomic determination of Megalocerine remains at GBY, whose similar size to that of 'Ubeidiya may point to the continuity of this species in the Levant.

Cervus cf. elaphus

Red deer remains from the site include two teeth from Layer V-5 (nos. 1,591, r; 1,743, l), a tooth from Layer V-6 (no. 1,482, dP^4, r), and an astragalus (no. 1,178, l) from the JB (Table 4.13).

Cervus remains from southern Levantine sites are rare, and its measurements vary only slightly (Rabinovich and Lister in preparation). Thus, on the basis of the available data, early and late acoronate and coronate red deer are indistinguishable from each other. Chronological considerations favor *C. e. acoronatus* as the probable candidate at

Table 4.13 Measurements (mm) of *Cervus* cf. *elaphus* teeth and skeletal elements (length and width of teeth measured from base of crown, crown height measured from bucca/linguall mid-point)

Layer	Skeletal element	Base length	Base width	CH1[a]	CH2[a]
V-5	LM[b] (no. 1,591, r)				11.42
V-5	M^1 (no. 1,743, l)			8.02	8.73
V-6	dP^4 (no. 1,482, r)			7.93	8.18
JB	P^2 (no. 1,111, r)	13.81	15.56	11.84	
		GL[c]	BD[c]	D1[c]	
JB	Astragalus (no. 11,788, l)	57.79	34.84	30.58	

[a]CH1—crown height of first lobe, CH2—crown height of second lobe
[b]LM—lower molar
[c]Measurements after von den Driesch (1976): GL—greatest length, BD—greatest breadth of the distal end, D1—greatest depth of the lateral half

GBY, although due to the absence of antlers, this proposal remains tentative.

Dama sp.

The most common species within the GBY cervid assemblage is *Dama*. The frequency of *Dama* skeletal elements is 285 in Layer V-5, 545 in Layer V-6, and 355 in the JB (Table 4.1). Since many of the bones are broken, the sample available for detailed species identification is small when compared to the overall frequency of cervid bones (Table 4.14). Included in this study are only those *Dama* elements from Area C and the JB, whereas the original species identification, undertaken by Rabinovich and Lister (in preparation), was based on all of the cervid material from GBY.

Table 4.14 *Dama* sp. skeletal-element frequencies in Area C and the JB based on NISP

Skeletal element	V-5 NISP	V-6 NISP	JB NISP
Antler	1	1	3
Antler fragment		1	1
Skull fragment	9	36	11
Maxilla fragment		2	1
Upper tooth	23	20	15
Mandible		1	1
Mandible fragment		24	15
Lower tooth	37	38	13
Tooth fragment	32	26	7
Atlas	1	5	2
Axis		2	2
Cervical vertebra	4	6	3
Thoracic vertebra	1	3	5
Lumbar vertebra	2	4	4
Vertebra fragment	21	33	26
Vertebral spine	2	10	10

Table 4.14 (continued)

Skeletal element	V-5 NISP	V-6 NISP	JB NISP
Sacrum		2	1
Rib	3	2	
Rib, proximal	20	7	3
Rib shaft		39	18
Scapula, proximal		1	
Scapula blade		3	1
Scapula, distal	2	5	3
Scapula fragment	2	7	5
Humerus, proximal		1	1
Humerus shaft	6	11	6
Humerus, distal	1	5	7
Radius, proximal	3	3	6
Radius shaft	7	26	17
Radius, distal	2	1	5
Ulna, proximal	2	2	5
Ulna shaft	3	1	2
Ulna, distal	1	1	1
Radius Ulna, distal		2	1
Carpal	2	7	
Metacarpal, proximal		8	1
Metacarpal shaft	4	17	12
Metacarpal, distal	3	1	2
Acetabulum	1	3	3
Ilium	2	8	3
Pubis	3	2	2
Ishium	2	1	
Pelvis fragment	1	3	5
Femur, complete			1
Femur, proximal	1		3
Femur shaft	8	17	17
Femur, distal		4	3
Tibia proximal			1
Tibia shaft	10	29	29
Tibia, distal	5	3	1
Astragalus	1	3	4
Calcaneum	1	3	2
Tarsal	3	6	8
Sesamoid	2	6	4
Metatarsal, proximal	8	14	13
Metatarsal shaft	11	21	20
Metatarsal, distal	1	5	2
Phalanx 1	1	1	1
Phalanx 1, proximal	3	11	5
Phalanx 1 shaft	2	2	1
Phalanx 1, distal	2	7	
Phalanx 2	1	2	
Phalanx 2, proximal	1	3	1
Phalanx 2 shaft		2	
Phalanx 2, distal	2	6	2
Phalanx 3	4	5	6
Phalanx 3, proximal	1	1	
Phalanx 3 shaft		1	
Phalanx 3, distal		1	
Metapodial shaft	5	5	2
Metapodial, distal		4	
Total	285	545	355

Identification of *Dama* at GBY is based on the morphological characteristics of antlers, postcranial elements, and body size (Rabinovich and Lister in preparation). Antlers from the site include shed and unshed specimen fragments and numerous tine fragments; none retained the upper part of the beam, which is essential for species identification. The following diagnostic antler elements were recovered; for their detailed description, see Rabinovich and Lister (in preparation): Layer V-5, base and pedicle (no. 1,625, l); Layer V-6, shed antler (no. 1,608, l), base and frontal: (nos. 1,459, r; 1,609, r); and the JB, shed alters (no. 754, r), palmation (no. 904, r), base and frontal (no. 219, l).

The *Dama* antlers from GBY (Fig. 4.5) are characterized by the absence of thick beams, a variable (probably short) brow tine, and some sort of palmation. Thus far, it appears that none of the GBY antlers exhibit typical characteristics of the *Dama dama mesopotamica* (Rabinovich and Lister in preparation), such as a developed second tine and flatness of the beam above the tine bur. Unfortunately, these two particular characteristics, among many others, were not preserved in the GBY specimens.

The length and width ratio of the burr base indicates that, compared to other Levantine sites, the GBY antlers were small- and medium-sized (Fig. 4.6).

Tooth size and their other descriptive characteristics are considered to be good species indicators (Fig. 4.7). The following are complete *Dama* specimens from the site:

Deciduous upper premolars: Layer V-6: nos. 1,465, P^2, r; 1,483, P^4, l; 1,713, P^4, l; the JB: no. 1,782, P^4, r.

Upper premolars: Layer V-5: nos. 2,667, P^3, l; 1,704, P^4, r; 1,805, P^4, l; 5,637, P^4, r; 5,807, P^4, l; Layer V-6: no. 1,560, P^3, r; 1,712, P^3, l; 1,728, P^4, r; 1,495, r, P^4–M^1.

Upper molars: Layer V-5: nos. 1,711, M^1, l; 1,735, M^3, l; 1,832, M^3, l; 2,103, M^3, l; Layer V-6: nos. 1,463, M^1, r; no. 1,602, M^1, r; 1,498, M^2, r; 1,566, M^3, r; 1,599, M^3, l; 1,699, M^3, l; 2,105, M^3, r; the JB: nos. 1,272, M^1, l; 2,164, M^1, l; 11,786, M^1, l; 2,086, M^2, r; 2,170, M^2, l; 2,173, M^2, l; 781, M^3, r; 2,168, M^3, l.

Deciduous lower premolars: Layer V-5: nos. 11,573, dP_2, r; 1,635, dP_4–M_3, l.

Lower premolars: Layer V-5: nos. 1,749, P_3, r; 2,104, P_4, l; Layer V-6: nos. 1,573, P_3, l; 1,582, P_3, r; the JB: nos. 2,151, P_2–P_4, l; 2,172, P_3, r; 898, P_3–P_4, l.

Lower molars: Layer V-5: nos. 1,462, M_2, r; 2,106, M_2, r; 2,107, M_3, r; 1,670, M_3, l; Layer V-6: nos. 1,540, M_1, r; 1,584, M_1, l; 1,647, M_2, l; 1,771, M_2, r; 1,488, M_3, l; the JB: nos. 901, M_1, r; 2,181, M_3, l.

Tables 4.15 and 4.16 present the measurements of the teeth listed above.

When the length of M^3 was plotted against its width, it was revealed to be particularly large, with a single exception similar to a tooth from Holon and from Hayonim D defined as *Dama dama* cf. *mesopotamica* (Lister 2007) (Table 4.5;

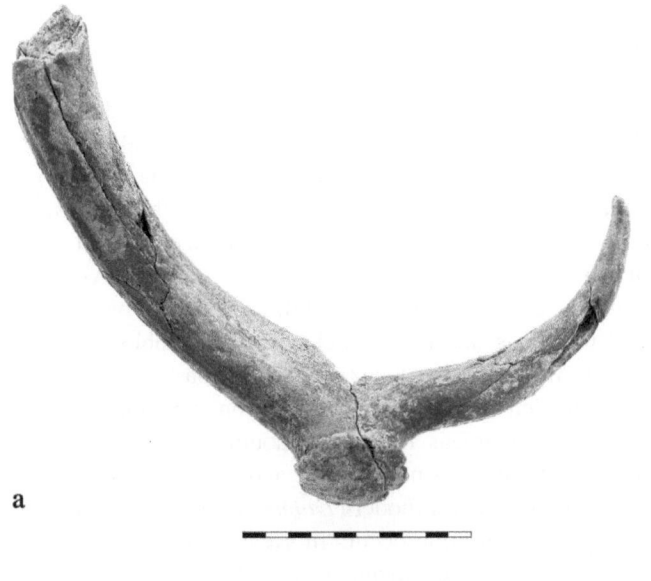

a

b

Fig. 4.5 a Shed *Dama* antler from the JB (no. 754, r) and **b** *Dama* antler fragment from the JB (no. 1,608, r)

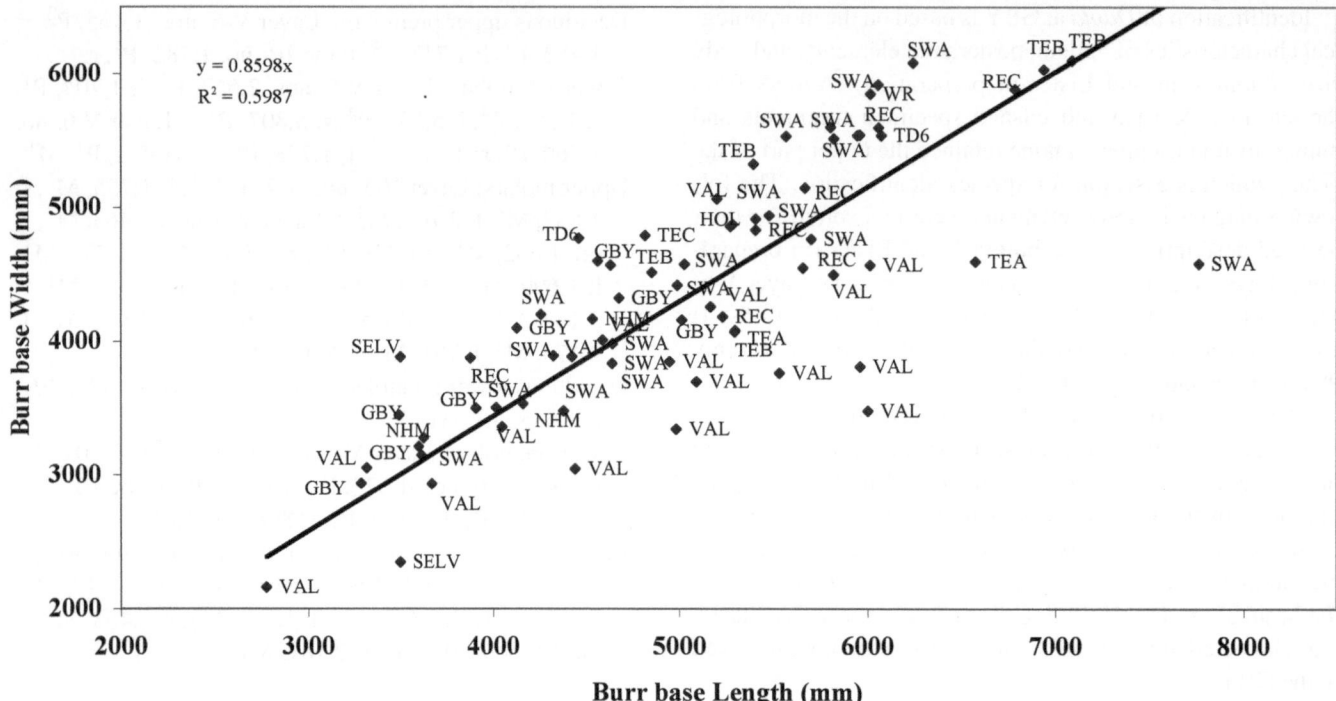

$y = 0.8598x$
$R^2 = 0.5987$

Fig. 4.6 Plot of *Dama* sp. antler size (burr base width and length) from the following sites: VAL—Vallonnet, SELV—Selvella, WR—West Runton, SWA—Swanscombe, TD6—Atapuerca TD6, Hol—Holon, TD—Tabun D, TEA—Tabun Ea, TEC—Tabun Ec, REC—*Dama mesopotamica* in Israel, NHM—Recent *Dama mesopotamica,* NHM, London. See Table 4.5 for details on each site

Fig. 4.8). This tendency towards relatively large teeth is also seen in the other upper teeth from GBY.

In addition to cranial elements, the following postcranial specimens were also used for a detailed taxonomic determination of *Dama*:

Scapula: Layer V-5: nos. 1,679, r; 2,082, r; Layer V-6: no. 1,663, l; the JB: no. 1,227, r.

Humerus: The JB: nos. 933, r; 539, l.

Metacarpals (distal parts): Layer V-5: nos. 1,531, 1,825, 2,099; Layer V-6: no. 1,592; the JB: nos. 943, 1,269.

Tibia: Layer V-6: no. 1,765, l.

Astragalus: Layer V-5: no. 5,829, r; Layer V-6: nos. 1,660, r; 1,661, l; the JB: nos. 829, r; 914, l; 2,079, r.

Calcaneum: The JB: nos. 112, r; 1,255, r.

Cubo-navicular: Layer V-6: no. 1,485, l; the JB: no. 2,084, r.

Metatarsal: Layer V-5: no. 1,644, r; Layer V-6: no. 1,665.

Phalanges: Layer V-5: nos. 1,678, 1,777; Layer V-6: nos. 1,662, 1,579, 1,666.

The postcranial elements, such as the scapula, distal metapodials, astragalus, cubo-navicular, and first phalanx, display *Dama* morphological characteristics, as described by Lister (1996a) and Pfeiffer (1999). Table 4.17 lists the measurements of *Dama* scapula, humerus, and metapodials, Table 4.18 presents the measurements of *Dama*

tarsal bones, and Table 4.19 presents the measurements of phalanges.

When considering the size of the *Dama* skeletal elements from the GBY, the small-sized distal humerus, as well as the remaining limbs, should be noted. The astragali from GBY also represent small specimens. Measurements of *Dama* postcranial skeletal elements from GBY support the possibility that this species featured quite small skeletal elements. However, the relatively large upper teeth come in contrast to this pattern. One may therefore conclude that the *Dama* sp. remains from GBY represent a small-sized animal in all but its teeth (Rabinovich and Lister in preparation). As stated above, the characteristics of the GBY *Dama* can be summarized as follows: absence of thick beams, a variable (probably short) brow tine, and some sort of palmation.

Rabinovich and Lister (in preparation) note that in comparison to specimens from other southern Levantine sites, the GBY *Dama* is smaller-sized and falls within the size range documented for modern *Dama mesopotamica* females. Though little variation occurs in skeletal elements between *Dama* species, the characteristics of the GBY *Dama* point to a species positioned between *Pseudodama*-like to modern *Dama* (as above).

Juvenile animals have been recorded within the sample of GBY *Dama*, originating from Layer V-5, Layer V-6, and the JB (Table 4.20). The presence of juvenile animals will be further discussed in Chapter 5.

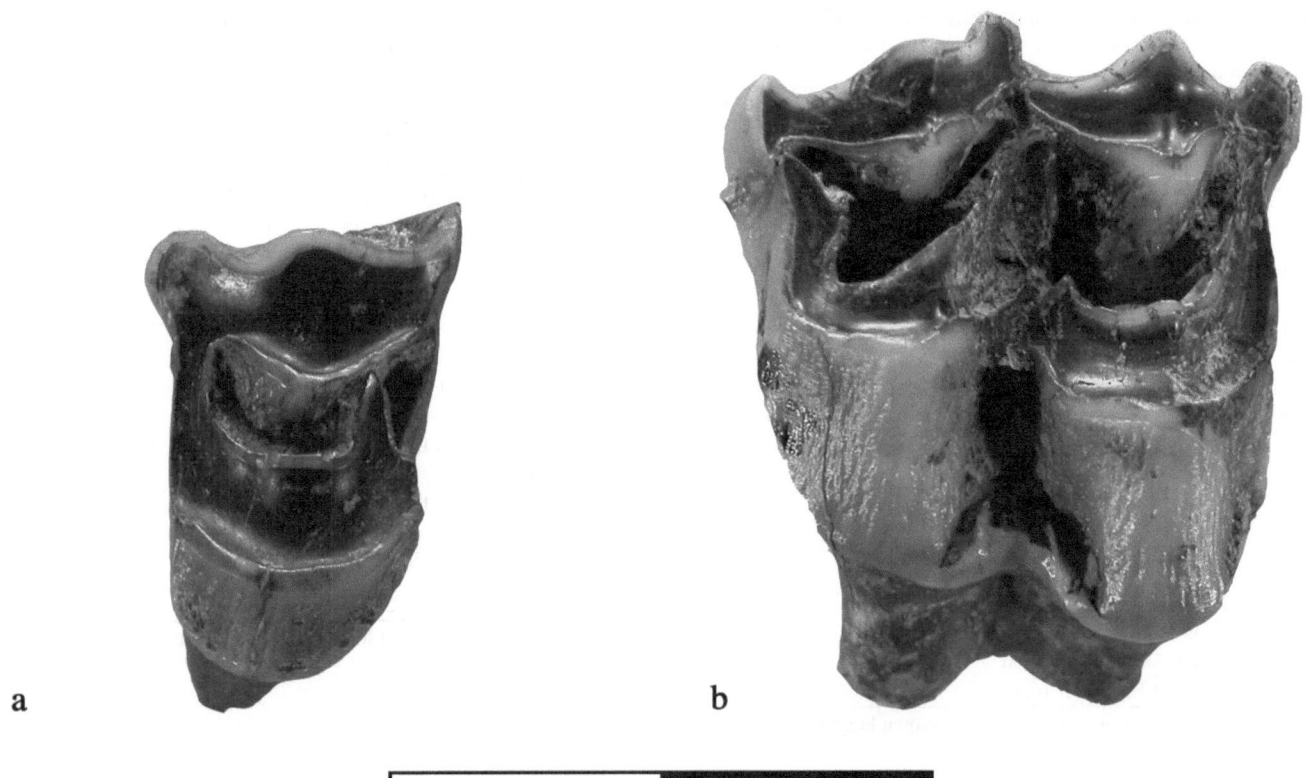

a

b

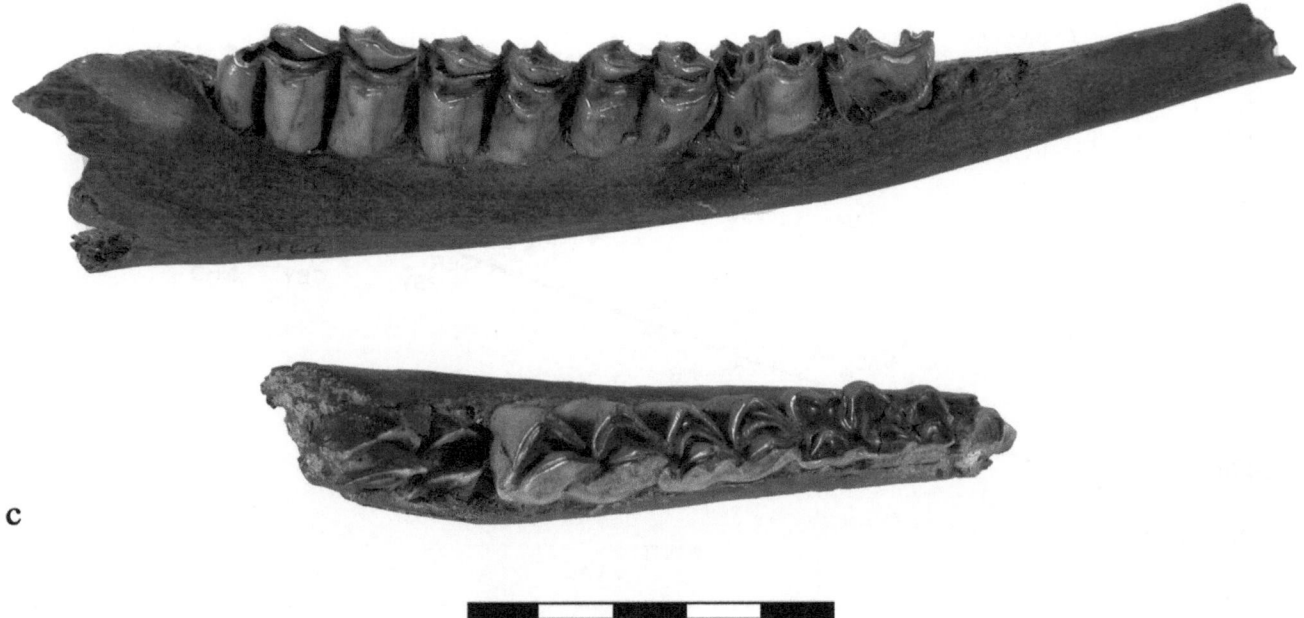

c

Fig. 4.7 *Dama* sp. teeth from Area C and the JB: upper teeth: **a** P[4] (no. 5,637, r; Layer V-5), **b** M[3] (no. 781, r; the JB), **c** two mandibles (nos. 1,635, l, Layer V-5; 1,461, r, Layer V-6)

Table 4.15 Measurements (mm) of *Dama* sp. upper teeth (length and width measured from base of crown, crown height measured from lingual mid-point)

Layer	Skeletal element	Base length	Base width	CH1[a]	CH2[a]
V-5	M^1 (no. 1,711, l)	15.74	18.16	4.98	5.54
V-5	M^3 (no. 2,103, l)	20.64	22.32	16.72	16.88
V-5	M^3 (no. 1,832, l)	20.47	21.33	16.82	17.03
V-5	M^3 (no. 1,735, l)	19.95	21.40	12.89	12.66
V-6	M^1 (no. 1,602, r)	15.97	18.97	5.86	6.42
V-6	M^1 (no. 1,463, r)	19.08	20.97	6.87	6.09
V-6	M^2 (no. 1,498, r)	19.87	21.92	8.26	7.50
V-6	M^3 (no. 1,699, l)	18.90	21.99	13.54	
V-6	M^3 (no. 1,566, r)	21.43	24.72	15.69	13.89
V-6	M^3 (no. 1,599, l)	20.10	21.42	10.01	9.67
V-6	M^3 (no. 2,105, r)	20.67	22.81	8.37	8.00
JB	M^1 (no. 1,272, l)	18.19	21.23	5.84	
JB	M^1 (no. 2,164, l)	16.21	19.97	13.19	15.00
JB	M^1 (no. 11,786, l)	17.77	20.24	14.12	13.04
JB	M^2 (no. 2,170, l)	19.66	22.30	12.43	12.39
JB	M^2 (no. 2,086, r)	19.78	20.81	9.80	9.46
JB	M^2 (no. 2,173, l)	20.46	19.01	11.17	
JB	M^3 (no. 781, r)	20.66	21.95	10.81	11.57
JB	M^3 (no. 2,168, l)	16.58	19.81	9.32	10.56

[a]CH1—crown height of first lobe, CH2—crown height of second lobe

Table 4.16 Measurements (mm) of *Dama* sp. lower teeth (length and width measured from base of crown, crown height measured from buccal mid-point)

Layer	Skeletal element	Base length	Base width	CH1[a]	CH2[a]
V-5	dP_4 (no. 1,635, l)	18.67	9.31	4.13	4.31
V-5	P_3 (no. 1,749, r)	13.19	8.48	9.87	
V-5	P_4 (no. 2,104, l)	11.58	5.27	Unerupted	
V-5	M_2 (no. 1,462, r)	19.04	12.32	13.61	14.03
V-5	M_2 (no. 2,106, r)	18.81	12.01	10.03	10.03
V-5	M_3 (no. 1,670, l)		12.47		18.74
V-6	P_3 (no. 1,573, l)	13.93	8.48	12.77	
V-6	P_3 (no. 1,582, r)	13.65	9.01	7.17	
V-6	M_1 (no. 1,540, r)	16.06	12.97	5.65	5.74
V-6	M_1 (no. 1,584, l)	16.68	13.16	7.00	7.92
V-6	M_2 (no. 1,647, l)	17.71	11.68	10.29	10.84
V-6	M_2 (no. 1,771, r)	16.99	13.01	5.91	6.22
V-6	M_3 (no. 1,488, l)	23.53	10.67	19.37	15.31
JB	P_3 (no. 2,172, r)	12.97	8.36	8.90	
JB	M_1 (no. 901, r)	16.98	10.66	11.91	11.90
JB	M_3 (no. 2,181, l)	24.78	12.81	9.54	8.16

[a]CH1—crown height of first lobe, CH2—crown height of second lobe

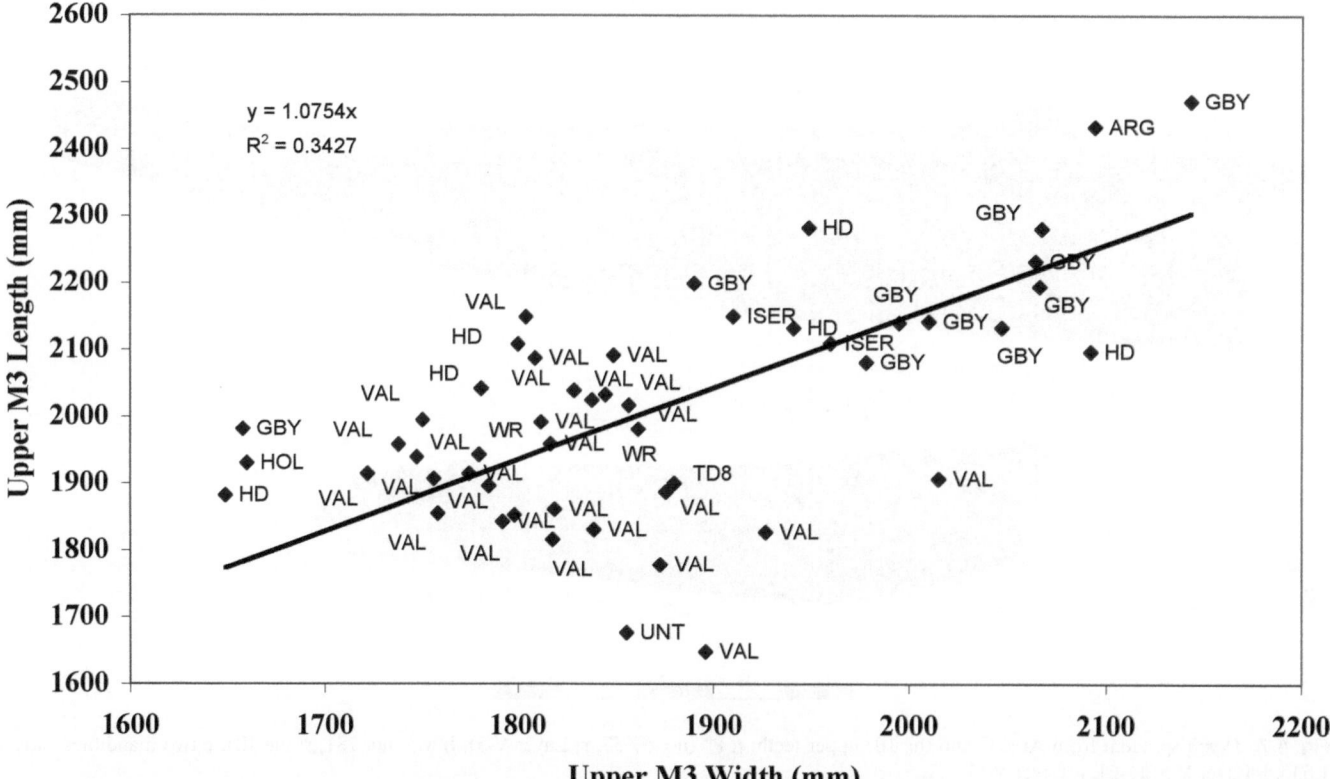

Fig. 4.8 Plot of *Dama* M^3 length and width from the following sites: VAL—Vallonnet, UNT—Untermassfeld, WR—West Runton, ISER—Isernia, TD8—Atapuerca; Hol—Holon, HD—Hayonim D. See Table 4.5 for details on each site

Table 4.17 Measurements (mm) of *Dama* sp. postcranial elements from Area C and the JB

Layer	Skeletal element	GL[a]	GLP[a]	LG[a]	BG[a]
V-5	Scapula (no. 1,679, r)		41.25	29.70	26.82
V-6	Scapula (no. 2,082, r)			37.82	34.86
V-6	Scapula (no. 1,663, l)	120.49	46.29	35.90	36.17
JB	Scapula (no. 1,227, r)		50.17	38.35	35.69
		GL[a]	BT[a]	BD[a]	H[b]
JB	HumD (no. 933, r)		38.67	41.85	23.50
JB	HumD (no. 539, l)		35.95	37.28	21.11
		BD[a]	MCHDW[b]	MCLC[b]	MCSC[b]
V-5	Metacapal (no. 1,825)	31.66	15.14	21.18	16.78
V-5	Metacapal (no. 1,531)		15.48	19.48	14.01
V-5	Metacapal (no. 2,099)	28.61	13.54	17.77	13.95
V-6	Metacapal (no. 1,592)		14.41	20.77	15.10
JB	Metacapal (no. 943)	29.10	13.92	18.66	14.30
JB	Metacapal (no. 1,269)	30.42	14.30	19.81	15.30
		BD[a]	MTHDW[b]	MTLC[b]	MTSC[b]
V-6	Metatarsal (no. 1,665)	35.56	16.03	22.94	16.62

[a]Measurements after von den Driesch (1976): GL—greatest length, GLP—greatest length of the Processus articularis, LG—length of the glenoid cavity, BG—breadth of the glenoid cavity, BT—greatest breadth of the trochlea, BD—greatest breadth of the distal end
[b]Measurements after Davis (1985): H—height of distal humerus, MCHDW—width of distal condyle metacarpal, MCLC—diameter or height of distal condyle (metacarpal), MCSC—width of distal trochlea (metacarpal), MTHDH—width of distal condyle (metatarsal), MTLC—diameter or height of distal condyle (metatarsal), MTSC—width of distal trochlea (metatarsal)

Table 4.18 Measurements (mm) of *Dama* sp. tarsal bones from Area C and the JB

Layer	Skeletal element	GL[a]	BD[a]	D1[a]	Notes
V-5	Astragalus (no. 5,829, r)	40.02	24.19	21.48	
V-6	Astragalus (no. 1,660, r)	40.48	21.80	24.45	
V-6	Astragalus (no. 1,661, l)	42.88	23.22	25.78	
JB	Astragalus (no. 914, l)	40.52	21.43	25.09	
JB	Astragalus (no. 2,079, r)	35.35	22.71	19.17	Eroded
JB	Astragalus (no. 829, r)	41.34	22.52	25.07	
		GL[a]	GB[a]	MB[a]	
JB	Calcaneum (no. 112, r)	92.50	19.88	23.43	
JB	Calcaneum (no. 1,255, r)	87.75	18.75	22.68	
V-6	Naviculocuboid (no. 1,485, l)	30.11	35.33		
V-6	Naviculocuboid (no. 2,084, r)	28.49	33.17		

[a]Measurements after von den Dreisch (1976): GL—greatest length, BD—greatest breadth of the distal end, D1—greatest depth of lateral half, GB—greatest breadth, MB—greatest depth of proximal end

Table 4.19 Measurements (mm) of *Dama* sp. tarsal bones from Area C and the JB

Layer	Skeletal element	GL[a]	BP[a]	BD[a]
V-5	Ph 1 (no. 1,678)	42.45	13.32	14.52
V-6	Ph 1 (no. 1,662)	48.05	13.27	16.38
V-5	Ph 2 (no. 1,777)	29.80	13.48	11.03
V-6	Ph 2 (no. 1,666)	32.19	14.71	12.61
V-6	Ph 2 (no. 1,579)	31.11	12.81	15.08

[a]Measurements after von den Dreisch (1976): GL—greatest length, BP—greatest breadth of the proximal end, BD—greatest breadth of the distal end

Table 4.20 *Dama* sp. skeletal elements attributed to juvenile animals from Area C and the JB

Layer	Skeletal element	Eruption and wear of tooth[a]		
	Deciduous teeth			
V-5	dP^4 (no. 7,951)			
V-5	dP$_2$ (no. 11,573, r)	Unworn		> 6 months
V-5	dP$_3$ (no. 6,672, r)			
V-5	dP$_3$ (no. 1,428, r)			
V-5	dP$_4$ (no. 1,618, l)			
V-5	dP$_4$ (no. 2,645, l)			
V-5	dP$_4$–M$_3$ (no. 1,635, l)	Approx. 18 months		
V-6	dP^2, dP^4 (no. 1,465, r)			
V-6	dP^4 (no. 1,483, l)			
	dP^4 (no. 1,713, l)			
JB	dP^4 (no. 13,193, l)			
JB	dP^4 (no. 1,782, r)			
JB	dP^4 (no. 14.723, l)			

Layer	Skeletal element	Fusion information[b]		
		Proximal	Distal	Fusion age in months
V-5	Tibia D (no. 2,770)		Unfused	18
V-6	Radius D (no. 1,519)		Unfused	24
V-6	Tibia D (no. 12,697)		Unfused	18
V-6	Metatars. D (no. 13,832)		Unfused	22–24
V-6	Metatars. D (no. 3,756)		Unfused	22–24
V-6	Metatars. D (no. 4,365)		Unfused	22–24
V-6	Metapod. D (no. 2,198)		Unfused	22–24
V-6	Calcaneum P (no. 1,541, r)	Unfused		23–24
V-6	Ph 1 (no. 12,544)	Unfused		12–14
V-6	Vertebra (no. 1,497)	Unfused	Unfused	30–42
V-6	Vertebral articular surface (no. 13,746)		Unfused	<30
JB	Radius D (no. 12,670, r)		Unfused	23–24
JB	Ulna P (no. 12,668, r)	Unfused		22
JB	Ulna P (no. 1,783, r)	Unfused		22
JB	Femur (no. 2,221, l)	Unfused	Unfused	18–24
JB	Pelvis (no. 2,195, r)	Acetabulum unfused		12–15
JB	Cervical vertebra (no. 2,220)	Unfused	Unfused	30–42
JB	Thoracic vertebra (no. 1,203)	Unfused	Unfused	84–108
JB	Thoracic vertebra (no. 831)		Unfused	84–108
JB	Vertebral articular surface (no. 11,820)		Unfused	<30

[a]After Chaplin and White (1969) and Brown and Chapman (1990, 1991)
[b]After Pohlmeyer (1985) and Kersten (1989)

Cervidae sp.

Numerous antler fragments from the site could not be assigned to a particular species. They mostly belonged to tines or beams (Layer V-5: N = 19, Layer V-6: N = 53, the

JB: N = 11). A few cervid tooth fragments are also included in this category (Layer V-5: N = 5, Layer V-6: N = 2). All the fragments fall within the size and morphological range of the site's cervids, and most are probably attributable to *Dama*.

4.4 Body-Size Groups (BSG)

Bones that lack diagnostic traits enabling further identification were attributed to a body-size group (see Table 3.1). Table 4.21 lists the number of skeletal elements attributed to the different body-size groups in each layer. These groups' importance in the interpretation of the site's faunal assemblages is further discussed in Chapter 5.

Table 4.21 Specimen frequencies in Body size groups (BSG A–F) in Area C and the JB (for details on each group, see Table 3.1)

Layer	BSGA	BSGB	BSGC	BSGD	BSGE	BSGF
V-5	10	12	49	201	105	24
V-6	15	15	86	329	211	53
JB	9	9	94	169	120	8

Body-Size Group A (BSGA): Elephant (> 1,000 kg)

Ten specimens from Layer V-5 have been assigned to this group: six tooth fragments with only part of their enamel preserved, two bone splinters, an atlas fragment, and a fragment of a heavily eroded tarsal (Table 4.22). Fifteen specimens from Layer V-6 were also assigned to this group: five skull fragments, a vertebra fragment, a caudal vertebra, two rib shafts, and six splinters. Nine elements from the JB include: two skull fragments, two tusk fragments, three rib fragments, a vertebra, and a long bone shaft. Because of the unique anatomical characteristics of the elephant (robusticity and particular bone shape), it was difficult to identify the thick, bulky pieces; the category of splinters has therefore been included.

Table 4.22 Skeletal-element distribution of BSGA (elephant >1,000 kg) from Area C and the JB

Skeletal element	Layer V-5	Layer V-6	JB
Skull fragment		5	2
Tusk fragment			2
Tooth fragment	6		
Vertebra	1	2	1
Rib		2	3
Tarsal	1		
Long bone fragment			1
Flat splinter		1	
Splinter	2	5	
Total	10	15	9

Body-Size Group B (BSGB): Hippopotamus, Rhinoceros (Approx. 1,000 kg)

Twelve specimens from Layer V-5 have been assigned to this group: a vertebral fragment, a first phalanx fragment, two radius shafts, two long bone shafts, and five flat splinters (Table 4.23). The fifteen specimens from Layer V-6 include a skull fragment, a tooth fragment, a vertebra fragment, a rib portion, a radius shaft, humerus shaft, two sesamoid bones, five long bone shafts, and two large flat splinters. The JB yielded a skull fragment, two rib shafts, two vertebrae, a long bone shaft, and three flat splinters. Based on the morphological characteristics and bone thickness, it seems that most of the specimens from this group probably originated from hippopotamus.

Table 4.23 Skeletal-element distribution of BSGB (hippopotamus, rhinoceros, approx. 1,000 kg) from Area C and the JB

Skeletal element	Layer V-5	Layer V-6	JB
Skull fragment		1	1
Tooth fragment		1	
Vertebrae fragment	1	1	2
Rib fragment	1	1	2
Humerus fragment		1	
Radius fragment	2	1	
Sesamoid		2	
Phalanx fragment	1		
Long bone shaft	2	5	1
Flat splinter	5	2	3
Total	12	15	9

Body-Size Group C (BSGC): Giant Deer, Red Deer, Boar, and Bovine (80–250 kg)

Elements from this group originated in Layer V-5 (N = 49), Layer V-6 (N = 86), and the JB (N = 94) (Table 4.24). Long bone shafts and flat splinters constitute almost 50% of this group in the assemblages from Area C and the JB. Other body parts from this group are skull and postcranial elements. The species assignment is less evident since candidates of several species may have contributed to this group—bovidae, cervidae, and suidae.

Table 4.24 Skeletal-element distribution of BSGC (giant deer, red deer, boar, bovine, 80–250 kg) from Area C and the JB

Skeletal element	V-5	Layer V-6	JB
Skull fragment	3	8	5
Mandible fragment	1	3	2
Tooth fragment	3	3	1
Vertebrae fragment	2	4	5
Rib fragment		8	3

Table 4.24 (continued)

Skeletal element	V-5	Layer V-6	JB
Scapula fragment		2	1
Humerus fragment		1	
Radius fragment		2	3
Ulna fragment	1		
Metacarpal fragment		1	1
Carpal fragment	1	2	
Sesamoid	1		
Pelvis fragment	2	4	4
Femur fragment	1		3
Tibia fragment		4	11
Tarsal			2
Metatarsal fragment			5
Metapodial fragment		2	3
Phalanx fragment	1		
Long bone shaft	23	30	35
Flat splinter	10	12	10
Total	49	86	94

Body-Size Group D (BSGD): Fallow Deer, Caprinae (40–80 kg)

This group consists of bones that could not be attributed to a particular species but are similar in size to *Dama* (Table 4.25). It includes 201 fragments from Layer V-5, 329 fragments from Layer V-6, and 169 fragments from the JB. Most of the bones are long bone shaft fragments or flat splinters.

Table 4.25 Skeletal-element distribution of BSGD (fallow deer, Caprinae, 80–40 kg) from Area C and the JB

Skeletal element	Layer V-5	Layer V-6	JB
Long bone fragment	180	277	155
Flat splinter	21	52	14
Total	201	329	169

Body-Size Group E (BSGE): Gazelle, Roe Deer (15–40 kg)

This group is the second most frequent in the studied assemblages (Layer V-5: N = 105, Layer V-6: N = 211, the JB: N = 120). Gazelle is probably the main contributor to this group. Its long bone shafts comprise half the specimens in all the assemblages from the site. Vertebrae and rib fragments comprise another significant part of the assemblages of Layers V-5 and V-6, followed by fore and hind limb shafts.

Body-Size Group F (BSGF): Hare, Red Fox (2–10 kg)

This group is comprised of skeletal elements of hares or foxes less than 3 cm long, primarily fragments of long bone shafts, vertebrae, and limbs (Layer V-5: N = 24, Layer V-6: N = 53, the JB: N = 8).

Chapter 5

Taphonomic Analysis

Abstract An important objective of the current research is the analysis of mammalian taphonomy in order to differentiate between biotic and non-biotic agents, and to thus reach a better understanding of the various processes that formed the Gesher Benot Ya'aqov (GBY) assemblage. This has been achieved by a zooarchaeological analysis of the assemblage, which has identified and characterized bone damage caused by pre- and/or post-depositional biotic agents, such as hominins, carnivores, and rodents. Among the non-biotic processes that modify thanatocoenosis assemblages are weathering and bone abrasion.

5.1 Bone Preservation

The mammalian thanatocoenosis discovered at GBY displays different stages of bone preservation. Since the bones were retrieved from waterlogged deposits, they were discolored by the anaerobic environment and characterized by different shades of color, mainly ranging from bluish-gray to brownish-black. Surface preservation also varies, although many of the bones are characterized by a shiny surface. Sediment type and texture are considered to be among the steering factors in differences in bone-surface preservation. Six different types of surface preservation have been identified at the site (see Chapter 3).

In general, many of the bones from Area C and the JB are brown and shiny. Bones from Layer V-5 often feature slight surface weathering, a distinct blackish color, and still-adhering mollusk fragments (Type 2). In Layer V-6 we found fragments with sharp breaks in a dull, metallic bluish-gray color (Type 4). Some bones are slightly weathered and have a yellowish surface (Type 3), and others exhibit a shiny, mottled brownish-black surface, with a seemingly fresh appearance to the breaks (Type 1). Signs of weathering and exfoliation are rarely noticed (Type 5), as are morphologically very weathered and disintegrated bones (Type 6). The two most common types in the GBY assemblage, Types 2 and 4, consist of slightly weathered bones, either blackish or a dull bluish-gray in color (Table 5.1).

We also examined whether bone preservation varies according to species, since during analysis there were some indications that preservation levels could change according to mammalian body size. Thus, a correlation of preservation types by layer and species was undertaken (Table 5.2). Preservation types documented for *Dama*—the most common species at the site—typically reflect bone preservation of medium-sized mammals. Preservation types 2 and 4 dominate at the site, as is also seen in the unidentified mammal bones (UNM), although weathered and exfoliated bone surfaces (Types 5 and 6) are slightly more abundant. A positive significant correlation in the distribution of the preservation types was found only between Layer V-6 and the JB ($r = 0.899$, $p = 0.015$).

Table 5.1 State of preservation of the mammalian fauna from Area C and the JB

State of preservation	Layer V-5		Layer V-6		JB	
	N	%	N	%	N	%
Shiny mottled brownish-black surface (Type 1)	11	2.2	39	4.1	40	9.0
Slightly weathered blackish surface, mollusk fragments (Type 2)	246	49.9	380	40.2	161	36.4
Slightly weathered yellowish surface (Type 3)	32	6.5	24	2.5	5	1.1
Slightly weathered, dull metallic bluish-gray color (Type 4)	163	33.1	426	45.0	224	50.7
Signs of weathering and exfoliation (Type 5)	27	5.5	45	4.8	6	1.4
Very eroded, morphology changed (Type 6)	14	2.8	32	3.4	6	1.4
Total	493	100	946	100	442	100

R. Rabinovich et al., *The Acheulian Site of Gesher Benot Ya'aqov Volume III: Mammalian Taphonomy.*
The Assemblages of Layers V-5 and V-6, Vertebrate Paleobiology and Paleoanthropology,
DOI 10.1007/978-94-007-2159-3_5, © Springer Science+Business Media B.V. 2012

Table 5.2 State of bone preservation by species from Area C and the JB

V-5	BSGA	BSGB	BSGC	BSGD	BSGE	BSGF	CER	DAMA	BOS SP	BOVINI	CAPR	GAZ	EQ	ELEP	UNM	Total
Type 1				5	1		1	1							3	11
Type 2	2	4	18	82	32	6		41		2			2	2	54	246
Type 3			2	27	2										1	32
Type 4		3	5	58	31	3		28	2		1	1	3		28	163
Type 5			4	12	2										9	27
Type 6	1	1		7			1							1	3	14
Total	3	8	30	191	68	9	2	70	2	2	1	1	5	3	98	493

V-6	BSGA	BSGB	BSGC	BSGD	BSGE	BSGF	CER	DAMA	BOS SP	BOVINI	CAPR	GAZ	EQ	ELEP	HIPO	SUS	CARN	UNM	Total
Type 1	2	1	2	15				16				1	1					1	39
Type 2	4	3	25	108	69	13	13	65		1		2		2	1	3	1	70	380
Type 3			3	12	4			4										1	24
Type 4	1	4	16	147	65	6	2	122	1		5	2						55	426
Type 5	1	1	1	18	8	3	1	1										11	45
Type 6	1		3	8	2		4											14	32
Total	9	9	50	308	148	22	20	208	1	1	5	5	1	2	1	3	1	152	946

JB	BSGA	BSGB	BSGC	BSGD	BSGE	BSGF	CER	DAMA	BOS SP	BOVINI	CAPR	GAZ	EQ	ELEP	HIPO	SUS	UNM	Total
Type 1	1		2	17	8	1	10										1	40
Type 2	2	1	8	48	37		27	4	1	1	1		2	2	2	2	23	161
Type 3				4			1											5
Type 4	1	5	28	101	30		36	5	1	1	1	2			1		12	224
Type 5		2	2	1										1				6
Type 6			3	1			1										1	6
Total	4	8	43	172	75	1	75	9	2	2	2	2	2	3	3	2	37	442

See Table 5.1 for explanation of Types 1-6.

Bone preservation of the different body-size groups is similar to that of their possible corresponding mammalian species (i.e., compare *Dama* with BSGD) (Table 5.2). The relative frequency of weathered bones (Types 5 and 6) is observed especially in the very few remains attributed to the larger species (e.g., elephant, BSGA, and BSGB).

Yet another pattern of weathering has been observed in several cases, appearing in the form of weathered surfaces located just along the edges of the bone (Layer V-5: NISP = 1, Layer V-6: NISP = 5, JB: NISP = 4). These included *Dama* and bones assigned to BSGD, as well as one case of a BSGE bone from Layer V-6 (no. 12,807). Such erosion occurred mainly on unfused vertebrae (Layer V-5: no. 1,619; Layer V-6: nos. 1,467, 1,719; the JB: nos. 820, 2,183, 11,797), on long bone epiphyses (Layer V-6: nos. 1,587, 12,807; the JB: no. 1,284), and on a rib (Layer V-6: no. 13,571). The reason for this kind of erosion is unknown, although its appearance in unfused elements may suggest a type of local rolling.

5.2 Striations

In addition to the evident weathering, the bone surfaces from all the GBY strata are characterized by striations. We defined striations as very narrow marks of various depths, with profiles that differ from the v-shape indicative of hominin-induced cut marks. An additional criterion for differentiating between cut marks and striations is the fact that the former is often located in parallel groups/clusters and restricted to specific parts of the skeleton (Fig. 5.1). Quite often these marks display a very distinct bifurcating morphology.

Various striations have been observed in the GBY assemblages. Among them are cut-mark-like damages, whose identification as such is uncertain, due to bone-surface corrosion (Fig. 5.2).

One such category has been termed "micro-striations," characterized by localized groups of parallel, fine shallow striations, usually overlying all other bone-surface modifications. We observed no pattern in the occurrence of these micro-striations with respect to a particular anatomical location (Fig. 5.3).

Another category of striations consists of flat-based striations, observed on several bone specimens. These modifications usually occurred in the form of a single striation that begins abruptly with a flat-based profile and terminates as a thin line (Fig. 5.4).

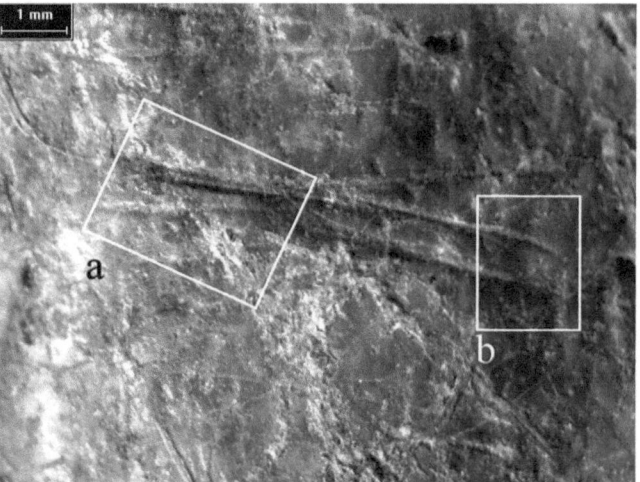

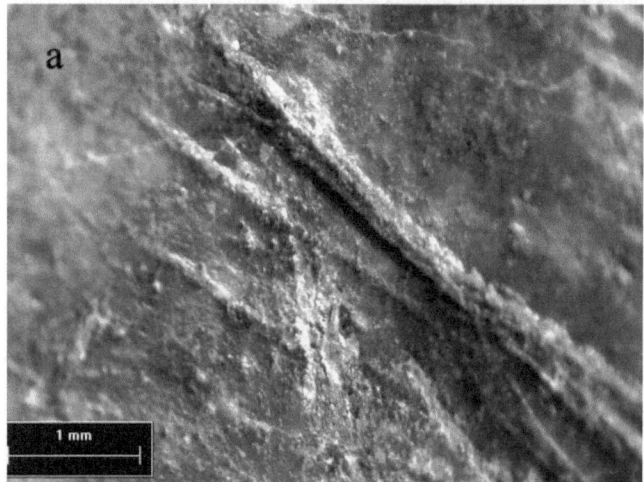

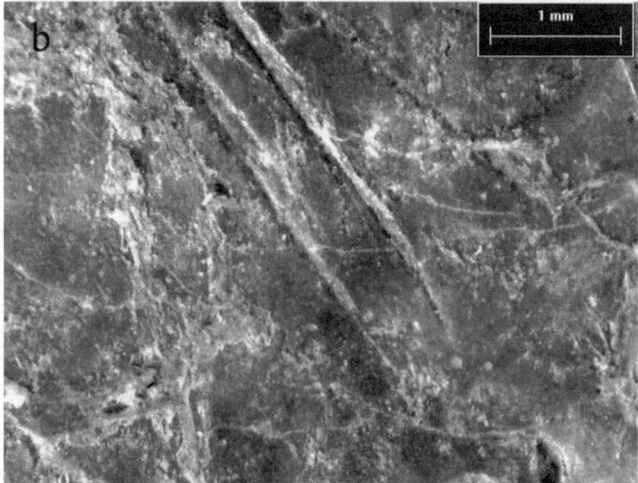

Fig. 5.1 Details (a and b) of a cut mark on a cervid atlas from Layer V-6

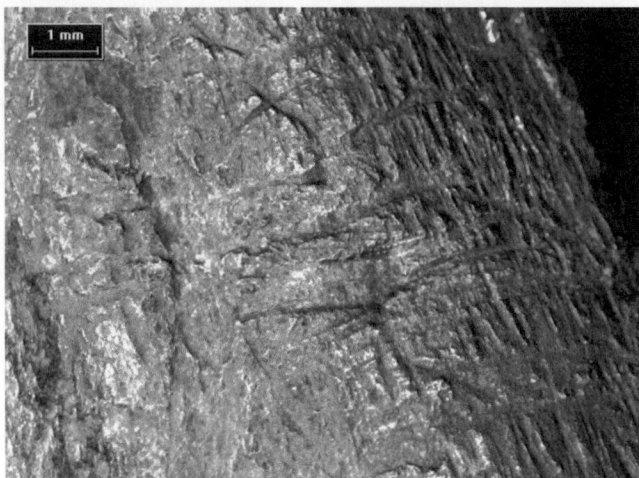

Fig. 5.2 Detail of a cut mark like striation on the corroded surface of a long bone from Layer V-5

Fig. 5.3 "Micro-striations" on a rib fragment from Layer V-5

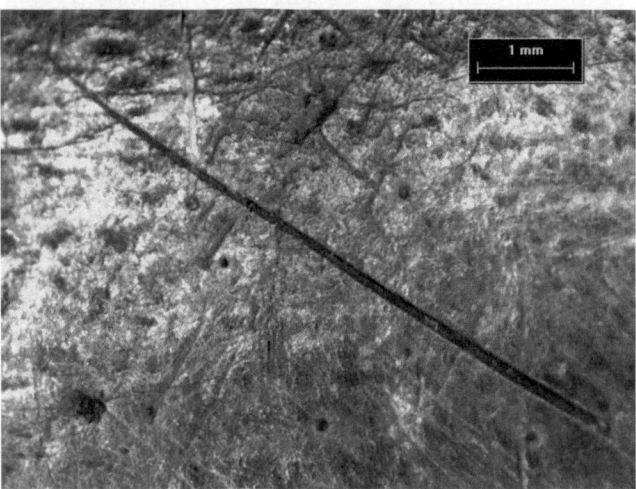

Fig. 5.4 Flat-based striations on a long bone from Layer V-5

A third category is comprised of u-shaped striations characterized by broad thick lines that often appear as gently curving or semi-circular. They are often crossed by short striations, usually at an approximately right angle (Fig. 5.5).

Apart from the above-mentioned categories, the GBY faunal assemblage includes bones whose surface was heavily marked by striations (Fig. 5.6). These consist mainly of very fine, shallow striations that, unlike the micro-striations, display no overall orientation.

Striations appear on most of the identified mammal species bones at GBY, as well as on unidentified bones. Striation frequencies differ in both individual species and bones of different body-size groups. It is interesting to note that striations are more common on bones of *Dama*, and on its corresponding body-size group and on Gazella and its corresponding body-size group. These are species that exhibit relatively many cut marks and evidence of marrow extraction, as will be discussed below.

When comparing striated-bone frequency per layer (Layer V-5: N = 105, Layer V-6: N = 220, the JB: N = 211) (Table 5.3) it becomes evident that Layer V-6 has more bones with striations, but that the frequency of striated bones in the JB is higher (18.99%; Table 5.3). It is also clear that a correlation exists between species frequency and striation frequency, with striations appearing mainly on bones of the more common taxon (*Dama*), and BSGC, BSGD, and BSGE.

Striation frequency per species, as presented in Table 5.4, is not consistent for the different excavation areas. The relative frequency of striated *Dama* bones in the JB (26.20%, NISP = 93) is considerably higher than that of Layer V-5 (14.39%, NISP = 41) and Layer V-6 (18.90%, NISP = 103). Body-size groups are also relatively more striated in the JB. A somewhat different picture is reflected in the less abundant species where most bones are striated (i.e., Layer V-6: *Sus scrofa*, one striated bone = 33.33%; the JB: *Megaloceros*, one striated bone = 100%).

We examined the distribution of striated bones per body-size group, lumping all identified species with their relevant body size. As can be observed in Fig. 5.7, a striking similarity is seen between the layers, with the most common body-size group BSGD showing the highest frequencies of striated bones in all layers. A positive significant correlation in the distribution of striated bones was found between Layers V-5 and V-6 (r = 0.883, p = 0.008), Layer V-5 and the JB (r = 0.955, p = 0.001), and Layer V-6 and the JB (r = 0.809, p = 0.028).

To conclude, we observed a general correlation between species- and bone-striated frequencies in the GBY faunal assemblages. Within this context, we also noted that bones

Fig. 5.5 U-shaped striations on a femur fragment from the JB

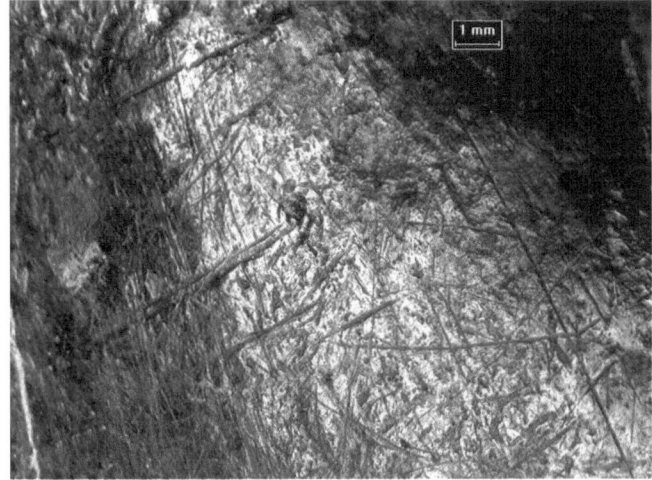

Fig. 5.6 Scattered striations on the surface of a humerus from the JB

of high fragmentation (grouped within the body-size groups) exhibit relatively higher striation frequencies. Differences in striation frequency per species can be observed throughout the studied assemblages.

5.3 Skeletal-Element Representation at GBY

Skeletal elements were quantified according to several criteria: the number of identified specimens (NISP), the minimum number of individual animals (MNI), the minimum number of skeletal elements (MNE) and the minimum number of animal units (MAU) (Binford 1981; Lyman 1994). Discussed below are the taxa and their assigned BSG.

5.3.1 Dama Skeletal-Element Representation

In this section we shall examine body-part distribution according to the *Dama* NISP (see Table 4.14), where it is evident that almost all skeletal elements are present and, as discussed in Chapter 4, complete bones are rare. A general comparison of body parts between the excavation areas reveals similar distribution of NISP (%NISP) of *Dama* bones (Layers V-5 and V-6: r = 0.712, p = 0.000; Layer V-5 and the JB: r = 0.727, p = 0.000; Layer V-6 and the JB: r = 0.758, p = 0.000). The few *Dama* sp. antlers

Table 5.3 Relative frequency of striated bones from Area C and the JB

	V-5	%NISP (N = 974)	V-6	%NISP (N = 1,868)	JB	%NISP (N = 1,111)
Sus scrofa			1	0.05		
Megaloceros					1	0.09
Dama sp.	41	4.21	103	5.51	93	8.37
Cervidae	1	0.10	1	0.05		
Bos sp.			2	0.11	4	0.36
Bovid			1	0.05		
Gazella cf. *gazella*			3	0.16	1	0.09
Caprini indet.					1	0.09
*E.*cf. *africanus*			1	0.05		
Equus sp.	2	0.21	2	0.11	2	0.18
Palaeoloxodon antiquus	1	0.10	1	0.05		
BSGA			1	0.05	1	0.09
BSGB	2	0.21	2	0.11	3	0.27
BSGC	8	0.82	15	0.80	25	2.25
BSGD	34	3.49	52	2.78	55	4.95
BSGE	12	1.23	21	1.12	22	1.98
BSGF			3	0.16		
UNM	4	0.41	11	0.59	3	0.27
Total	105	10.78	220	11.78	211	18.99

Table 5.4 Relative frequency of striated bones by species from Area C and the JB

	V-5 Striated bones	V-5 % Striated bones of SP NISP	V-6 Striated bones	V-6 % Striated bones of SP NISP	JB Striated bones	JB % Striated bones of SP NISP
Sus scrofa			1	20		
Megaloceros					1	100.00
Dama sp.	41	14.39	103	18.90	93	26.20
Cervidae	1	5.26	1	1.89		
Bos sp.			2	28.57	4	23.53
Bovid			1	9.09		
Gazella cf. *gazella*			3	37.50	1	8.33
Caprini indet.					1	100.00
*E.*cf. *africanus*			1	16.67		
Equus sp.	2	16.67	2	14.29	2	10.00
Palaeoloxodon antiquus	1	8.33	1	4.76		
BSGA			1	6.67	1	11.11
BSGB	2	16.67	2	13.33	3	33.33
BSGC	8	16.33	15	17.44	25	26.60
BSGD	34	16.92	52	15.81	55	32.54
BSGE	12	11.43	21	9.95	22	18.33
BSGF			3	5.66		
UNM	4	1.82	11	2.35	3	1.24

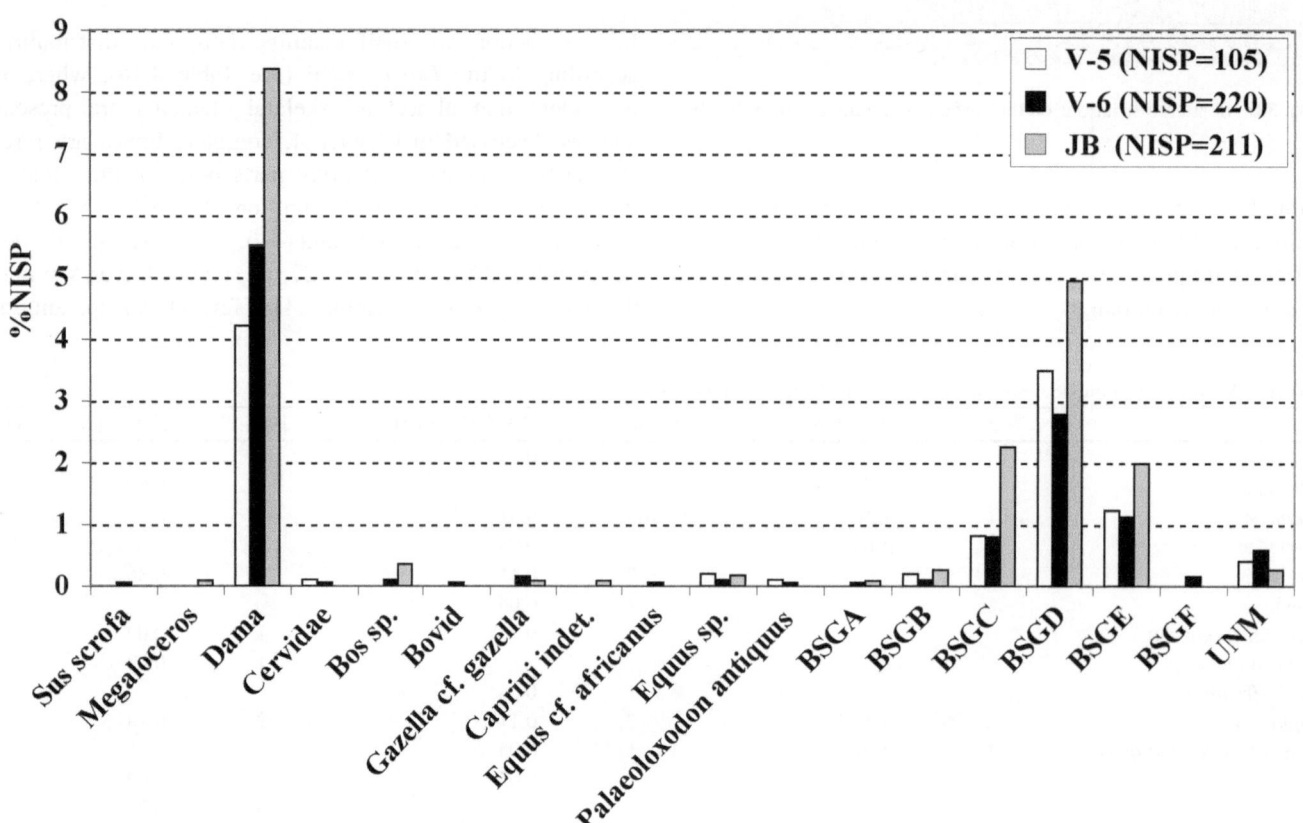

Fig. 5.7 Relative frequency of the striated bones per body-size group and layer

from the site—albeit broken—provided information on sex and seasonality (see Chapter 4). Skull remains include fragments of parietal, orbit, occipital, and bulla tympanica from Layers V-5 and V-6, and orbit and bulla tympanica from the JB. Maxilla fragments were uncovered only in Layer V-6 and the JB, although maxilla teeth appear in all the assemblages. Mandible fragments are present in similar frequencies throughout the assemblages (Layer V-5: NISP = 9, 3.15%; Layer V-6: NISP = 24, 4.40%; the JB: NISP = 15, 4.22%), lower teeth and teeth fragments are most abundant in Layer V-5 (Fig. 5.8). Vertebrae and ribs comprise around 20% of the *Dama* remains from the site (Layer V-5: NISP = 54, 18.94%; Layer V-6: NISP = 115, 21.10%; the JB: NISP = 74, 20.84%). Vertebrae fragments were defined as atlas, axis, cervical, thoracic, lumbar, and sacrum, although most are incomplete and broken vertebrae bodies. Rib shafts are numerous.

When examining fore and hind limb remains from the site, it becomes clear that the most common element is shaft fragments, as is further emphasized by the numerous long bone shaft fragments assigned to BSGD, which probably also belonged to *Dama*. The ratio of the fore limbs' (humerus, radius, and ulna) proximal to distal parts (prox/dist) varies among the assemblages with a higher survivorship of the proximal part (Layer V-5: prox/dist = 1.25, Layer V-6: prox/dist = 0.67, the JB: prox/dist = 0.86).

The ratio of the proximal to distal parts of the hind limbs (femur, tibia) reflects a higher survivorship of the distal part in both Layers V-5 and V-6 (Layer V-5: prox/dist = 0.20, Layer V-6: prox/dist = 0.00), and is equally distributed in the JB (prox/dist = 1). The survivorship of metacarpal parts varies between Layer V-6, which contains assemblages with a high representation of proximal parts (prox/dist = 8), to Layer V-5, which reveals a complete absence of proximal parts, and the JB, with a ratio of 0.5. Proximal metatarsal parts are more numerous than distal parts in all the assemblages (see Table 4.14, Fig. 5.8). Proximal, distal, and shaft parts of phalanges have also been found in each assemblage; this phenomenon is further discussed in Section 5.5.

The MNI was calculated in those cases where the most common-sided anatomical part served as the basis for estimation, after taking into account information on body side, age, and sex. Due to the incompleteness of the bones, few specimens were available for MNI calculation. In order to prevent duplicates, only bones that were more than half complete were included in the MNI calculation.

5.3.1.1 Ageing and Sexing Considerations in MNI Estimation

For the ageing of *Dama*, tooth-eruption data and a tooth-loss scheme were taken from Brown and Chapman (1990, 1991),

and a tooth-wear scheme from Chaplin and White (1969). Deciduous teeth are lost by 30 months of age; while the first molar erupts at about four or five months (Brown and Chapman 1990). Thus, the presence of isolated deciduous teeth is only a very general age indicator.

The sole complete mandible from Layer V-5 (no. 1,635, l; Fig. 4.7b) features a still in-place dP_4 and an erupting M_3, suggesting an age of approximately 18 months. A P_4 (no. 2,104, l) that had not erupted from the mandible suggests an age of approximately 22 months. An unworn dP_2 (no. 11,573, r) belonged to an animal less than six months old.

Based on Kersten's (1989) age classes for the Epipaleolithic fallow deer (*Dama mesopotamica*) from Ksar Akil and on the *Dama* sided dentition from Layer V-5, we propose that Layer V-5 consists of at least 3 young adults (12–30 months old), represented by deciduous premolars (i.e., dP_4; Table 5.5), and at least a single aged specimen represented by an eroded M_3 (no. 1611, r). Only one unfused bone, a distal tibia (no. 2,770), has been found in this layer, which according to long bone epiphyseal data (Pohlmeyer 1985; Kersten 1989), may represent an animal aged younger than a year and a half.

Sided teeth and postcranial elements from Layer V-5 point to the presence of three adults and a single juvenile (Table 5.5). Based on the above data, the bones probably belonged to young adults. Information on the sexing was retrieved from two specimens, a male skull fragment with an antler pedicle (no. 1,625, l) and a female distal scapula (no. 1,679, r).

The reconstruction of the MNI in Layer V-6 is based on both cranial and postcranial bones, including a complete adult mandible (no. 1,461, r), aged six–nine years, and at least two specimens with a deciduous upper premolar. Due to the deciduous teeth, the mandible of the adult specimen cannot be assigned to the same animal, and we thus propose the presence of two juveniles and at least three adults based on two right-sided isolated P_3s (nos. 1,582, 10,131) and the above-mentioned mandible (Table 5.5).

Most of the unfused elements from Layer V-6 are of animals younger than 22 months, such as a distal radius (no. 1,519), distal metapodials (nos. 13,832, 3,756, 4,365, 2,198), and a proximal calcaneum (no. 1,541, r), as well as an unfused distal tibia of an animal younger than 18 months (no. 12,697) and an unfused proximal phalanx of an animal younger than 12–14 months (no. 12,544). The rest are unfused vertebrae body, but since the anterior and posterior extremities of the vertebrae fuse quite late in an animal's life-span, they are not suitable as age indicators (see Table 4.20).

Based on an estimation of their overall size, two atlas fragments from Layer V-6 were identified as female (no. 1,499) and male (no. 1,714). A presence of a male is also indicated by an antler fragment (no. 1,459, l). Combining

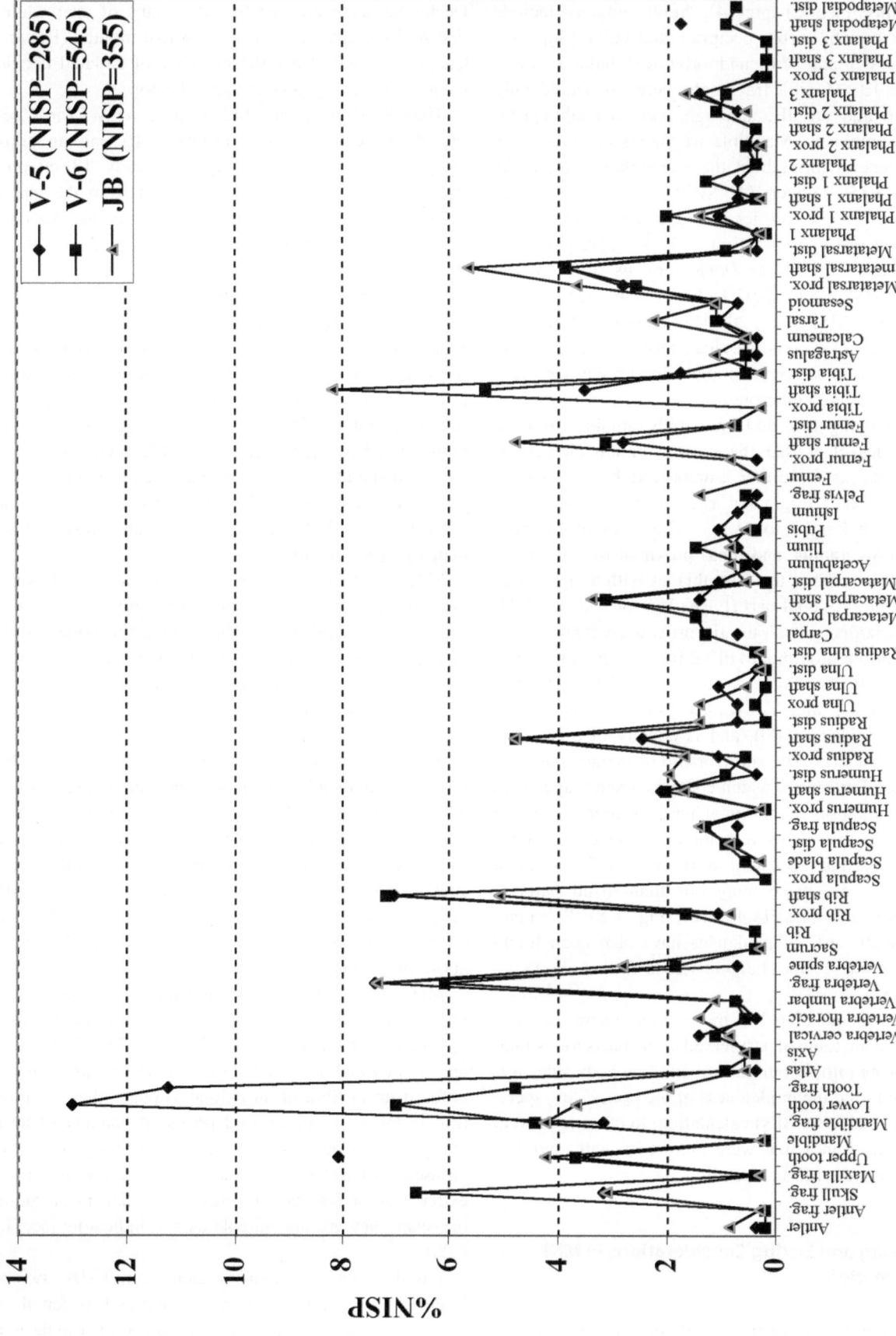

Fig. 5.8 *Dama* %NISP from Area C and the JB

Table 5.5 Sided *Dama* sp. elements from Area C and the JB

	Skeletal element	L	R
V-5			
Dama sp.			
	Skull fragment	1	
	P^4	3	
	M^3	3	
	dP_3		2
	dP_4	3	
	P_3	2	2
	M_2	1	2
	M_3	2	1
	Scapula D		2
	Metatars. P	1	3
V-6			
Dama sp.			
	Antler		1
	Premaxilla	1	
	dP^2–dP^4		1
	dP^4	2	
	P^3	2	2
	P^4	2	2
	M^1		2
	M^3	2	2
	P_2–M_3		1
	P_2	1	1
	P_3	2	2
	M_1	2	1
	M_2	1	1
	M_3	1	1
	Mandible fragment	2	1
	Mandible (coronoid process and condyle)	3	4
	Scapula D	2	
	Radius P	1	1
	Ulna D		1
	Metacarp. P	1	1
	Pelvis (ischiatic spine)	1	2
	Tibia (crest)	2	1
	Metatars. P	2	2
	Astragalus	1	1
JB			
Dama sp.	Antler	1	1
	dP^4	2	1
	M^1	3	1
	M^2	2	1
	M^3	1	1
	P_2–P_4	1	
	P_2		1
	P_3–P_4	1	
	P_3	1	1
	P_4	1	1
	M_1–M_3		1
	M_1	1	
	M_3	1	
	Mandible (coronoid process and condyle)	1	2
	Scapula D		2
	Radius P		2
	Ulna P	1	2
	Pelvis	2	1

Table 5.5 (continued)

Skeletal element	L	R
Femur	3	2
Tibia	3	2
Metatars. P	2	2
Naviculocuboid (T4)	1	2
Calcaneum		2
Astragalus	1	3

the information on age and body sides, we propose that Layer V-6 yielded at least two juveniles and three adults, which included at least one male and one female.

The upper sided teeth from the JB include dP^4s of at least two specimens (nos. 13,193, l; 14,723, l; 1,782, r; see Table 4.20), in addition to isolated molars. Also found were lower-jaw premolars (no. 898, P_3–P_4, l; no. 2,151, P_2–P_4, l) from an animal aged three–four years, and worn lower-row molars of another lower jaw (no. 2,181, l), which indicate a mature animal aged nine–ten years. Loose lower-jaw teeth represent at least one other animal.

Unfused postcranial elements from the JB belonged to animals younger than 12–15 months (unfused acetabulum, no. 2195, r) and younger than 18–24 months (e.g., distal radius, no. 12,670, r; proximal ulna, nos. 12, 668, r; 1,783, r; unfused femur, no. 2,221, l). As previously mentioned, the vertebrae cannot be used for ageing, even though they comprise a major part of the postcranial unfused data from the JB (Table 4.20). The presence of males is indicated by two antler pieces, a base with a frontal bone (no. 219, l) and a shed antler (no. 754, r). The above-mentioned unfused acetabulum is probably of a female.

Based on the above sided elements, together with the ageing data, we argue for the presence of at least two juveniles and two adults in the JB. The coexistence of both shed and unshed antlers in the same assemblage will be further discussed below.

A quantitative analysis of *Dama* bones resulted in an MNI of four animals (a single juvenile and three adults) in Layer V-5, five (two juveniles and three adults) in Layer V-6, and four (two juveniles and two adults) in the JB. The NISP to MNI ratio indicates that the bones in all the assemblages are highly fragmented (Layer V-5: 285:4, Layer V-6: 545:5, the JB: 355:4), a phenomenon that is further accentuated by the inclusion of bones attributed to BSGD (Table 4.25). This group consists of bones that could not be attributed to a particular species but are similar in size to *Dama*. Of the three GBY assemblages, BSGD includes mainly long bone fragments (Layer V-5: 89.55%, Layer V-6: 84.19%, the JB: 91.71%). The distribution of BSGD is similar between Layer V-5 and Layer V-6, and not significantly different when compared to the JB (Layers V-5 and V-6, r = 1.00, Layer V-6 and the JB, r = 0.866, p = 0.333).

BSGD long bone fragments from Layers V-5 and V-6 measure 3.0–3.4 cm in length, 1.0–1.4 cm in width, and 0.5 cm in thickness. The size of the long bone fragments from the JB is somewhat larger, with a length of nearly 4 cm, a width of 1.49 cm, and a thickness of 0.56 cm (Table 5.6). The longest flat splinters (4.35 cm) originated in the JB, while those from Layers V-5 and V-6 are shorter (2.7–2.8 cm). The length and width of flat splinters from all the assemblages are similar, measuring 1.3–1.5 and 0.4–0.5 cm, respectively (Table 5.6). In comparison to the above BSGD fragments, *Dama* long bones (i.e., limb and leg bones) from all the assemblages are larger in length, width, and thickness. However, there is much overlap in the size of all three dimensions (Table 5.7); thus, the attribution of an element to a particular species- or body-size is not only size-dependent: the smaller the fragment, the less characteristic, and therefore less easily assigned to a particular anatomical element.

Furthermore, portions of antler tines and beams and some fragmented cervidae teeth (see Chapter 4) are probably also of *Dama*. It is unclear what caused the GBY antlers to break into numerous pieces (Layer V-5: N = 19, Layer V-6: N = 53, the JB: N = 11); deer antlers are known to be an extremely tough material, durable under applied stress (Evans et al. 2005) and an efficient raw material for projectile points (Pokines 1998). We shall try to address this topic when discussing the site's taphonomic history (see Chapter 7).

The MNE was obtained by counting the bone epiphyses per body side as well as long bone shaft fragments whose anatomical position could be determined (Lam et al. 1998). Tables 5.8, 5.9, and 5.10 present the calculated values of the MNE and MAU. Since both utilize sided parts for calculation (Lyman 1994: 106), the *Dama* assemblage is reduced. Vertebrae, ribs, and shafts become underrepresented during these calculations. We have calculated the MAU in two ways: the first is based on the sided-MNE (termed "MAU-sided") and the second, on the NISP.

A significant positive correlation was found in Layer V-5 between the MAU-sided counts and the NISP-based MAU counts (r = 0.723, p = 0.001). After adding shaft limbs to the calculation, this correlation weakened (r = 0.593, p = 0.001) since most shaft limbs are not sided (Table 5.8). A similar trend is seen in Layer V-6 (r = 0.696, p = 0.000). This correlation is less evident than when shaft counts are included (r = 0.571, p = 0.000) (Table 5.9). In the JB, the inclusion of shafts in the correlation between the MAU-sided counts and NISP-based MAU counts did not result in a pronounced difference (without shafts: r = 0.777, p = 0.000; with shafts: r = 0.668, p = 0.000), suggesting the somewhat less-fragmented nature of the *Dama* remains in this assemblage (Table 5.10). This is also suggested by the bone-fragmentation analysis, as fragments from the JB tend to be larger than those from the other layers.

5.3.2 Skeletal-Element Representation of Additional Species and Their Probable BSG

Elephant remains from Area C and the JB are represented primarily by teeth and tusk fragments (see Tables 4.6; Fig. 5.9). The tendency of elephant teeth and tusks towards high fragmentation has resulted in numerous splinters and these elements' subsequent overrepresentation in the GBY assemblages (Layer V-5: 72.73%, Layer V-6: 50%, the JB: 47.06%). Moreover, this specific trait of the elephant teeth and bones has resulted in particular exfoliation and weathering patterns: tooth disintegration expressed by breakages between the enamel plates, followed by weathering into smaller fragments, and tusk breakage into cuboids particles, while the long bones disintegrate into thick pieces (Rabinovich, pers. comm.).

Combining the elephant remains with BSGA yields data on the presence of trunk elements that are comprised of vertebrae and ribs (Layer V-5: 4.55%, Layer V-6: 13.89%, the JB: 23.53%), as well as carpal and tarsal bones (Layer V-5: 9.10%, Layer V-6: 5.56%, the JB: 11.76%) (Fig. 5.9). The MNI cannot be estimated when sided elements are so few and the NISP is also low. For this reason, other calculated values, such as the MNE and MAU, can hardly be used.

The only complete tooth, that of a 26-year-old animal, was found in the JB (no. 753, r). Based on the complete bones found, we conclude that both young adults and adults are present in all three assemblages. Detailed measurements of the bones have yielded additional information on this taxon's remains (Table 5.11). The large range of its measurements is due to the incorporation of small tusk and larger trunk elements. The higher value of elephant and BSGA-elements from the JB should be attributed to the incorporation of a complete tooth.

Skeletal representation of hippopotamus and BSGB is limited. Teeth outnumber body parts in all three assemblages (Layer V-5: 25%, Layer V-6: 50%, the JB: 35.71%), followed by long bone shafts (Layer V-5: 12.50%, Layer V-6: 17.86%, the JB: 7.14%) and flat splinters (Layer V-5: 31.25%, Layer V-6: 7.14%, the JB: 21.43%). In addition to a few trunk fragments, fore limbs and leg remains are also assigned to BSGB (Fig. 5.10).

Based on Laws (1968) tooth-ageing criteria, all the GBY assemblages represent juveniles: in Layer V-5 and the JB there is evidence of animals less than three years old; and in Layer V-6, of an animal approximately one year old. The MNI, based on the hippopotamus elements and BSGB, is of a juvenile and an adult from each assemblage. Considering the low number of bones, this MNI presents a high value when compared to those of other species represented by a higher number of skeletal remains, such as *Dama*.

Table 5.6 Measurements (mm) of BSGD skeletal elements (fallow deer, Caprinae, 40–80 kg) from Area C and the JB

Length		Width		Thickness	
Long bone fragments V-5					
Mean	33.32	Mean	14.08	Mean	5.23
Median	31.38	Median	13.69	Median	5.18
Standard deviation	12.18	Standard deviation	4.01	Standard deviation	1.25
Range	101.64	Range	24.40	Range	6.01
Minimum	13.06	Minimum	6.26	Minimum	2.54
Maximum	114.70	Maximum	30.66	Maximum	8.55
Count	179	Count	179	Count	176
Long bone fragments V-6					
Mean	33.77	Mean	13.14	Mean	5.16
Median	30.68	Median	12.53	Median	4.78
Standard deviation	14.74	Standard deviation	4.48	Standard deviation	2.00
Range	117.46	Range	35.90	Range	16.60
Minimum	10.91	Minimum	4.53	Minimum	1.94
Maximum	128.37	Maximum	40.43	Maximum	18.54
Count	275	Count	275	Count	275
Long bone fragments JB					
Mean	39.49	Mean	14.97	Mean	5.63
Median	38.75	Median	14.68	Median	5.48
Standard deviation	13.23	Standard deviation	4.99	Standard deviation	1.76
Range	84.48	Range	23.66	Range	10.86
Minimum	15.72	Minimum	7.00	Minimum	2.07
Maximum	100.20	Maximum	30.66	Maximum	12.93
Count	154	Count	154	Count	153
Flat splinters V-5					
Mean	27.71	Mean	13.92	Mean	5.45
Median	27.30	Median	15.43	Median	5.08
Standard deviation	9.09	Standard deviation	4.23	Standard deviation	1.86
Range	36.63	Range	11.79	Range	6.64
Minimum	10.21	Minimum	6.50	Minimum	3.14
Maximum	46.84	Maximum	18.29	Maximum	9.78
Count	15	Count	15	Count	15
Flat splinters V-6					
Mean	28.74	Mean	15.11	Mean	4.81
Median	28.53	Median	14.96	Median	4.94
Standard deviation	11.67	Standard deviation	6.04	Standard deviation	1.16
Range	50.71	Range	2.884	Range	5.07
Minimum	11.31	Minimum	8.00	Minimum	1.93
Maximum	62.02	Maximum	36.84	Maximum	7.00
Count	32	Count	32	Count	31
Flat splinters JB					
Mean	43.57	Mean	13.70	Mean	4.34
Median	33.03	Median	13.98	Median	4.10
Standard deviation	19.00	Standard deviation	3.47	Standard deviation	1.63
Range	50.23	Range	10.36	Range	5.29
Minimum	28.60	Minimum	8.71	Minimum	2.36
Maximum	78.83	Maximum	19.07	Maximum	7.65
Count	7	Count	7	Count	7

Table 5.7 Measurements (mm) of *Dama* sp. long bones from Area C and the JB

Length		Width		Thickness	
V-5					
Mean	48.48	Mean	18.18	Mean	6.86
Median	48.20	Median	17.54	Median	5.51
Standard deviation	21.98	Standard deviation	5.38	Standard deviation	4.76
Range	97.51	Range	25.63	Range	26.37
Minimum	14.09	Minimum	8.45	Minimum	2.34
Maximum	111.60	Maximum	34.08	Maximum	28.71
Count	80	Count	80	Count	78
V-6					
Mean	49.96	Mean	17.49	Mean	6.24
Median	45.98	Median	17.04	Median	5.24
Standard deviation	24.00	Standard deviation	6.29	Standard deviation	3.63
Range	129.63	Range	42.52	Range	25.00
Minimum	1.14	Minimum	3.00	Minimum	2.06
Maximum	130.77	Maximum	45.52	Maximum	27.06
Count	179	Count	179	Count	179
JB					
Mean	58.03	Mean	20.74	Mean	7.84
Median	52.00	Median	20.20	Median	6.19
Standard deviation	26.88	Standard deviation	6.37	Standard deviation	4.86
Range	199.31	Range	33.94	Range	36.76
Minimum	15.69	Minimum	8.94	Minimum	2.45
Maximum	215.00	Maximum	42.88	Maximum	39.21
Count	157	Count	157	Count	156

Table 5.8 NISP, MNE, and MAU values of *Dama* sp. skeletal elements from Layer V-5

	NISP	MNE L	MNE R	MNE not sided	Skeletal-element frequency	MAU sided	MAU NISP
Antler	1				2	0.00	0.50
Skull fragment	9	1			2	0.50	4.50
Maxilla fragment	0				2	0.00	0.00
Upper tooth	23						
dP^4	1				2	0.00	0.50
P^3	2	1	2		2	1.50	1.00
P^4	6	3	2		2	2.50	3.00
M^1	1	1			2	0.50	0.50
M^3	3	3			2	1.50	1.50
Mandible fragment	9	4	3		2	3.50	4.50
Lower tooth	37						
I$_1$–I$_3$	6	3	2		3	1.67	2.00
dP$_2$	1		1		2	0.50	0.50
dP$_3$	2		2		2	1.00	1.00
dP$_4$	3	3			2	1.50	1.50
P$_3$	5	2	2		2	2.00	2.50
P$_4$	3	2			2	1.00	1.50
M$_1$	1		1		2	0.50	0.50
M$_2$	2	1	2		2	1.50	1.00
M$_3$	3	2	2		2	2.00	1.50
Tooth fragment	32						
Atlas	1			1	1	1.00	1.00
Axis	0			0	1	0.00	0.00
Cervical vertebra	4			4	5	0.80	0.80
Thoracic vertebra	1			1	13	0.08	0.08
Lumbar vertebra	2			2	7	0.29	0.29

Table 5.8 (continued)

	NISP	MNE L	MNE R	MNE not sided	Skeletal-element frequency	MAU sided	MAU NISP
Vertebra fragment	21			21			
Vertebrae spine	2			2			
Rib, proximal	3			3	26	0.12	0.12
Rib shaft	20						
Scapula, distal	2		2		2	1.00	1.00
Scapula fragment	2						
Humerus, proximal	0				2	0.00	0.00
Humerus shaft	6	1	1		2	1.00	3.00
Humerus, distal	1				2	0.00	0.50
Radius, proximal	3	1	1		2	1.00	1.50
Radius shaft	7	1	1		2	1.00	3.50
Radius, distal	2				2	0.00	1.00
Ulna, proximal	2		1		2	0.50	1.00
Ulna shaft	3				2	0.00	1.50
Ulna, distal	1				2	0.00	0.50
Carpal	2	1			12	0.08	0.17
Metacarpal, proximal	0				2	0.00	0.00
Metacarpal shaft	4				2	0.00	2.00
Metacarpal, distal	3				2	0.00	1.50
Acetabulum	1				2	0.00	0.50
Ilium	2		1		2	0.50	1.00
Pubis	3				2	0.00	1.50
Ishium	2		1		2	0.50	1.00
Pelvis fragment	1						
Femur, proximal	1		1		2	0.50	0.50
Femur shaft	8		2		2	1.00	4.00
Femur, distal	0				2	0.00	0.00
Tibia, proximal	0				2	0.00	0.00
Tibia shaft	10	1	1		2	1.00	5.00
Tibia, distal	5	1	1		2	1.00	2.50
Astragalus	1		1		2	0.50	0.50
Calcaneum	1				2	0.00	0.50
Tarsal	3	1			6	0.17	0.50
Sesamoid	2						
Metatarsal, proximal	8	1	3		2	2.00	4.00
Metatarsal shaft	11				2	0.00	5.50
Metatarsal, distal	1				2	0.00	0.50
Phalanx 1	1			1	8	0.13	0.13
Phalanx 1, proximal	3			3			
Phalanx 1 shaft	2			2			
Phalanx 1, distal	2			2			
Phalanx 2	1			1	8	0.13	0.13
Phalanx 2, proximal	1			1			
Phalanx 2 shaft	0			0			
Phalanx 2, distal	2			2			
Phalanx 3	4			4	8	0.50	0.50
Phalanx 3, proximal	1			1			
Metapodial shaft	5						

Skeletal elements of *Sus scrofa* include teeth and leg and limb bones. The three specimens from Layer V-5 comprise two fused phalanges and a single tooth fragment. Layer V-6 yielded a fused proximal radius (no. 1,587) of an animal older than 12 months and an unfused lateral phalanx (no. 7,980) of an animal approximately 12 months old. The remains from the JB comprise a single adult animal tooth (no. 13,108, I^3, l). The paucity of *Sus scrofa* remains allow only limited age reconstruction. It is possible that some of its bones are actually found in BSGC, which includes animals weighing 80–250 kg. We shall thus first examine the remains of other species potentially found in BSGC, such as

Table 5.9 NISP, MNE, and MAU values of *Dama* sp. skeletal elements from Layer V-6

	NISP	MNE L	MNE R	MNE not sided	Skeletal-element frequency	MAU Sided	MAU NISP
Antler	1		1		2	0.50	0.50
Antler fragment	1						
Skull fragment	36		1		2	0.50	18.00
Maxilla fragment	2	1			2	0.50	1.00
Upper tooth	20						
dP^2	1		1		2	0.50	0.50
dP^4	3	2	1		2	1.50	1.50
P^2	2	2			2	1.00	1.00
P^3	4	2	2		2	2.00	2.00
P^4	2		2		2	1.00	1.00
M^1	3		3		2	1.50	1.50
M^2	1		1		2	0.50	0.50
M^3	4	2	2		2	2.00	2.00
Mandible	1		1		2	0.50	0.50
Mandible fragment	24	7	6				
Lower tooth	38						
I_1–I_3	15	9	5		3	4.67	5.00
dP_2					2	0.00	0.00
dP_3					2	0.00	0.00
dP_4					2	0.00	0.00
P_2	3	1	3		2	2.00	1.50
P_3	5	3	2		2	2.50	2.50
P_4					2	0.00	0.00
M_1	3	2	1		2	1.50	1.50
M_2	2	1	1		2	1.00	1.00
M_3	2	1	1		2	1.00	1.00
Tooth fragment	26						
Atlas	5			5	1	5.00	5.00
Axis	2			2	1	2.00	2.00
Cervical vertebra	6			6	5	1.20	1.20
Thoracic vertebra	3			3	13	0.23	0.23
Lumbar vertebra	4			4	7	0.57	0.57
Vertebra fragment	33			33			
Vertebrae spine	10			10			
Sacrum	2				1	0.00	2.00
Rib	2			2	26	0.08	0.08
Rib, proximal	9			9			
Rib shaft	39			39			
Scapula, proximal	1						
Scapula, blade	3	2					
Scapula, distal	5	3			2	1.50	2.50
Scapula fragment	7						
Humerus, proximal	1				2	0.00	0.50
Humerus shaft	11	2	1		2	1.50	5.50
Humerus, distal	5				2	0.00	2.50
Radius, proximal	3	2	1		2	1.50	1.50
Radius shaft	26		1		2	0.50	13.00
Radius, distal	1				2	0.00	0.500
Ulna, proximal	2		1		2	0.50	1.000
Ulna shaft	1				2	0.00	0.500
Ulna, distal	1		1		2	0.50	0.500
Radius ulna, distal	2						
Carpal	7	2	2		12	0.33	0.583
Metacarpal, proximal	8	1	2		2	1.50	4.000
Metacarpal shaft	17				2	0.00	8.500
Metacarpal, distal	1				2	0.00	0.500

Table 5.9 (continued)

	NISP	MNE L	MNE R	MNE not sided	Skeletal- element frequency	MAU Sided	MAU NISP
Acetabulum	3	1			2	0.50	1.500
Ilium	8	1					
Pubis	2						
Ishium	1	1	3				
Pelvis fragment	3						
Femur, proximal	0				2	0.00	0.00
Femur shaft	17	1	1		2	1.00	8.50
Femur, distal	4	2			2	1.00	2.00
Tibia, proximal					2	0.00	0.00
Tibia shaft	29	3	1		2	2.00	14.50
Tibia, distal	3	2			2	1.00	1.50
Astragalus	3	1	1		2	1.00	1.50
Calcaneum	3		2		2	1.00	1.50
Tarsal	6	3	2		6	0.83	1.00
Sesamoid	6						
Metatarsal, proximal	14	3	5		2	4.00	7.00
Metatarsal shaft	21				2	0.00	10.50
Metatarsal, distal	5				2	0.00	2.50
Phalanx 1	1			1	8	0.13	0.13
Phalanx 1, proximal	11			11			
Phalanx 1 shaft	2			2			
Phalanx 1, distal	7			7			
Phalanx 2	2			2	8	0.25	0.25
Phalanx 2, proximal	3			3			
Phalanx 2 shaft	2			2			
Phalanx 2, distal	6			6			
Phalanx 3	5			5	8	0.63	0.63
Phalanx 3, proximal	1			1			
Phalanx 3 shaft	1			1			
Phalanx 3, distal	1			1			
Metapodial shaft	5						
Metapodial, distal	4						

Megaloceros, Cervus cf. *elaphus*, and Bovini indet., and only then continue to examine BSGC.

Three bones from Layer V-6 and the JB represent giant deer (*Megaloceros*). Their presence is of importance from the perspective of biostratigraphy and paleoecology, but cannot be used for the taphonomic reconstruction of the GBY assemblages. The same holds true for the red deer; a total of five bones were identified, comprising a tooth from Layer V-5, a tooth from Layer V-6, and two bones from the JB. The tooth from V-6 is of a juvenile animal (dP^4, no. 1,482, r).

Elements attributed to BSGC are present in each assemblage (Layer V-5: N = 49, Layer V-6: N = 86, the JB: N = 94; see Table 4.24). The attribution of the remains to a specific species within BSGC is more difficult since both bovids and cervids may be present in this group. In each assemblage, long bone shafts and flat splinters constitute at least 50% of this group. The remaining body parts in this group are skull fragments, trunk elements, and limb shafts. By definition, a BSG includes less complete elements, and the same can be observed here (Fig. 5.11). A significant positive correlation was found between skeletal-element frequency of BSGC between Layer V-6 and the JB (r = 0.657, p = 0.001), with a less evident correlation observed between Layers V-5 and V-6 (r = 0.467, p = 0.033), suggesting a similar skeletal distribution in BSGC.

Two incomplete elements of *Bos* sp. (previously attributed to *Pelorovis*; see Chapter 4) were found in Layer V-5; one is a proximally fused metatarsal, an element that fuses before birth. A proximal metatarsal, a tarsal, and upper- and lower-jaw teeth were found in Layer V-6. The presence of a dP^4 (no. 2,102, l) suggests an animal younger than 30 months (Grigson 1982). The sided teeth indicate the probable presence of two animals, as the dP^4 and the P^2 (no. 10,160, l) display different levels of wear. Although P^{2}s erupt at approximately 18–24 months (Grigson 1982), the tooth wear suggests a much older specimen.

Bos sp. remains in the JB are more abundant, with 17 skull and postcranial elements attributed to this species. The skull elements include a horn core fragment, a mandible body,

Table 5.10 NISP, MNE, and MAU values of *Dama* sp. skeletal elements from the JB

	NISP	MNE L	MNE R	MNE not sided	Skeletal-element frequency	MAU sided	MAU NISP
Antler	3	1	1		2	1.00	1.50
Antler fragment	1						
Skull fragment	11				2	0.00	5.50
Maxilla fragment	1		1		2	0.50	0.50
Upper tooth	15						
dP^2					2	0.00	0.00
dP^4	3	2	1		2	1.50	1.50
P^2					2	0.00	0.00
P^3					2	0.00	0.00
P^4					2	0.00	0.00
M^1	5	4	1		2	2.50	2.50
M^2	3	2	1		2	1.50	1.50
M^3	4	1	1		2	1.00	2.00
Mandible	1	1			2	0.50	0.500
Mandible fragment	15	2	7				
Lower tooth	13						
I$_1$–I$_3$	4	2	1		3	1.00	1.33
dP$_2$					2	0.00	0.00
dP$_3$					2	0.00	0.00
dP$_4$					2	0.00	0.00
P$_2$					2	0.00	0.00
P$_3$	3	2	1		2	1.50	1.50
P$_4$	3	2	1		2	1.50	1.50
M$_1$	2	2			2	1.00	1.00
M$_2$	1	1			2	0.50	0.50
M$_3$	2	2			2	1.00	1.00
Tooth fragment	7						
Atlas	2			2	1	2.00	2.00
Axis	2			2	1	2.00	2.00
Cervical vertebra	3			3	5	0.60	0.60
Thoracic vertebra	5			5	13	0.39	0.39
Lumbar vertebra	4			4	7	0.57	0.57
Vertebra fragment	26			26			
Vertebrae spine	10			10			
Sacrum	1			1	1	1.00	1.00
Rib, proximal	3			3	26	0.12	0.12
Rib shaft	18			18			
Scapula blade	1	1					
Scapula, distal	3		2		2	1.00	1.50
Scapula fragment	5						
Humerus, proximal	1				2	0.00	0.50
Humerus shaft	6				2	0.00	3.00
Humerus, distal	7	2	2		2	2.00	3.50
Radius, proximal	6	1	3		2	2.00	3.00
Radius shaft	17	1	2		2	1.50	8.50
Radius, distal	5	1			2	0.50	2.50
Ulna, proximal	5	1	4		2	2.50	2.50
Ulna shaft	2				2	0.00	1.00
Ulna, distal	1				2	0.00	0.50
Radius ulna, distal	1				2	0.00	0.50
Carpal	0				12	0.00	0.00
Metacarpal, proximal	1				2	0.00	0.50
Metacarpal shaft	12				2	0.00	6.00

Table 5.10 (continued)

	NISP	MNE L	MNE R	MNE not sided	Skeletal-element frequency	MAU sided	MAU NISP
Metacarpal, distal	2				2	0.00	1.00
Acetabulum	3	2			2	1.00	1.50
Ilium	3	1	1				
Pubis	2	2					
Ishium							
Pelvis fragment	5						
Femur	1	1			2	0.50	0.50
Femur, proximal	3	2	1		2	1.50	1.50
Femur shaft	17	3			2	1.50	8.50
Femur, distal	3				2	0.00	1.50
Tibia, proximal	1				2	0.00	0.50
Tibia shaft	29	5	2		2	3.50	14.50
Tibia, distal	1		1		2	0.50	0.50
Astragalus	4	1	3		2	2.00	2.00
Calcaneum	2		2		2	1.00	1.00
Tarsal	8	1	4		6	0.83	1.33
Sesamoid	4						
Metatarsal, proximal	13	4	3		2	3.50	6.50
Metatarsal shaft	20	1			2	0.50	10.00
Metatarsal, distal	2				2	0.00	1.00
Phalanx 1	1			1	8	0.13	0.13
Phalanx 1, proximal	5			5			
Phalanx 1 shaft	1			1			
Phalanx 1, distal							
Phalanx 2					8	0.00	0.00
Phalanx 2, proximal	1			1			
Phalanx 2 shaft							
Phalanx 2, distal	2			2			
Phalanx 3	6			6	8	0.75	0.75
Phalanx 3, proximal							
Phalanx 3 shaft							
Phalanx 3, distal							
Metapodial shaft	2						

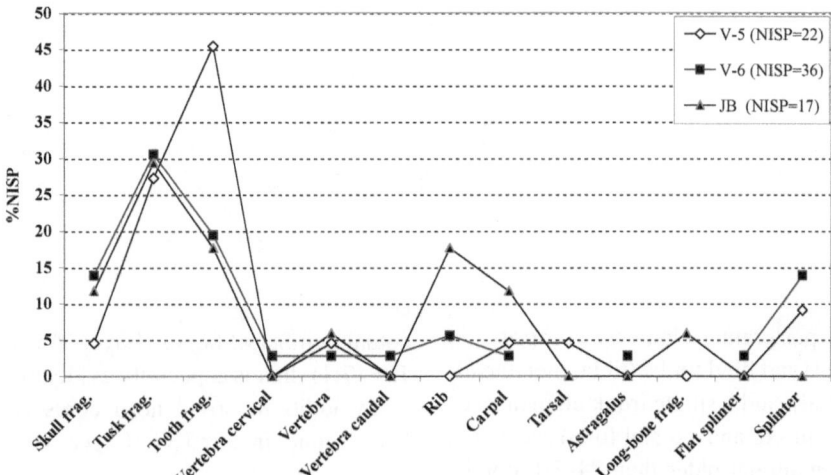

Fig. 5.9 Skeletal-element representation of *Palaeoloxodon antiquus* and BSGA (elephant > 1,000 kg) from Area C and the JB

Table 5.11 Measurements (mm) of *Palaeoloxodon antiquus* skeletal elements and BSGA (elephant > 1,000 kg) from Area C and the JB

Length		Width		Thickness	
V-5					
Mean	33.16	Mean	15.82	Mean	8.37
Median	19.09	Median	10.31	Median	4.82
Standard deviation	36.12	Standard deviation	12.01	Standard deviation	8.46
Range	162.66	Range	47.49	Range	31.40
Minimum	7.34	Minimum	4.27	Minimum	2.40
Maximum	170.00	Maximum	51.76	Maximum	33.80
Count	21	Count	21	Count	21
V-6					
Mean	49.98	Mean	32.13	Mean	15.39
Median	36.98	Median	20.15	Median	6.33
Standard deviation	45.70	Standard deviation	29.98	Standard deviation	21.53
Range	193.83	Range	119.86	Range	93.52
Minimum	9.00	Minimum	4.34	Minimum	1.78
Maximum	202.83	Maximum	124.20	Maximum	95.30
Count	34	Count	34	Count	33
JB					
Mean	75.75	Mean	44.64	Mean	18.28
Median	51.83	Median	36.50	Median	10.47
Standard deviation	70.46	Standard deviation	39.26	Standard deviation	21.32
Range	258.11	Range	155.06	Range	82.19
Minimum	16.54	Minimum	7.19	Minimum	2.81
Maximum	274.65	Maximum	162.25	Maximum	85.00
Count	17	Count	17	Count	17

Fig. 5.10 Skeletal-element representation of *Hippopotamus* and BSGB (hippopotamus, rhinoceros < 1,000 kg) from Area C and the JB

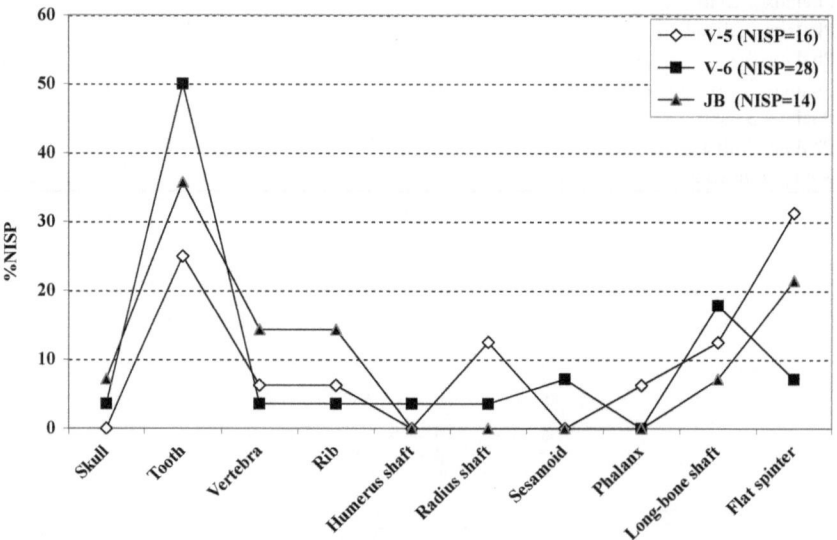

and isolated teeth. The postcranial elements comprise the fore (ulna and metacarpal) and hind limbs (tibia, metatarsal), lower leg fragments (tarsal), and a single trunk element (vertebrae). All elements are fused, and a distal fused metatarsal (no. 2,080) belongs to an animal older than 24–30 months. Based on the sided teeth, at least one animal is present at JB.

The NISP to MNI ratio of the *Bos* sp. indicates that all the assemblages feature high MNI counts in comparison to the actual number of bones (Layer V-5: 2:?,[1] Layer V-6: 6:2, the JB: 17:1); this is especially evident in Layer V-6.

In addition, bovid horn cores (*processus cornus*) have been found in Area C (Layer V-5: NISP = 2, Layer V-6:

---

[1] A question mark denotes when the available data was insufficient to specify the MNI.

Fig. 5.11 Skeletal-element
representation of BSGC (giant
deer, red deer, boar, bovine,
80–250 kg) from Area C
and the JB

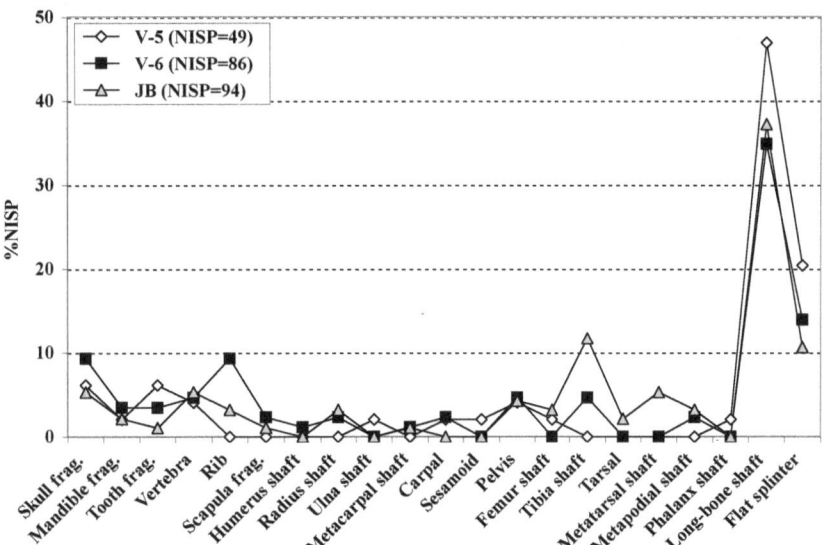

NISP = 11, the JB: NISP = 6), including a single fragment of a horn base from the JB with a fragment of the
frontal bone (no. 774). Although lacking identifiable marks
necessary for species identification, the horn core fragments
are assigned by their size and thickness to medium- and
large-sized bovids. The fragments display a wide size range,
especially in the JB, where an 11.8-cm-long fragment was
found (Table 5.12).

Additional bovid remains are a distal radius (Layer V-5),
a distal scapula (Layer V-6), and a proximal ulna (the JB)
of the Caprini indet. These remains are important biostrati-
graphic landmarks but cannot be used for the taphonomic
identification.

A few gazelle bones were found in each assemblage
including cranial and postcranial elements. Four gazelle

bones were found in Layer V-5: a tooth fragment, fore limbs
(a distal ulna and a proximal metacarpal), and a sesamoid
bone; and eight fragments were found in Layer V-6: five
cranial elements (a horn core fragment, a hyoid fragment,
two mandible fragments, and a tooth fragment) and three
postcranial elements (a scapular blade, a femoral shaft, and
a fragmented astragalus). The surface appearance of a horn
core fragment (no. 4,366) attributes it to a young gazelle, as
male gazelle horn cores are characterized by a spongy surface
until the age of two years (Davis 1980).

Twelve gazelle bones were identified in the JB: four cranial elements (M^2, two M_2s, and a mandible fragment) and
eight postcranial elements (two atlas fragments, a proximal
radius, a pelvis fragment, a proximal metatarsal, a distally
unfused metapodial, a first phalanx, and a proximal portion

Table 5.12 Measurements (mm) of Bovid horn core fragments from Area C and the JB

Length		Width		Thickness	
V-6					
Mean	35.76	Mean	21.34	Mean	8.18
Median	37.00	Median	20.29	Median	8.88
Standard deviation	6.51	Standard deviation	8.53	Standard deviation	3.43
Range	20.90	Range	25.88	Range	10.10
Minimum	25.41	Minimum	10.27	Minimum	3.45
Maximum	46.31	Maximum	36.15	Maximum	13.55
Count	11	Count	11	Count	11
JB					
Mean	66.97	Mean	30.00	Mean	10.53
Median	67.03	Median	31.53	Median	9.24
Standard deviation	34.14	Standard deviation	15.07	Standard deviation	3.45
Range	92.93	Range	35.26	Range	9.36
Minimum	25.41	Minimum	10.27	Minimum	7.03
Maximum	118.34	Maximum	45.53	Maximum	16.39
Count	6	Count	6	Count	6

Fig. 5.12 Skeletal-element representation of BSGE (gazelle, roe deer, 15–40 kg) from Area C and the JB

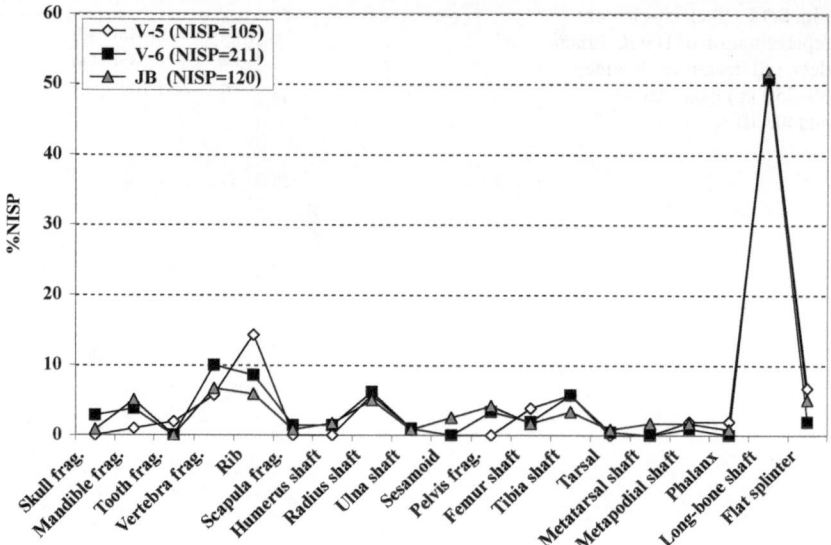

of a second phalanx). The two atlas fragments (nos. 1,016, 1,033) most probably belonged to the same animal. The distally unfused metapodial (no. 11,904) is of an animal younger than 10–16 months (Davis 1980). The M^2 (no. 1,842, l) is worn and therefore suggests an age older than three years (Rabinovich 1998).

Thus, based on the sided and the age of the gazelle elements, the NISP to MNI ratio of the three assemblages reveal high MNI counts when compared to the number of actual bones (Layer V-5: 4:?[2], Layer V-6: 8:2, the JB: 12:2). In Layer V-6 there is evidence of an animal younger than two years, but no further information could be retrieved from this layer. There is evidence of the presence of a juvenile and an adult animal in the JB.

Gazelle is most likely the main contributor to BSGE (Layer V-5: NISP = 105, Layer V-6: NISP = 211, the JB: NISP = 120). Long bone shafts comprise half the specimens in all assemblages. Vertebrae and rib fragments comprise another significant component of the finds in Layers V-5 (20%) and V-6 (18.48%), followed by fore (Layer V-5: 6.67%, Layer V-6: 9.95%, the JB: 8.33%) and hind limbs (Layer V-5: 9.52%, Layer V-6: 10.9%, the JB: 9.17%) (Fig. 5.12). Indeed, a significant positive correlation was found between the skeletal frequencies of BSGE in the three assemblages (Layers V-5 and V-6: r = 0.592, p = 0.008; Layer V-5 and the JB: r = 0.604, p = 0.006; Layer V-6 and the JB: r = 0.782, p = 0.000), suggesting similar skeletal representation in each assemblage.

Equus sp. and *E.* cf. *africanus* have also been identified in the GBY assemblages. The remains of *Equus* sp. from Layer V-5 (NISP = 12) comprise teeth (a molar fragment and two incisors), postcranial elements from the fore (a distal scapula and a residual metacarpal) and hind limbs (a pubis fragment, an ilium shaft, a tibia shaft, and three residual metatarsals), and a sesamoid bone. Based on sided elements and ageing, an MNI of one adult has been determined. *E.* cf. *africanus* is represented by an upper premolar, a carpal, a tarsal, and a second phalanx.

The *Equus* sp. remains from Layer V-6 (NISP = 14) include a mandible, teeth (three lower teeth and one tooth fragment), and the following postcranial elements: a scapula fragment, a radius, two pelvis parts, a metacarpal, two metatarsals and a sesamoid. *Equus* cf. *africanus* is represented by six upper and lower teeth, a humerus, and a metacarpal (see Table 4.10).

Based on tooth eruption and epiphysis fusion, four of the skeletal elements from Layer V-6 (a mandible, a deciduous incisor, a radius, and a pelvis) were attributed to an *Equus* cf. *africanus* foal (see Chapter 4). Thus, the MNI counts from Layer V-6 include an *Equus* cf. *africanus* foal and adult (no. 2,008, eroded M_3, l) and an *Equus* sp. adult. *Equus* sp. remains from the JB (NISP = 20) are comprised of upper and lower teeth, a mandibular symphysis, limbs (a radius shaft, a proximal femur shaft, two distal femur shafts, and a proximal metatarsal), pelvis fragment, two first phalanx fragments, and a sesamoid. The two lower deciduous premolars (nos. 2,098, 2,030) and the mandibular symphysis were assigned to a foal. Thus, one foal and one adult are represented in the JB.

Equus cf. *africanus* (NISP = 11) is represented by three upper and three lower teeth, a proximal ulna, three carpals (pisiform, semi-lunar, and scaphoid), and a first phalanx. Based on two left P^3s, an MNI of two has been calculated. Thus, based on sided and aged elements, there is a single adult *Equus* sp. in both Layer V-5 and V-6. In Layer V-6, *Equus* cf. *africanus* is represented by a single juvenile and a

---

[2] See note 1 above.

Fig. 5.13 Skeletal-element representation of *Equus* sp. and *E.* cf. *africanus* from Area C and the JB

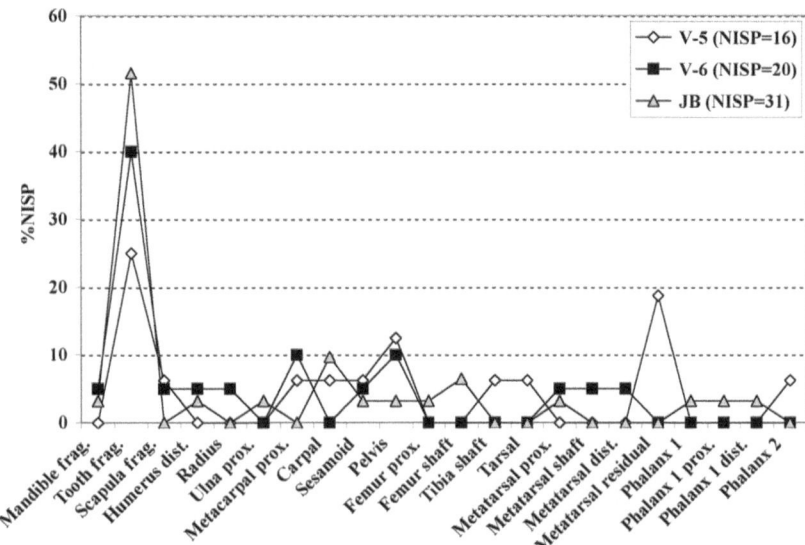

single adult, and by two adults in the JB. The equids' NISP to MNI ratio from the three assemblages reveal higher MNI counts in comparison to the actual number of bones.

Because of the small sample size available for the analysis of equids, we have lumped together all equid skeletal elements. Teeth were found to be the most frequent component, while other skeletal elements are minimally represented, usually by one to three bones (Fig. 5.13). The absence of postcranial elements may be due to the high degree of bone fragmentation, which hinders identification of these elements. In such a scenario, we would assume that these elements are found in BSGC. However, since the delicate remains of a one-year-old foal survived, this scenario seems rather unlikely.

5.3.3 Skeletal-Element Representation Versus Density Values

The skeletal elements of the various species from the GBY assemblages were correlated with bone-density values obtained through photon densitometry (Lyman 1994: table 7.6). We correlated the %NISP of *Dama* with the bone-density values of deer (Lyman 1994). Excluded from analysis was an unfused femur (no. 2,222, l), the only complete limb bone from the JB, in order to not overestimate the femur representation; i.e., including this bone will erroneously add to each femur category—proximal, shaft, and distal. It should be noted that due to the highly fragmented nature of the GBY assemblages, many bone fragments will be excluded from analysis. For the metapodials we used an average of the bone-density values of metacarpals and metatarsals. No significant correlation was found between

Dama skeletal-element frequency and bone density (Layer V-5: r = 0.255, p = 0.122; Layer V-6: r = 0.266, p = 0.65; the JB: r = 0.232, p = 0.125; Fig. 5.14), indicating that the survival of skeletal elements was not governed by bone density.

Only a few incomplete hippopotamus and BSGB bones are found in the GBY assemblages. Density values of hippopotamus bones have thus far not been obtained. One should bear in mind that as a semi-aquatic animal, the hippopotamus perhaps features a somewhat different mineral density value than terrestrial animals (Gray et al. 2007).

Only a few remains have been attributed to *Sus*, *Megaloceros*, *Cervus* cf. *elaphus*, and Bovini indet., and since some of these different species may possibly be included in BSGC, we refrained from density analysis. The sample size of *Bos* sp., which excludes teeth, is too small to allow any comprehensive and comparative analysis of the skeletal elements and average mineral densities.

We have correlated the %NISP of BSGE with the bone-density values of deer (Lyman 1994: table 7.6) and found no significant correlation (Fig. 5.15). As in the case of *Dama*, we thus concluded that the survival of skeletal elements was not mediated by bone density.

Teeth dominate the GBY equid assemblages, with the remaining elements represented by only a few bones. Equids have a thick bone wall (Outram and Rowley-Conwy 1998), a trait that might have influenced their density values. However, after correlating the GBY %NISP of equids with bone-density values of *Equus* sp. (Lam et al., 1999: table 1), no significant correlation was found between equid skeletal-element frequency and bone density (Layer V-5: r = 0.000, p = 1.000; Layer V-6: r = 0.364, p = 0.336; the JB: r = −0.236, p = 0.512), indicating that the survival of skeletal elements was not dependent on bone density (Fig. 5.16).

Fig. 5.14 Skeletal-element representation of *Dama* plotted against mineral-density values of deer bones (Lyman 1994: table 7.6)

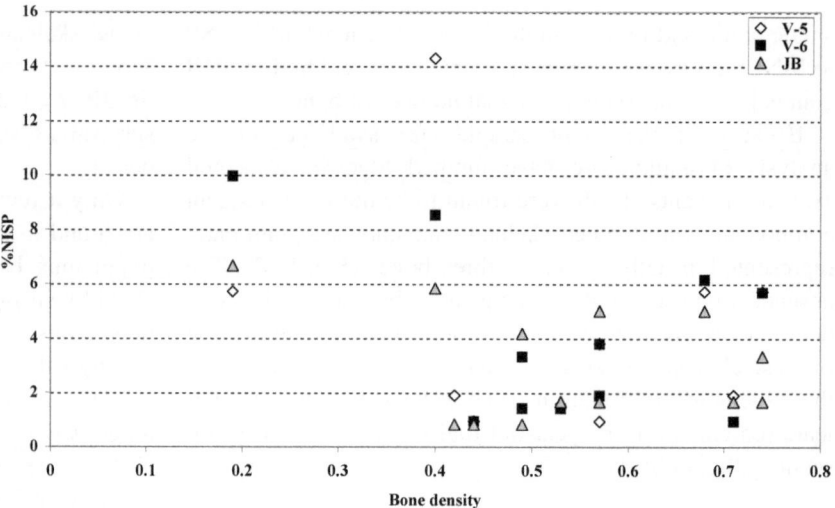

Fig. 5.15 Skeletal-element representation of BSGE plotted against mineral-density values of deer bones (Lyman 1994: table 7.6)

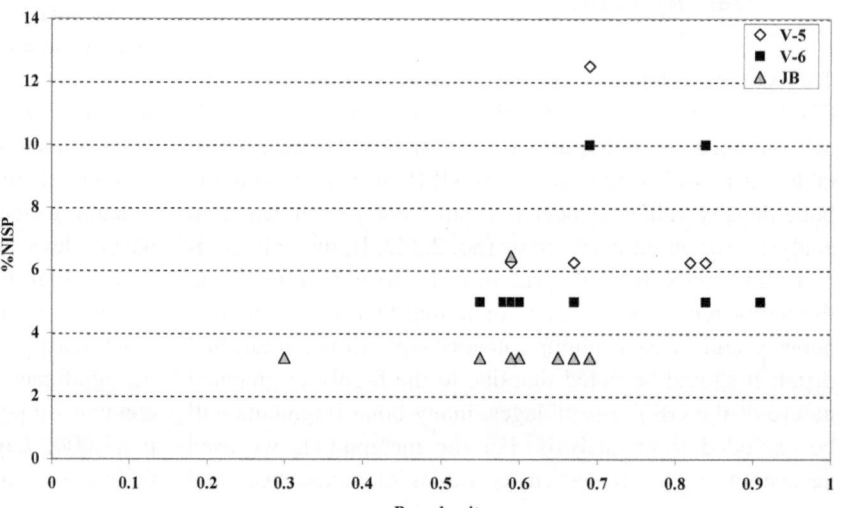

Fig. 5.16 Skeletal-element representation of Equids plotted against mineral-density values of *Equus* (Lam et al. 1999: table 1)

5.3.4 Skeletal-Element Representation and Winnowing

The site of GBY consists of a series of occupation levels bedded on a lake margin, making necessary the need to examine the susceptibility of the faunal elements to environmental agents, such as fluvial transport, assuming that the water body changed seasonally and that the remains along its shores were removed or transported. Although there is no immediate sedimentological evidence of a fluvial agent within the excavated depositional sequence that contains Layers V-5 and V-6, we refer to studies describing site-formation processes in a fluvial regime (see Chapter 2). Thus, we have adoped Voorhies' (1969) categories that group mammalian skeletal elements according to their susceptibility to fluvial transport (in Lyman 1994: table 6.5, Groups I–III). The groups are organized in increasing order from the non-fluvial winnowed bones (Group I: rib, vertebra, sacrum, and sternum), to the gradually removed (Group I & II; scapula, phalanges, and ulna; Group II: femur, tibia, humerus, metapodial, pelvis, and radius), winnowed (Group II & III: mandible ramus), to the lag deposit (Group III: skull and mandible). Since Group II includes the mammalian limb bones, we have included in our analysis the long bone shafts as well.

Dama remains, comprising the most common species at the site, are presented in Fig. 5.17, which summarizes the *Dama* %NISP per assemblage organized according to Voorhies (1969). Group I, a non-fluvial winnowed assemblage, comprises approximately 20% of the *Dama* NISP (Layer V-5: 18.94%, Layer V-6: 21.20%, the JB: 20.84%). The two groups that tend to be the most affected by fluvial transport, Group I & II and Group II, feature the highest bone survival rates in the GBY assemblages, while fewer elements

populate the last two groups that are considered as lag deposit indicators.

Elephant and BSGA, the largest species in the GBY assemblages, are over-represented by tooth and tusk fragments (Fig. 5.18). They also include skeletal elements that tend to be immediately moved by fluvial transport, such as vertebrae and astragalus (Fluvial transport index (FTI) > 75 according to Frison and Todd 1986). An inverse significant correlation was found between the FTI and the bone frequencies in the JB (r = –0.894, p = 0.041), suggesting that the elephant remains were not affected by fluvial transport (Fig. 5.18).

Examination of the hippopotamus and BSGB skeletal elements grouped according to Voorhies reveals the survival of bones in Group I, the non-fluvial winnowed assemblage. Groups I & II and II feature the highest bone survival rate of all the GBY assemblages, and only a few elements from Layer V-6 and the JB are assigned to Group III (Fig. 5.19). No significant correlation was found between the assemblages.

Examination of the BSGC skeletal elements grouped according to Voorhies reveals a low bone survival rate in Group I (Layer V-5: 4.08%, Layer V-6: 13.95%, the JB: 8.51%), a higher survival rate in Group I & II (Layer V-5: 24.49%, Layer V-6: 16.28%, the JB: 11.70%), and a particularly high rate in Group II (Layer V-5: 53.06%, Layer V-6: 51.16%, the JB: 69.15%). Fewer skeletal elements are found in Group III (Fig. 5.20). This is a similar distribution to that of *Dama* (Fig. 5.17), with the exception of relatively fewer elements attributed to Group I. A significant similarity in survival rates was found only between BSGC skeletal elements from Layer V-6 and the JB (r = 1.000).

No representatives of *Bos* sp. are found among the almost complete absence of trunk elements in Group I (except for

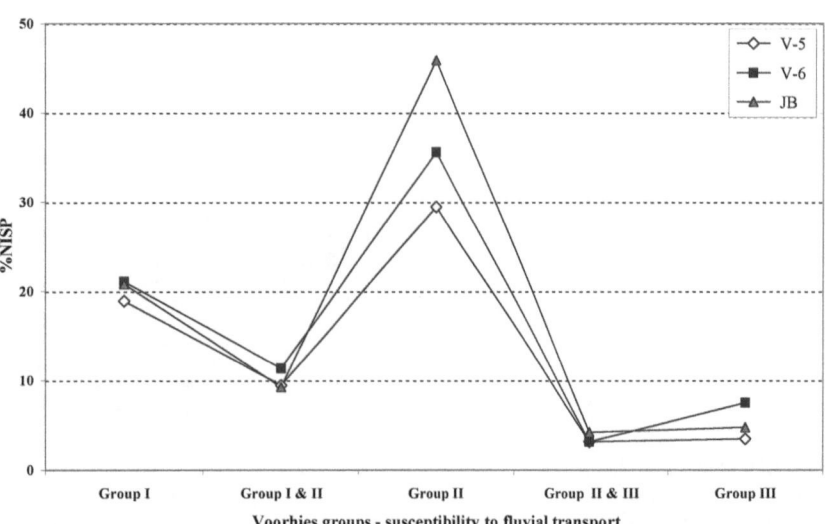

Fig. 5.17 Susceptibility of *Dama* to fluvial transport according to Voorhies groups

Fig. 5.18 Correlation between elephant and BSGA skeletal elements and FTI (Fluvial Transport Index, Frison and Todd 1986)

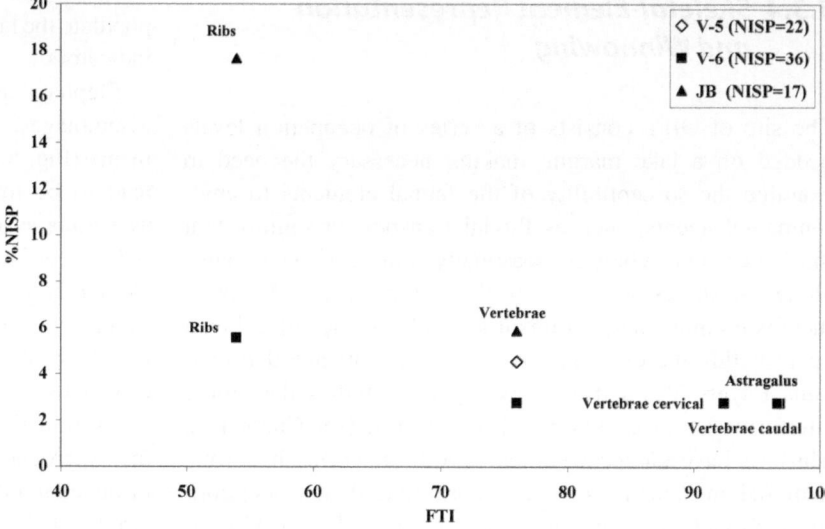

Fig. 5.19 Susceptibility of *Hippopotamus* and BSGB to fluvial transport according to Voorhies groups

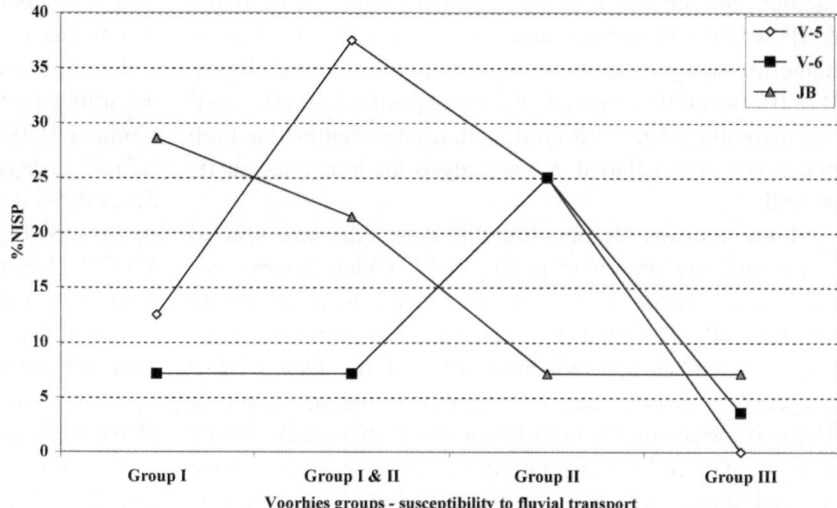

Fig. 5.20 Susceptibility of BSGC to fluvial transport according to Voorhies' groups

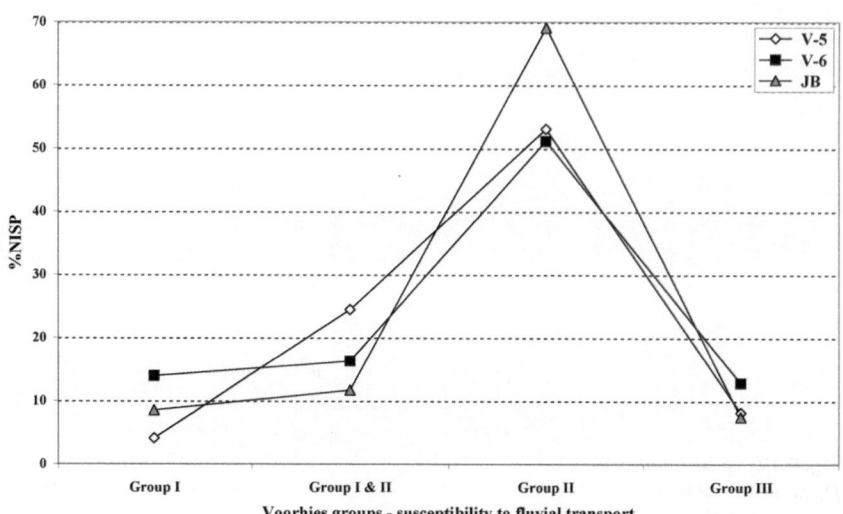

Fig. 5.21 Susceptibility of
BSGE to fluvial transport
according to Voorhies' groups

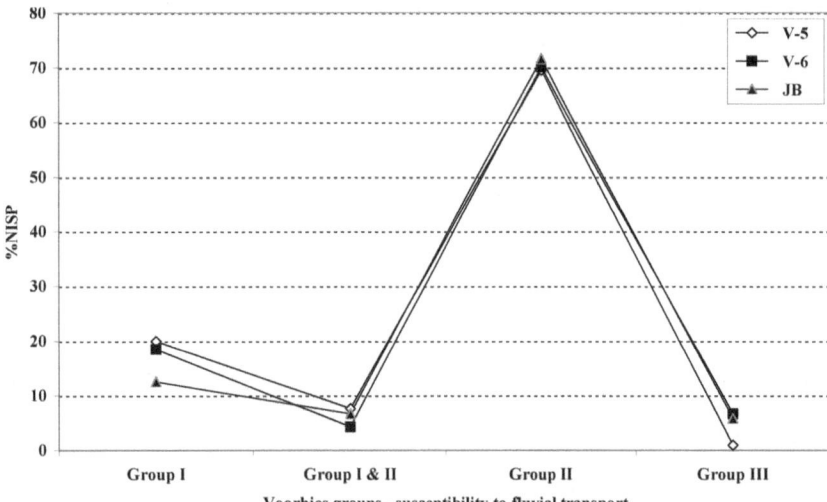

Voorhies groups - susceptibility to fluvial transport

one vertebra from the JB). The most abundant elements
(excluding teeth) are found in Group II. Bovid horn core fragments are lighter than a bulky complete skull and might be
attributed to Group I or to Group I & II, together with other
moveable elements.

Examination of BSGE skeletal elements grouped according to Voorhies reveals a relatively high bone survival rate
in Group I (Layer V-5: 20%, Layer V-6: 18.48%, the JB:
12.5%), and a lower rate in Group I & II (Layer V-5: 7.62%,
Layer V-6: 4.26%, the JB: 6.67%). The highest survival rate
is assigned to Group II (Layer V-5: 69.52%, Layer V-6:
70.14%, the JB: 71.67%), and fewer in Group III (Layer V-5:
0.95%, Layer V-6: 6.63%, the JB: 5.83%) (Fig. 5.21). This is
a similar distribution to that of *Dama* (see Fig. 5.17). Layer
V-5 and the JB feature a similar survival rate (r = 1.000),
while Layer V-6 is not significantly similar to the other
assemblages (r = 0.800, p = 0.200).

Equid skeletal-element frequencies grouped according to
Voorhies indicates that Group I is missing and that the
elements are distributed among Group I & II (Layer V-5:
12.5%, Layer V-6: 5%, the JB: 12.90%), Group II (Layer
V-5: 43.75%, Layer V-6: 35%, the JB: 22.50%), and Group
III (Layer V-6: 5%, the JB: 3.22%).

5.3.5 Nutritional Values and Skeletal-Element Representation

The nutritional value potential of the GBY assemblages was
examined. As *Dama* bones are the most abundant species,
we examined the correlation between MGUI, meat, marrow,
grease, and *Dama* element frequency. The data available for
this comparison is body parts (see Binford's normed utility indices (1978) in Lyman 1994: table 7.1). However, the
GBY body-part representation does not correlate with the

available list (as above), in which shafts are not represented
and scapulae, pelvises, and mandibles receive only one
value. Since the GBY assemblages contain incomplete skeletal elements, mainly represented by shafts (as discussed
above), we assume that the nature of the assemblages is best
represented by NISP counts. Thus, when nutritional-value
calculations are concerned, we must adapt the available element to the existing information (Table 5.13). No significant
correlation was found between MGUI, meat, marrow, and
grease, and the %NISP.

In the absence of utility indices for elephant and hippopotamus, the largest species at the site, we have applied
other measures in quantifying the possible nutritional potential of these animals. Even though an elephant, being a large
animal weighing over 1,000 kg, can theoretically provide
most of the meat-fat hominin nutritional needs, there are no
indications from the southern Levant that this large game was
heavily exploited, or even the most abundant animal in the
Paleolithic record (e.g., Rabinovich et al. 2005; Stiner 2005;
Monchot and Horwitz 2007; on Europe, see Gaudzinski
et al. 2005). The potential weight of a one-year-old hippopotamus is about 167 kg, and that of a three year-old is
483 kg; adults can reach more than 1,000 kg (Transboundary
Species Project: appendix 2). In terms of edible weight,
this species is thus of a high nutritional value even at a
young age.

Due to the meager remains of *Sus, Megaloceros, Cervus*
cf. *elaphus*, and Bovini indet. at the site, we cannot correlate the quantitative abundance of these species' skeletal
elements with utility indices. For the sake of the argument,
had we compared the skeletal-element representation with
bison, after excluding the long bone shafts and flat splinters,
we would have been left with very few candidates for such
an analysis (Fig. 5.11).

Bos sp. represents a large bovid that could have highly
contributed to the hominin diet, but its remains are rare at

Table 5.13 Nutritional values of deer (Lyman 1994: table 7.1) correlated with %NISP of *Dama* sp. from Area C and the JB

	MGUI	Meat	Marrow	Grease	V-5 %NISP	V-6 %NISP	JB %NISP
Skull	8.74	9.05	1		3.16	6.61	3.10
Antler	1.02		1		0.35	0.18	0.85
Mandible	30.26	31.1	5.74	12.51		0.18	0.28
Atlas	9.79	10.1	1	13.11	0.35	0.92	0.56
Axis	9.79	10.1	1	12.93		0.37	0.56
Cervical	35.71	37	1	17.46	1.40	1.10	0.85
Thoracic	45.43	47.2	1	12.26	0.35	0.55	1.41
Lumbar	32.05	33.2	1	14.82	0.72	0.73	1.13
Vertebral centrum	37.76	39.13	1	14.84	7.37	6.06	7.32
Rib	49.77	51.6	1	7.5	7.02	7.16	5.07
Scapula	43.47	44.7	6.4	7.69	0.70	0.92	0.85
Humerus P	43.47	28.9	29.69	75.46	0.00	0.18	0.28
Humerus D	36.52	28.9	28.33	27.84	0.35	0.92	1.97
Radius P	26.64	14.7	43.64	37.56	1.05	0.55	1.69
Radius D	22.23	14.7	66.11	32.7	0.70	0.18	1.41
Ulna P	26	14.7	43.64	37.56	0.70	0.37	1.41
Ulna D	22	14.7	66.11	32.7	0.35	0.18	0.28
Carpal	15.53	5.2	1	36.47	0.70	1.28	
Metacarpal P	12.18	5.2	61.68	16.71		1.47	0.28
Metacarpal D	10.5	5.2	67.08	42.47	1.05	0.18	0.56
Pelvis	47.89	49.3	7.85	29.26	0.70	1.47	0.85
Femur P	100	100	33.51	26.9	0.35		0.85
Femur D	100	100	49.41	100		0.73	0.85
Tibia P	64.73	25.5	43.78	69.37			0.28
Tibia D	47.09	25.5	92.9	26.05	1.75	0.55	0.28
Astragalus	31.66	11.2	1	32.47	0.35	0.55	1.13
Calcaneum	31.66	11.2	21.19	46.96	0.35	0.55	0.56
Metatarsal P	29.93	11.2	81.74	17.88	2.81	2.57	3.66
Metatarsal D	23.93	11.2	100	43.13	0.35	0.92	0.56
Phalanx 1	13.72	1.7	30	33.27	1.05	2.02	1.41
Phalanx 2	13.72	1.7	22.15	24.77	0.35	0.55	0.28
Phalanx 3	13.72	1.7	1	13.59	1.40	0.92	1.69

Spearman's rho correlations	MGUI	Meat	Marrow	Grease
V-5 %NISP	0.024 (p = 0.907)	−0.012 (p = 0.954)	−0.072 (p = 0.722)	−0.334 (p = 0.103)
V-6 %NISP	0.149 (p = 0.432)	0.017 (p = 0.930)	−0.236 (p = 0.209)	−0.294 (p = 0.129)
JB %NISP	0.058 (p = 0.757)	0.144 (p = 0.447)	−0.324 (p = 0.075)	−0.220 (p = 0.251)

GBY; only a few elements were found in the JB, where no significant correlation between the MGUI and %NISP was noted. As BSGE is mainly composed of long bone shaft fragments, correlation with MGUI indices was not undertaken.

We have examined equid (e.g., *Equus* sp. and *E*. cf. *africanus*) skeletal elements from the site vis-à-vis the General Utility Index (GUI; Outram and Rowley-Conwy 1998: table 5). According to this empirical data, more meat exists on a horse's trunk and the upper limbs than on its lower limbs (as above). With this conclusion in mind, it should be noted that trunk elements are very rare in the GBY equid assemblages. Figure 5.22 demonstrates that the GBY equid assemblages are characterized by a higher frequency of skeletal elements with a low meat and

marrow GUI, originating from the lower legs, in comparison to fewer skeletal elements with higher GUI values, such as the trunk and upper limbs. The presence of both low and high GUI skeletal elements at the site should be further explained; however, this cannot be demonstrated statistically.

Absolute horse marrow nutritional yields are low when compared to other species, with the lowest values found in the horse's lower limbs (Outram and Rowley-Conwy 1998). Since the *Equus* cf. *africanus* identified at GBY is a primitive form of an ass, caution should be exercised when applying horse nutritional values to the remains. Comparison of the marrow standardized cavity volume for horse (Outram and Rowley-Conwy 1998: table 4) to the standardized mean wet weights for zebra (Blumenschine

Fig. 5.22 Nutritional values of equids from Area C and the JB (GUI—General Utility Index from Outram and Rowley-Conwy 1998: table 5)

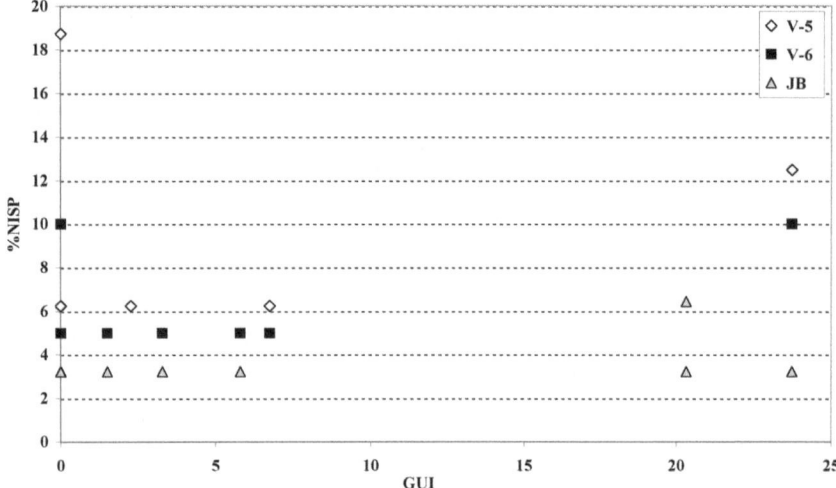

and Madrigal 1993: table 2) has revealed several differences that were previously observed by Outram and Rowley-Conwy (1998: 847). Furthermore, no significant correlation was found neither between the marrow values of the two species nor between the two marrow values (horse and zebra) and the limb %NISP of the GBY assemblages (Table 5.14).

Table 5.14 Marrow cavity volume for horse, zebra mean wet weights, and %NISP of equid limbs from Area C and the JB

	Horse[a]	Zebra[b]	V-5 %NISP	V-6 %NISP	JB %NISP
Humerus	86.3	54	0	5	3.22
Radius	50.5	52	0	5	0
Metacarpal	16	43	6.25	10	0
Femur	114	65	0	0	9.67
Tibia	59.7	100	6.25	0	
Metatarsal	16.8	81	18.75	15	3.22

[a]Outram and Rowley-Conwy (1998: table 4)
[b]Blumenschine and Madrigal (1993: table 2)

5.3.6 Skeletal-Element Representation: Conclusion

The general picture that emerges from the mammalian skeletal-element representation at GBY reveals that mammals were affected neither by density-mediated factors nor by winnowing. We could not find any significant nutritional correlation between the skeletal remains and their supposed relative nutritional value. Many reasons may explain this phenomenon, but at this stage we prefer to address the presence of the remains at face value together with the remainder of the archaeological data, and to view it as a sufficient indicator of the interest hominins

expressed in the animals. A certain similarity is observed between the layers, quite often between Layer V-6 and the JB, the significance of which will be further discussed in Chapter 7.

5.4 Animal-Induced Damage

The taphonomic history of biotic modifications on mammalian bones at GBY includes animal influence. Damage marks include tooth scratches, gnaw marks, and traces of digestion. Rodents, small- (i.e., fox), medium- (i.e., wolf), and large-sized carnivores (i.e., hyena) have been identified as responsible for these traces (Fig. 5.23). Rodent modifications occur on less than 0.5% of the total NISP of each assemblage (Table 5.15). Carnivore modifications occurr on less than 1.5% of the total NISP from Layer V-5 (NISP = 14, 1.43%) and the JB (NISP = 12, 1.08%). At 2.41% (NISP = 45) in Layer V-6, the frequency of these traces is higher than in the other layers (Table 5.15). Small and medium carnivores are the major animal agents that left their mark on the fauna of Area C and the JB.

Over 50% (NISP = 9, 52.94%) of the carnivore damage marks in Layer V-5 could be attributed to an agent. Fox- and wolf-sized carnivores are the main agents responsible here (damage caused by fox-sized carnivores: NISP = 3; damage caused by wolf-sized carnivores: NISP = 6). Rodents also scratched and gnawed the bones (NISP = 3). Also seen are four cases of gnaw marks of unknown origin, as well as a single scratched bone (Table 5.16).

Layer V-6 yielded damage marks inflicted by all three carnivores sizes, as well as by rodents. We could not assign a particular agent to the damage traces in 11 cases (20.75%) (Table 5.17). Fox-sized carnivores left 26.41% (NISP = 14) of the visible damage in the shape of tooth scratches, gnaw marks, and tooth pitting. Scratches and gnaw marks seen on 12 bones (22.64%) were caused by a medium-sized carnivore, along with 2 digested bones, a second *Dama* phalanx, and an unidentified mammal metatarsal shaft fragment. Tooth scratches, gnaw marks, and tooth pitting attributed to large-sized carnivores were recognized in six cases (11.32%).

Within the JB assemblage, damage was caused by small- (NISP = 3, 18.75%), medium- (NISP = 3, 18.75%), and large-sized carnivores (NISP = 4, 25%). Rodents damaged an additional four bones, and in two other cases the damaging agent was not identified (Table 5.18).

It should be noted that carnivores affected bones of several mammalian species at GBY (Table 5.15). This applies to skeletal elements of *Palaeoloxodon antiquus,* as an astragalus from Layer V-6 (no. 9274, r) displays gnaw marks from a medium-sized carnivore. In addition, a rib shaft fragment (the JB: no. 5,187) attributed to BSGA exhibits a large carnivore tooth scratch. An *Equus* sp. mandible (Layer V-6: no. 2,033, r) on the medial face above the ventral border features a rectangular pit measuring 9.51 mm in length and 4.71 in

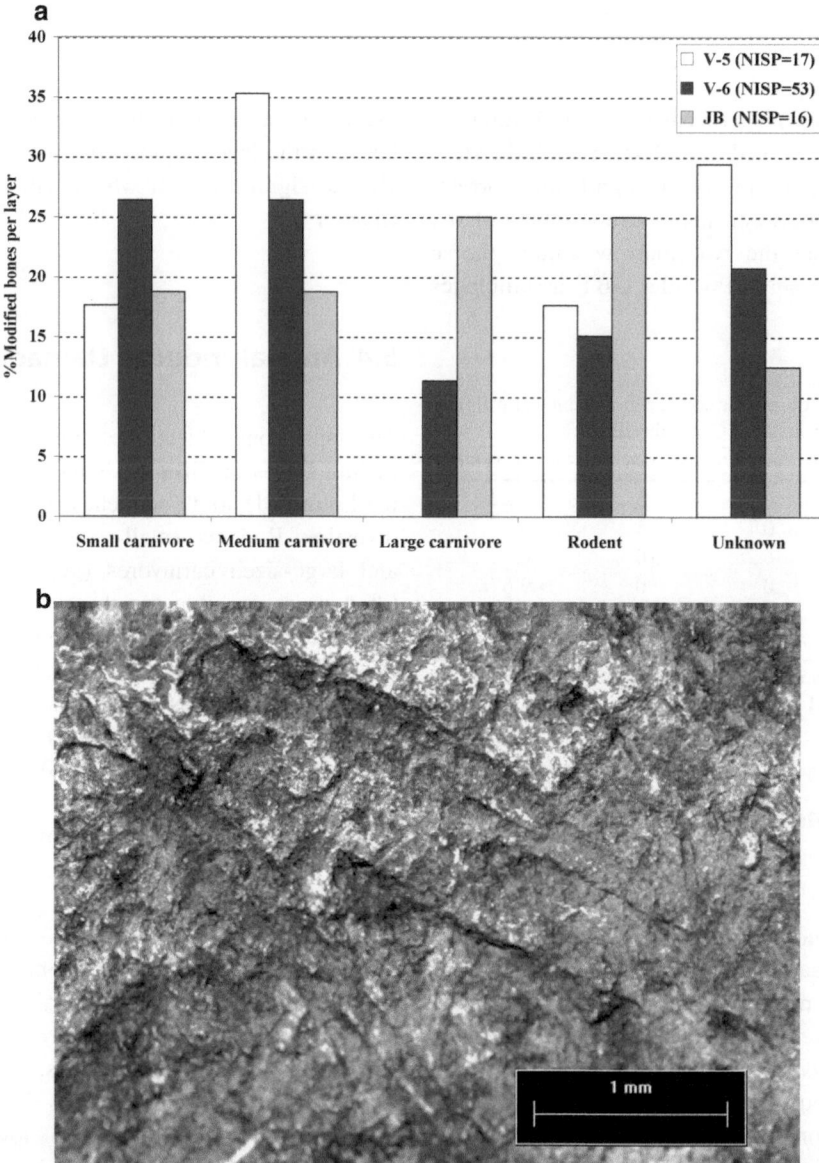

Fig. 5.23 **a** Animal-induced modification by agent from Area C and the JB; **b** tooth scratches on a *Dama* distal scapula from Layer V-5 (no. 1,679); **c** tooth scratches on a *Dama* distal tibia shaft from Layer V-6 (no. 1,765)

Fig. 5.23 (continued)

Table 5.15 Animal-induced damage by species (rodents, carnivores, and others) from Area C and the JB

	V-5		V-6		JB	
	Rodents modifications	Carnivores modifications	Rodents modifications	Carnivores modifications	Rodents modifications	Carnivores modifications
Dama sp.	3	7	4	18	1	5
Equus sp.				1		1
Palaeoloxodon antiquus				1		
BSGA						1
BSGB						
BSGC			2	3		2
BSGD		3	2	15	3	2
BSGE		3		6		1
UNM		1		1		
Total	3	14	8	45	4	12
% animal-induced modifications	0.3	1.43	0.42	2.41	0.36	1.08

width, probably caused by a large carnivore. A first phalanx (no. 12,973) of *Equus* sp. from the JB was heavily gnawed by a medium-sized carnivore.

Five bones attributed to animals weighing 80–250 kg (BSGC) from Layer V-6 display tooth scratches along their shafts. Three were caused by small- (no. 1,812), medium- (no. 1,507), and large-sized (no. 12,753) carnivores, as well as one by a rodent (no. 1,926), which also features rodent tooth marks on the surface. In addition, a mandible fragment (no. 12,980) was found in Layer V-6 with gnaw marks from an unknown agent. From the JB, a large-sized carnivore scratched the surface of a BSGC metacarpal shaft fragment (no. 12,977), and an unknown agent damaged a radius shaft fragment (no. 4,428).

Animal modifications also appear on bones within the 15–40 kg weight range (BSGE) (Layer V-5: NISP = 3, Layer V-6: NISP = 6, the JB: NISP = 1). Medium-sized carnivores are responsible for most of the marks in Layer V-5. Examples are gnaw marks on an inner anterior articular cavity and tooth scratches on an atlas fragment (no. 2,936). Small- and medium-sized carnivores and rodents damaged BSGE bones in Layer V-6 and a small-sized carnivore gnawed a radius shaft fragment (no. 4,265) from the JB.

Dama, the most abundant species within the GBY faunal assemblage, was only minimally affected by animal modification (Layer V-5: NISP = 10, 3.5% of *Dama* NISP; Layer V-6: NISP = 22, 4.03% of *Dama* NISP; the JB: NISP = 6, 1.69% of *Dama* NISP). *Dama* bones from Layer V-5 were scratched and gnawed by small- and medium-sized carnivores and rodents (Table 5.16). More than a third of the modified *Dama* bones from Layer V-6 were damaged by small- and medium-sized carnivores. Large-sized carnivores scratched three bones (13.63%). Damage was also caused by

rodents (NISP = 4, 18.18%), as well as by unknown agents (NISP = 7, 31.81%). Six *Dama* bones from the JB displayed signs of modification caused by carnivores, rodents, and unknown agents (Table 5.18).

Damage marks are also seen on bones attributed to animals weighing 40–80 kg (BSGD) (Layer V-5: NISP = 3, 1.49% of BSGD; Layer V-6: NISP = 17, 5.16% of BSGD; the JB: NISP = 5, 2.95% of BSGD). Most of the marks are found on long bone shaft fragments. The bones were affected by fox- (Layer V-6: NISP = 8, the JB: NISP = 1), wolf- (Layer V-5: NISP = 2, Layer V-6: NISP = 4), and hyena-sized (Layer V-6: NISP = 1, the JB: NISP = 1) carnivores. Also seen are a few cases in which rodents have damaged the bones (Layer V-5: NISP = 1, Layer V-6: NISP = 2, the JB: NISP = 3). The remaining bones from BSGD were modified by unknown agents.

If the frequency of carnivore traces is viewed as an indication of intensity, fox-sized carnivores should then be considered the main agent responsible for thanatocoenosis modification in Layer V-6. However, large carnivores, such as hyena, tend to feed on bones. Their general impact on a faunal assemblage is thus not necessarily expressed in the number of gnawed bones and bone fragments they leave after feeding.

The categories summarized by Pickering (2002) serve as the most reliable criteria for differentiating between the faunal assemblages accumulated by hyenas and hominids. We can thus observe that based on the rarity of carnivore species (Table 4.1), the extensive fragmentation of the limb bones, the low abundance of carnivore damage, the rarity of digested bones, and the absence of coprolites, carnivores exerted minimal influence as collectors and destroyers of the faunal assemblages from Area C.

Table 5.16 Animal-induced damage from Layer V-5

Tooth scratches

Small-sized carnivore	Anatomical position	Medium-sized carnivore	Anatomical position	Rodent	Anatomical position	Unknown agent
SCD *Dama* (no. 1,679, r)	Anterior and posterior borders	TBSH *Dama* (no. 11,985)	Shaft	RIBP *Dama* (no. 1,482)	Shaft and proximal	FSH *Dama* (no. 1,753, r)
PEL ISH *Dama* (no. 5,412, r)	Ischium	MTSH *Dama* (no. 1,441)	Shaft			
		LBSH BSGD (no. 6,842)	Over shaft			
		LBSH BSGE (no. 5,996)	Edge			

Gnaw marks

Small-sized carnivore	Anatomical position	Medium-sized carnivore	Anatomical position	Rodent	Anatomical position	Unknown agent
PH1PH *Dama* (no. 5,626)	Body	LBSH BSGD (no. 5,693)	Edge	FPH *Dama* (no. 1,693, r)	Femur neck	MTSH *Dama* (no. 6,280)
		VATLP BSGE (no. 2,936)	Inner anterior articular cavity	LBSH BSGD (no. 6,038)	Shaft	RIB *Dama* (no. 2,845)
						RIB BSGE (no.13,726)
						SPL UNM (no. 3,133)

Table 5.17 Animal-induced damage from Layer V-6

Tooth scratches

Small-sized carnivore	Anatomical position	Medium-sized carnivore	Anatomical position	Large-sized carnivore	Anatomical position	Rodent	Anatomical position	Unknown agent	Anatomical position
FSH *Dama* (no. 1,559)	Shaft	SCDH *Dama* (no. 1,516, 1)	Glenoid cavity rim	SKFH *Dama* (no. 12,649)		MANF *Dama* (no.1,478, 1)	Condyle edge	FSH *Dama* (no.1,632)	Shaft
TBSH *Dama* (no. 12,779)	Inner shaft	HSH *Dama* (no. 12,823)	Shaft	SKFH *Dama* (no. 12,650)		TBSH *Dama* (no. 2,252)	Shaft	MTPH *Dama* (no. 1,577, r)	Proximal
MCSH BSGC (no. 1,812)	Shaft	RDSH *Dama* (no. 12,508)	Shaft	TBD *Dama* (no. 1,765, 1)	Distal shaft	SPL FLAT BSGC (no. 1,926[a])		VEL BSGD (no. 1,487)	Body
LBSH BSGD (no. 15,119)	Shaft	RIBSH *Dama* (no. 4,341)	Shaft	TBSH BSGC (no. 12,753)	Shaft	LBSH BSGD (no. 12,501)	Shaft		
LBSH BSGD (no. 1,808)	Shaft	LBSH BSGC (no. 1,507)	Shaft			LBSH BSGD (no. 1,589)	Shaft		
LBSH BSGD (no. 6,778)	Shaft	LBSH BSGD (no. 13,298)	Shaft			LBSH BSGE (no. 12,505)	Shaft		
LBSH BSGD (no. 14,006)	Shaft	LBSH BSGD (no. 1,837)	Shaft						
LBSH BSGD (no. 3,142)	Shaft								
LBSH BSGD (no. 12,813)	Shaft								
LBSH BSGE (no. 12,209)	Edge								

Table 5.17 (continued)

Tooth marks

Small-sized carnivore	Anatomical position	Medium-sized carnivore	Anatomical position	Large-sized carnivore	Anatomical position	Rodent	Anatomical position	Unknown agent	Anatomical position
LBSH BSGD (no. 12,100)	Shaft			MAND *Equus* (no. 2,033, r)	Medial face above ventral border	SPL FLAT BSGC (no. 1,926[a])			
LBSH BSGE (no. 7,961)	Edge								

Gnaw marks

Small-sized carnivore	Anatomical position	Medium-sized carnivore	Anatomical position	Large-sized carnivore	Anatomical position	Rodent	Anatomical position	Unknown agent	Anatomical position
FSH *Dama* (no. 12,701)	Edge	AST elephant (no. 9,274, r)	Edge	LBSH BSGD (no. 13,294)	Edge	SCB *Dama* (no. 13,934,1)	Blade	FSH *Dama* (no. 12,778)	Shaft
SPL BSGD (no. 13,391)		LBSH BSGD (no. 12,321)	Edge			FSH *Dama* (no. 12,718)		TBSH *Dama* (no. 1,518)	Shaft
		LBSH BSGD (no. 14,097)	Edge					TBSH *Dama* (no. 1,524,1)	Shaft
		LBSH BSGE (no. 9,255)	Edge					TBSH *Dama* (no. 12,797)	Shaft
		LBSH BSGE (no. 14,039)	Edge					TBSH *Dama* (no. 1,794)	Shaft
		Digested						MANF BSGC (no. 12,980)	Body
		PH2D *Dama* (no. 1,684)						LBSH BSGD (no. 6,462)	
		MTSH UNM (no. 14,131)						RDSH BSGE (no. 6,124)	Shaft

[a] Appears under both tooth scratches and gnaw marks (no. 1,926)

Table 5.18 Animal-induced damage from the JB

Tooth scratches

Small-sized carnivore	Anatomical position	Medium-sized carnivore	Anatomical position	Large-sized carnivore	Anatomical position	Rodent	Anatomical position	Unknown agent	Anatomical position
SKFH *Dama* (no. 13,245)		SCD *Dama* (no. 2,196, r)	Blade	RIBSH BSGA (no. 5,187)	Shaft	LBSH BSGD (no. 4,754)	Shaft	RDSH BSGC (no. 4,428)	Shaft
LBSH BSGD (no. 15,078)	Shaft			MCSH BSGC (no. 12,997)	Shaft	LBSH BSGD (no. 12,522)	Shaft		
				LBSH BSGD (no. 2,254)	Shaft				

Tooth marks

Small-sized carnivore	Anatomical position	Medium-sized carnivore	Anatomical position	Large-sized carnivore	Anatomical position	Rodent	Anatomical position	Unknown agent	Anatomical position
				VTRS *Dama* (no. 15,194)	Body	RDS *Dama* (no. 12,670, r)	Proximal		

Gnaw marks

Small-sized carnivore	Anatomical position	Medium-sized carnivore	Anatomical position	Large-sized carnivore	Anatomical position	Rodent	Anatomical position	Unknown agent	Anatomical position
RDSH BSGE (no. 4,265)	Edge	TBD *Dama* (no. 15,123)	Distal			LBSH BSGD (no. 4,431)	Shaft	RIBS *Dama* (no. 1,778)	Shaft
		PH1PH *Equus* (no. 12,973)	Shaft						

5.5 Hominin-Induced Damage

The vast array of hominin activity at GBY is reflected in particular by the stone artifacts, as well as by the pitted stones used for cracking nuts and the spatial distribution of evidence for fire. The faunal assemblage is no exception, indicating that hominins interacted with animal carcasses. The most striking example of this behavior is the modifications on *Dama* bones from the three assemblages. From this context, we were able to reconstruct an entire butchering sequence for *Dama* (Rabinovich et al. 2008a).

Our analysis of the anthropogenic bone-damage patterns at GBY is based on an interpretation of cut marks and hominin-induced fractures as the by-product of exploitation activities guided by knowledge of animal anatomy (Binford 1981; Lyman 1994). The GBY assemblages of well-preserved *Dama* bones bearing hominin-made marks and the scarcity of carnivore markings on these bones permit a comprehensive study of hominin butchery patterns.

5.5.1 Cut Marks and Indications of Marrow Extraction

Hominin-induced modifications were frequently observed on bones from GBY. Among these are cut- and percussion marks. Cut marks occur on bones of most species and of most body-size groups (Table 5.19). In addition, the faunal assemblages display a high number of limb and long bone shaft fragments with conical fractures and percussion marks, as well as bone flakes. These features are interpreted as resulting from marrow extraction by hominins.

Bone breakage for marrow extraction leaves diagnostic and undiagnostic traces. The former most likely appears at the point where the blow was struck either on the outer or inner bone face. Had the bone been broken with an anvil, a more complete breakage pattern (i.e., scars, flakes) would occur. Splitting is also found in association with percussion points indicating blows that were inflicted perpendicularly to the shaft axis.

Cut marks were observed in each assemblage, with fewer seen in Layer V-5 than Layer V-6 and the JB (Layer V-5: NISP = 17, 1.74%; Layer V-6: NISP = 87, 4.66%; the JB: NISP = 64, 5.74%). In each layer *Dama* is the dominant species displaying cut marks. Percussion marks and resulting bone flakes also occur in all three assemblages (Layer V-5: NISP = 36, 3.69%; Layer V-6: NISP = 131, 7.03%; the JB: NISP = 84, 7.54%). Percussion marks are noted in most species and body-size groups (Table 5.19).

The following section contains a discussion of the hominin-induced modifications per species at GBY, which culminates with a detailed description of the *Dama* butchery sequence. Concluding remarks will incorporate a reconstruction of the taphonomic history of all the mammalian species at the site.

Megafauna

Early and Middle Pleistocene sites have revealed hardly any evidence of direct butchery of large animals with thick skins ("pachyderm") (Crader 1983; Delagnes et al. 2006). It is thus not surprising that the elephant bones from GBY do not

Table 5.19 Hominin-induced marks: cut marks and percussion marks from Area C and the JB

	V-5		V-6		JB	
	Percussion marks	Cut marks	Percussion marks	Cut marks	Percussion marks	Cut marks
Bos sp.					2	3
Dama	11	9	32	50	35	47
Sus					2	
Cervus. cf. *elaphus*						1
E. cf. *africanus*				1		
Equid	2			2		
Megaloceros			2	1		
BSGB	2	1	2	1		1
BSGC	2	1	9	6	10	4
BSGD	13	2	57	10	23	4
BSGE	5	3	19	11	4	4
BSGF		1	2			
UNM	1		8	5	8	
Total	**36**	**17**	**131**	**87**	**84**	**64**
% cut marks		1.74		4.67		5.75
% breakage	3.69		7.03		7.54	

display undisputed cut marks. In contrast, long bones from species weighing approximately 1,000 kg (BSGB) clearly display both cut- and percussion marks. Layer V-5 yielded a long bone shaft displaying cut marks, percussion marks, and striations (no. 1,696), as well as a radius shaft with percussion marks (no. 1,736).

In Layer V-6, cut marks are seen along the shaft of a long bone (no. 12,952). Two long bone shafts exhibit flake scars (nos. 12,487 and 12,702), and one also displays striations (no. 12,702). Cut marks are also observed on the proximal part of a rib from the JB (no. 1,197). These marks might indicate the defleshing and marrow processing of the extremities of hippopotamus.

Equids

A tibia shaft (no. 5,887) and an ilium shaft (no. 1457) from Layer V-5 exhibit evidence of bone breakage. A mandible belonging to a juvenile equid (less than 12 months old) from Layer V-6 (no. 2,033; see Chapter 4) displays cut marks along both the inner and outer bone face, suggesting tongue extraction. In addition, cut marks are seen on the distal medial lateral condyle of a humerus (no. 2,023), possibly indicating disarticulation of the elbow joint. Cut marks also occur on a proximal lateral residual metatarsal (no. 2,010) from Layer V-6, indicating disarticulation of equid carcasses.

Bos sp.

From the JB, a mandibular fragment (no. 2,157), a distal part of a calcaneum (no. 896), and a second phalanx (no. 884) provide evidence of cut marks. The mandibular fragment displays cut marks on its symphyseal surface and on the medial face of the mandibular body, suggesting tongue extraction. We conclude that these cut marks resulted mainly from carcass disarticulation.

Also observed is the longitudinal splitting of metacarpals and metatarsals (the JB: nos. 919, 1,164), produced by heavy blows to the proximal part of the bone. It is highly likely that such marks resulted from hominin marrow extraction.

Megaloceros sp.

The volar face of a metacarpal shaft from Layer V-6 (no. 1,490) features cut marks near the epiphysis. Two metapodials (metatarsal proximal, no. 1,669; metacarpal shaft, no. 1,490) from Layer V-6 display percussion marks.

Cervus cf. *elaphus*

One astragalus (no. 11,788) from the JB displays cut marks on the medial face.

Body-Size Group C (BSGC: Giant Deer, Red Deer, Boar, and Bovine, 80–250 kg)

Only a few bones from BSGC exhibit cut marks: a flat splinter from Layer V-5 (no. 1,3348), six from Layer V-6, and four from the JB. In Layer V-6, cut marks are seen on two mandibular fragments (nos. 2,097, 12,980). One of the latter (no. 2,097) also displays signs of marrow extraction on its lateral face. Additional cut marks appear on two long bone shafts (nos. 13,305 and 13,307), a rib shaft (no. 13,338), a distal radius-ulna (no. 2,097), and the anterior face of an ilium fragment (no. 12,726). Cut marks from the JB are found on a vertebral spine (no. 1,219), as a result of filleting, and on a tibia shaft (no. 909) and a distal scapula (no. 757), as a result of carcass disarticulation.

Bone flakes and percussion marks were observed in all three layers as a result of intentional bone breakage (Layer V-5: NISP = 2, Layer V-6: NISP = 9, the JB: NISP = 10).

Body-Size Group E (BSGE: Gazelle, Roe Deer, 15–40 kg)

Bones from BSGE display both cut- and percussion marks. Cut marks are seen on three elements from Layer V-5: two tibia shafts (nos. 1,445, 2905) and one long bone shaft (no. 7,876). Cut marks from Layer V-6 are observed on the anterior face of a mandibular symphysis (no. 12,425) and on the medial and lateral face of a mandibular fragment (no. 12,970). Cut marks are also seen on a radius shaft (no. 12,692), an ilium (no. 2,247), and long bone shafts (nos. 2,233, 4,028, 12,505, 12,625, 13,328) from Layer V-6. Cut marks from the JB are found on a rib

Table 5.20 Cut marks on BSGE specimens (gazelle, roe deer, 15–40 kg) from Area C and the JB

Layer	Skeletal element	Number of elements with cut marks	Anatomical position of cut marks
V-5	Tibia shaft	2	Lateral face
V-5	Long bone shaft	1	Shaft
V-6	Skull squamosal	1	Squamosal
V-6	Skull fragment	1	Surface
V-6	Mandibular fragment (r)	1	Mandibular angle: medial and lateral face
V-6	Mandinbular symphisis	1	Medial face
V-6	Radius shaft	1	Shaft
V-6	Ilium shaft (l)	1	Ilium shaft on lateral face
V-6	Long bone shaft	5	Shaft
JB	Rib	1	Shaft
JB	Pubis	1	Shaft
JB	Long bone shaft	2	Shaft

(no. 862), a pubis fragment (no. 5,986) and two long bone shafts (nos. 801 and 1,333). Due to their location, orientation, and shape, we conclude that these marks resulted from both the disarticulation and filleting of gazelle-sized animals (Table 5.20).

In addition, bone flakes and percussion marks have been observed in each assemblage on long bone shafts and on radius, femur, and tibia shafts (Layer V-5: NISP = 5, Layer V-6: NISP = 19, the JB: NISP = 4). The data presented above suggests butchering of complete animals, due to evidence of disarticulation, filleting, and marrow extraction.

Body-Size Group F (BSGF: Hare, Red Fox, 2–10 kg)

Cut marks from BSGF are found only on one long bone shaft (Layer V-5: no. 5,633). Two long bone shafts from Layer V-6 display signs of breakage in the form of flake-like features. The general fragmentation pattern of bones from this body size group is consistent among all specimens.

Dama sp. and Body-Size Group D

Cut marks are seen on 2.26% (NISP = 11) of *Dama* and *Dama*-sized bones from Layer V-5 (Table 5.21), 6.85% (NISP = 60) of those from Layer V-6 (Table 5.22), and 9.73% (NISP = 51) from the JB (Table 5.23). One-third of the cut marks are found on the trunk (Layer V-5: NISP = 4, 36.36%; Layer V-6: NISP = 19, 31.66%; the JB: NISP = 18, 35.29%) (Fig. 5.24), followed by upper limbs (Layer V-5: NISP = 3, 27.27%; Layer V-6: NISP = 14, 23.33%; the JB: NISP = 14, 27.45%), lower limbs (Layer V-5: NISP = 1, 9.09%; Layer V-6: NISP = 11, 18.33%; the JB: NISP = 10, 19.60%), long bone shafts (Layer V-5: NISP = 2, 18.18%; Layer V-6: NISP = 10, 16.66%; the JB: NISP = 4, 7.84%), and skull elements (Layer V-5: NISP = 1, 9.09%; Layer V-6: NISP = 6, 10%; the JB: NISP = 5, 9.80%).

Cut-mark frequency per skeletal element does not correlate with the relative abundance of these elements. For example, the JB yielded one sacrum, which constitutes 0.3% of the total NISP for the layer. However, since this sacrum bears cut marks, the frequency of cut marks on sacra in JB is 100%. Moreover, a significant negative correlation was found between the %NISP and %cut marks per element in Layer V-6 (r = –0.0819, p = 0.000) and the JB (r = –0.823, p = 0.000). Comparison of the cut-mark frequency per element between the GBY assemblages has revealed a significant correlation between Layer V-6 and the JB (r = 0.710, p = 0.001).

Tables 5.21, 5.22, and 5.23 describe each bone from the GBY assemblages that displays cut marks, along with each cut's anatomical position and our interpretation of the activity that caused the cuts.

Table 5.19 demonstrates that cut marks are more frequent on bones attributed to *Dama* than any other species at the site, and that breakage is more frequent on BSGD bones from Layers V-5 and V-6 than other body-size groups. This stems from the frequency of the long bone shaft in BSGD.

In addition to cut marks, the *Dama* assemblages include percussion marks and bone flakes, evidence of marrow extraction. Extensive long bone fragmentation resulted in a high number of shaft fragments with percussion marks (Table 5.24). These fragments have been assigned to BSGD (Layer V-5: 5%, Layer V-6: 13%, the JB: 11.61%). At times, percussion marks are found on one or both faces of the same bone. In such cases, the fractures may have resulted from return shocks, which also may have produced bone flakes (Layer V-5: NISP = 4; Layer V-6: NISP = 21; the JB: NISP = 5). More than 80% of percussion marks occur on *Dama* limb shafts (Layer V-5: 81.81%, NISP = 9; Layer V-6: 84.37%, NISP = 27; the JB: 85.71%, NISP = 30) (Fig. 5.25). The rest are found on long bone shafts attributed to the BSGD, but percussion marks are rarely observed on the proximal or distal part of a limb (Table 5.24; Fig. 5.25). The systematic occurrence of percussion marks and their location points to consistent marrow exploitation.

Percussion-mark frequency per skeletal element does not correlate with the relative abundance of *Dama* elements. No significant correlation was found between the %NISP per element and the %percussion marks per element in Layers V-5 and V-6. A weak correlation was found in the JB (r = 0.553, p = 0.050).

Marrow and grease nutritional values of limb and leg bones (Lyman 1994: table 7.1) were correlated with the %percussion marks per *Dama* element. In order to represent the limb shafts, we have used the lowest marrow and grease value available for each element; i.e., marrow of the humerus shaft = 28.33, which, in fact, is the value of the distal humerus (Lyman 1994). However, no significant correlation was found between the %percussion marks per *Dama* element and the marrow and grease values. Plotting of the marrow and the grease values (from lowest to highest) against the %percussion marks per element reflects extensive use of *Dama* limb and leg bones (Figs. 5.26 and 5.27). Elements with low and high marrow values were exploited.

Complete first and second *Dama* phalanges are not common in the GBY assemblages (Tables 5.8, 5.9, and 5.10). Longitudinally split first and second phalanges are found in Layers V-5 (NISP = 1, 4.16%) and V-6 (NISP = 4, 4.49%). This fragmentation type indicates an exploitation method similar to that of long bones, a striking phenomenon when considering the minimal amount of marrow available in phalanges. The practice of phalanx splitting is known from Late Pleistocene sites in the Levant (e.g., Bar-Yosef et al.

Table 5.21	Anatomical position, frequency, and interpretation of cut marks in Layer V-5

Skeletal element	Anatomical position of cut marks[a]	Number of elements with cut marks	Interpretation
Mandible	Symphyseal surface, medial	1	Disarticulation
Atlas	Cranioventral rim	1	Dislocation of joint between skull and atlas
Rib	Lateral	1	Removal of meat between ribs or removal of periosteum
Scapula	Glenoid cavity, lateral	1	Disarticulation of humerus
Humerus	Lower midshaft, cranial	1	Defleshing of trunk
Pelvis	Pubis	1	Dismemberment of hip region
Femur	Midshaft, lateral	1	Filleting; cutting and removal of periosteum
Tibia	Lower midshaft, cranial (dorsal)	1	Filleting; cutting and removal of periosteum
Phalanx 1	Shaft, dorsal	1	Disarticulation of joint between Ph 1 and Ph 2

[a]Cut marks appear also on long bone shaft fragments (N = 2) and on one flat splinter

Table 5.22 Anatomical position, frequency, and interpretation of cut marks in Layer V-6

Skeletal element	Anatomical position of cut marks[a]	Number of elements with cut marks	Interpretation
Skull	Indet.	1	Skinning of carcass
Mandible	Body of mandible, lateral	4	Defleshing
Mandible	Mandibular ramus, medial	1	Removal of internal masticatory muscle
Atlas	Cranioventral rim	1	Dislocation of joint between skull and atlas
Thoracic vertebra	Articular process	1	Disarticulation of ribs from vertebral column
Lumbar vertebra	Corpus	1	Defleshing of spinal column
Vertebra indet.	Corpus	2	Defleshing of spinal column
Rib	Lateral	6	Removal of meat between ribs or removal of periosteum
Scapula	Glenoid cavity, lateral	1	Disarticulation of humerus
Scapula	Corpus, lateral	2	Defleshing of scapula
Humerus	Lower midshaft, lateral	1	Disarticulation of joint between humerus and ulna
Humerus	Lower midshaft, medial	1	Defleshing of trunk
Humerus	Lower midshaft, cranial	1	Defleshing of trunk
Radius	Midshaft, lateral	2	Filleting; cutting and removal of periosteum
Metacarpal	Proximal, midshaft (cranial)	2	Disarticulation of metacarpal
Pelvis	Acetabulum	2	Dismemberment of hip region
Pelvis	Pubis	1	Dismemberment of hip region
Pelvis	Ilium, greater sciatic notch	1	Dismemberment of hip region
Pelvis	Body of ischium, lateral	1	Dismemberment of hip region
Femur	Midshaft, lateral	3	Filleting; cutting and removal of periosteum
Tibia	Midshaft, caudal (plantar)	1	Filleting; cutting and removal of periosteum
Tibia	Midshaft, cranial (dorsal)	1	Filleting; cutting and removal of periosteum
Tibia	Lower midshaft, cranial (dorsal)	3	Defleshing
Tibia	Lower midshaft, caudal (plantar)	1	Defleshing
Astragalus	Proximal rim, medial	1	Disarticulation of tarsal joint
Naviculocuboid	Dorsal	1	Disarticulation of tarsal joint
Metatarsal	Midshaft, cranial (dorsal)	1	Skinning of carcass
Phalanx 1	Proximal, dorsal	2	Disarticulation of joint between Ph 1 and Ph 2
Phalanx 1	Shaft, dorsal	2	Disarticulation of joint between Ph 1 and Ph 2
Phalanx 2	Proximal, dorsal	1	Disarticulation of joint between Ph 1 and Ph 2

[a]Cut marks appear also on a metapodial shaft, on long bone shaft fragments (N = 8), and on flat fragments (N = 2)

Table 5.23 Anatomical position, frequency, and interpretation of cut marks in the JB

Skeletal element	Anatomical position of cut marks[a]	Number of elements with cut marks	Interpretation
Skull	Indet.	1	Skinning of carcass
Mandible	Basal rim of symphysis	1	Skinning of carcass
Mandible	Mandibular ramus, lateral	2	Cutting of masseter muscle during defleshing
Mandible	Symphyseal surface, medial	1	Disarticulation
Axis	Rim of facies articularis	1	Disarticulation of joint between axis and other cervical vertebrae
Cervical vertebra	Corpus, ventral	1	Defleshing of neck
Thoracic vertebra	Spinous process	3	Defleshing of spinal column
Thoracic vertebra	Articular process	2	Disarticulation of ribs from vertebral column
Lumbar vertebra	Corpus, dorsal	1	Defleshing of spinal column
Lumbar vertebra	Spinous process	2	Defleshing of spinal column
Sacrum	Cranial	1	Disarticulation of joint between lumbar vertebrae and sacrum
Rib	Head	1	Disarticulation of ribs from vertebral column
Rib	Lateral	1	Removal of meat between ribs or removal of periosteum
Scapula	Posterior border of the blade	1	Disarticulation of humerus
Scapula	Corpus, lateral	1	Defleshing of scapula
Humerus	Lower midshaft, lateral	1	Disarticulation of joint between humerus and ulna
Radius	Distal epiphysis, caudal (palmar)	1	Disarticulation of joint between radius and carpals
Radius	Midshaft, lateral	1	Cutting and removal of periosteum
Ulna	Proximal medial	1	Cutting of lower arm muscles during defleshing
Metacarpal	Midshaft, cranial (dorsal)	1	Skinning of carcass
Metacarpal	Distal, cranial (dorsal)	1	Disarticulation of joint between metacarpus and phalanges
Pelvis	Ilium, lateral	2	Defleshing of hip region
Pelvis	Ilium, dorsal	1	Defleshing of hip region
Femur	Midshaft, caudal	1	Defleshing of hind leg
Femur	Midshaft indet.	1	Defleshing of hind leg or cutting and removal of periosteum
Femur	Lower midshaft, caudal	1	Defleshing of hind leg
Femur	Distal epiphysis, caudal	1	Disarticulation of knee joint
Tibia	Midshaft, caudal (plantar)	3	Filleting; cutting and removal of periosteum
Tibia	Upper midshaft, lateral	1	Filleting; cutting and removal of periosteum
Tibia	Lower midshaft, cranial (dorsal)	1	Filleting; cutting and removal of periosteum
Tibia	Lower midshaft, caudal (plantar)	1	Filleting; cutting and removal of periosteum
Astragalus	Proximal rim, medial	1	Disarticulation of tarsal joint
Naviculocuboid	Dorsal	1	Disarticulation of tarsal joint
Naviculocuboid	Lateral	1	Disarticulation of tarsal joint
Metatarsal	Proximal caudal (plantar)	2	Disarticulation of metatarsus
Phalanx 1	Proximal dorsal	1	Disarticulation of joint between Ph 1 and Ph 2
Phalanx 1	Shaft dorsal	1	Disarticulation of joint between Ph 1 and Ph 2
Phalanx 2	Lower midshaft, dorsal	1	Disarticulation of joint between Ph 1 and Ph 2

[a]Cut marks appear also on long bone shaft fragments (N = 4)

Fig. 5.24 Cut mark distribution on *Dama* and BSGD according to body area

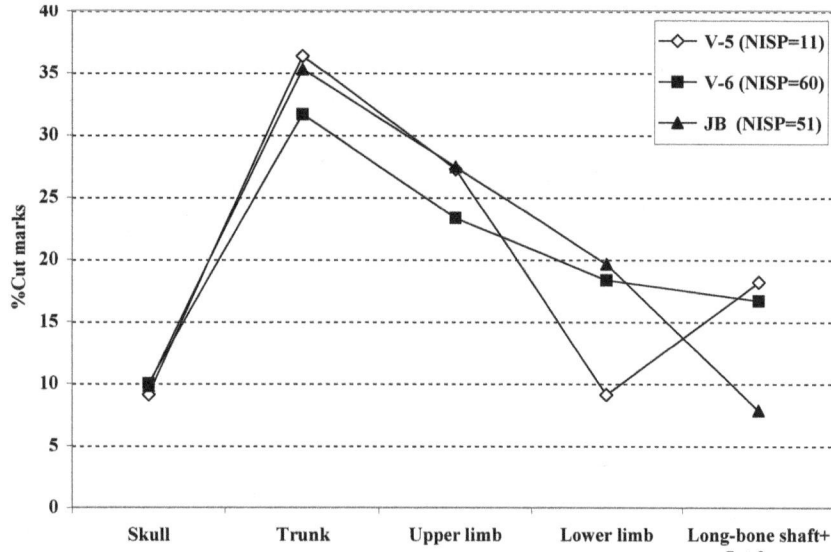

Table 5.24   Anatomical position of percussion-mark frequency in Area C and the JB

Skeletal element	Anatomical position of percussion marks	Number of elements with percussion marks		
		V-5	V-6	JB
Humerus	Midshaft, caudal	0	0	1
Humerus	Lower midshaft, caudal	0	0	1
Radius	Proximal, indet.	0	0	1
Radius	Midshaft, lateral	0	2	1
Radius	Midshaft, medial	1	2	1
Radius	Midshaft indet.	1	0	4
Metacarpal	Midshaft, dorsal	0	1	1
Metacarpal	Midshaft, palmar	1	2	1
Metacarpal	Distal, dorsal	0	0	1
Femur	Midshaft, lateral	0	3	1
Femur	Midshaft, caudal	1	1	4
Femur	Lower midshaft, caudal	0	0	2
Femur	Midshaft, indet.	0	3	2
Femur	Distal	0	1	0
Tibia	Lower midshaft, dorsal	0	3	1
Tibia	Upper midshaft, plantar	0	3	2
Tibia	Lower midshaft, plantar	0	0	2
Tibia	Midshaft, indet.	3	1	2
Tibia	Distal	0	1	0
Metatarsal	Proximal, dorsal	1	1	3
Metatarsal	Upper midshaft, dorsal	0	3	0
Metatarsal	Midshaft, dorsal	2	1	2
Metatarsal	Midshaft, plantar	0	0	2
Phalanx	Proximal end and shaft	0	1	0
Phalanx	Midshaft	0	2	0
Phalanx	Distal	1	1	0
Long bone shaft fragment	Indet.	9	36	18
Bone flake	Indet.	4	21	5
Total		24	89	58

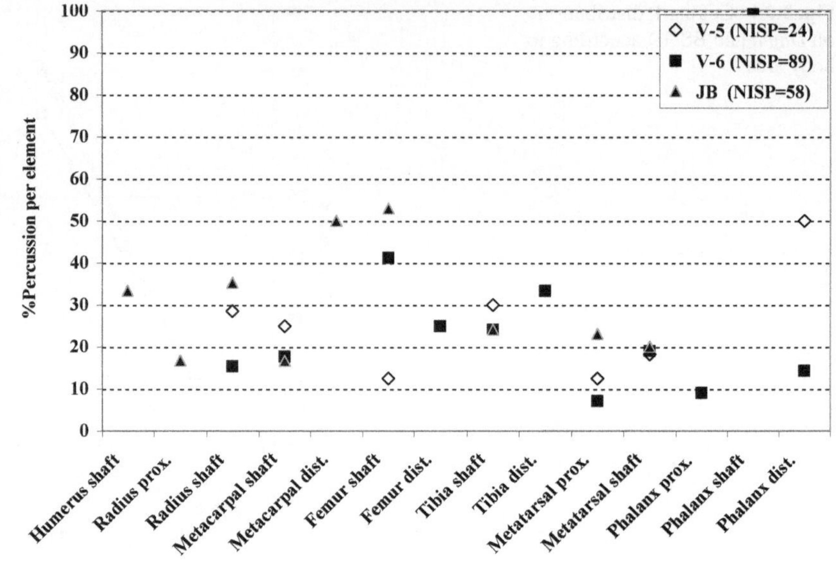

Fig. 5.25 Percentage of
percussion marks on *Dama* bones
from Area C and the JB

Fig. 5.26 Ranked marrow
indices (Lyman 1994: table 7.1)
and percentage of percussion
marks per *Dama* limb and leg
bones from Area C and the JB

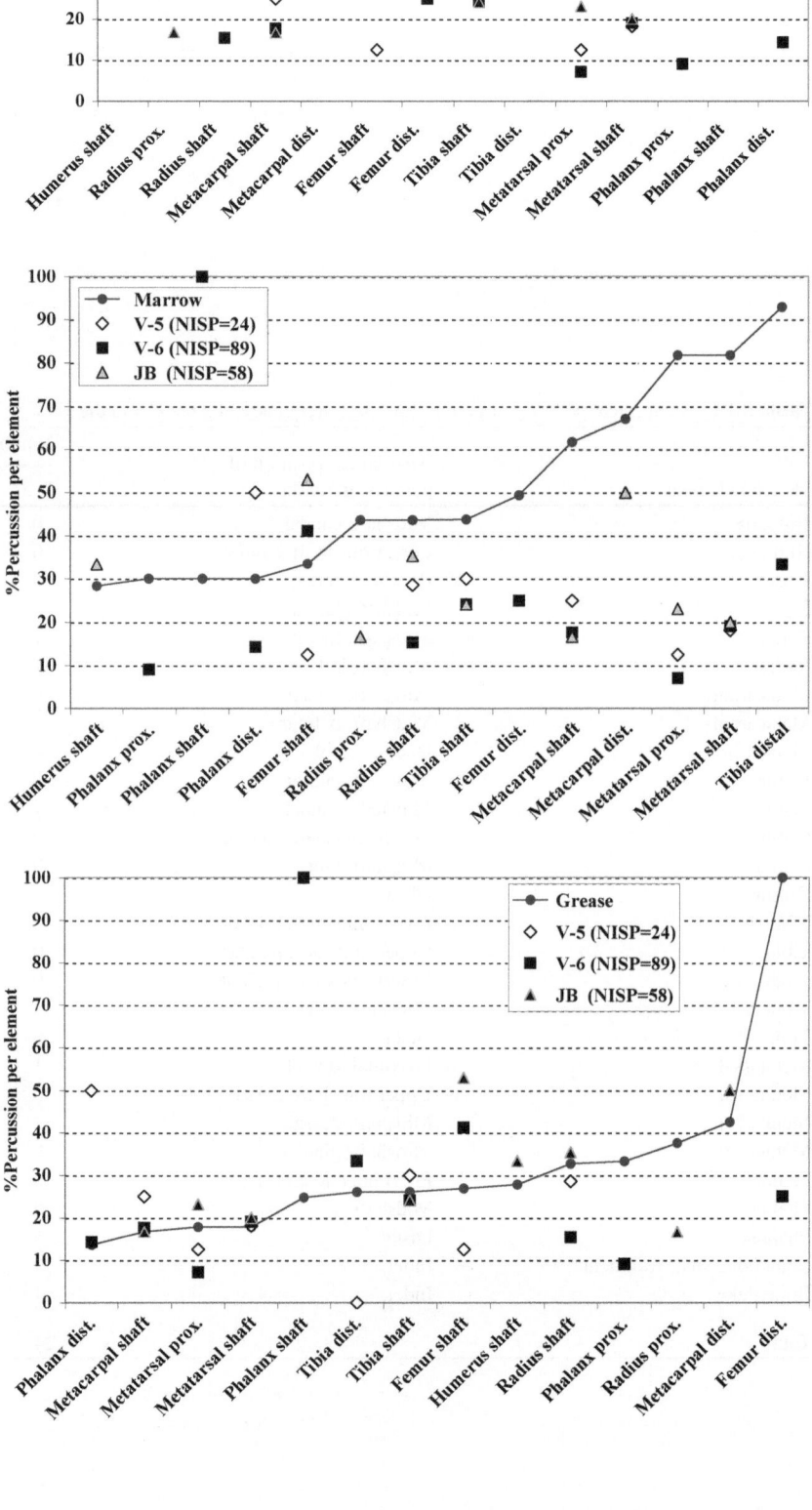

Fig. 5.27 Ranked grease indices
(Lyman 1994: table 7.1) and
percentage of percussion marks
per *Dama* limb and leg bones
from Area C and the JB

1992; Rabinovich 1998; Munro and Bar-Oz 2005) and elsewhere (e.g., Morin 2004).

On the basis of the data presented (Tables 5.21, 5.22, 5.23, and 5.24; Figs. 5.24–5.38), we propose the following sequence of *Dama* carcass processing at GBY:

(1) Skinning, evidenced by cut marks on crania, metapodials, and phalanges (Fig. 5.28).

(2) Disarticulation, evidenced by cut- and hack marks on first cervical vertebrae (atlases) and mandibles (to sever the head from the body), and by articulation of limb elements (to separate the fore limb and hind limb elements) (Figs. 5.29, 5.30, and 5.31).

(3) Defleshing and filleting, evidenced by cut marks on mandibles, vertebrae, ribs, scapulae, humeri, radii, ulnae, pelvises, femora, and tibiae (Figs. 5.32 and 5.33).

(4) Marrow extraction, evidenced by percussion marks on humeri, radii, metapodials, femora, tibiae, and phalanges (Figs. 5.34 and 5.35).

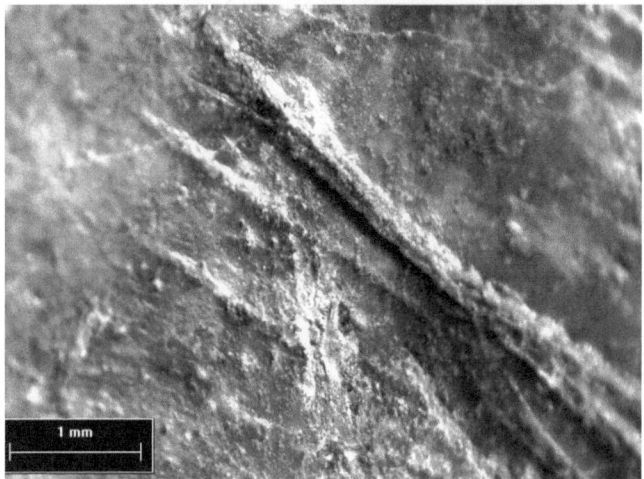

Fig. 5.29 Cut marks on a *Dama* sp. atlas from Layer V-6 (no. 1,714). *Top* – general view, *bottom* – detail

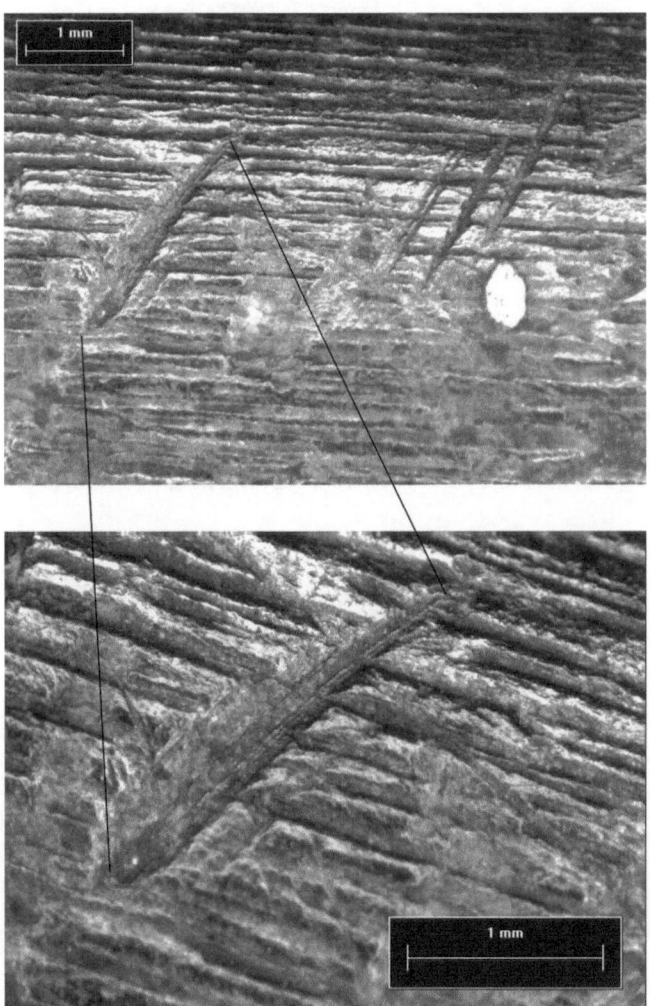

Fig. 5.28 Cut marks on a metapodial *Dama*-sized shaft from Layer V-6 (no. 12,827)

The high frequency of hominin-induced damage marks on *Dama* skeletal elements suggests a sequence of butchery processes that was repeated again and again in the same pattern. Furthermore, the presence in one area of all the major *Dama* bones and of all the hominin-induced damage types displayed on these bones leads to the conclusion that the entire carcass processing sequence took place in situ. This conclusion is further supported by the occurrence of clustered burned flint microartifacts, whose presence in the same archaeological horizons as the processed *Dama* bones is a sign of in-situ deposition (Goren-Inbar et al. 2004). Repeated evidence of such processing of complete carcasses indicates regular hunting by hominins, and suggests that these hominins possessed detailed knowledge of their prey's anatomy and thus exploited game in an efficient, systematic manner (Rabinovich et al. 2008a).

To ascertain the hominin patterns of *Dama* exploitation at GBY, we compared the butchery processes observed at

Fig. 5.30 Cut marks on a *Dama* sp. astragalus from Layer V-6 (no. 829)

Fig. 5.31 Cut marks on *Dama* sp. second phalanx from the JB (no. 1,327)

the site with those of *Dama mesopotamica* exploitation in Upper Paleolithic Aurignacian occurrences (Strata D) at Hayonim Cave in the western Galilee, Israel (Belfer-Cohen and Bar-Yosef 1981). Despite the great chronological gap between the two sites (Hayonim Layer D is dated ca. 28,000 BP; Goring-Morris and Belfer-Cohen 2003: appendix), the observed patterns of *Dama* butchering are similar (Fig. 5.36).

The excellent preservation of the bones and a minimal occurrence of carnivore damage marks at Hayonim Layer D allowed for a detailed study of animal-exploitation patterns by *Homo sapiens*. Only a few hack marks appear on vertebrae of *Dama mesopotamica* at Hayonim Layer D; their anatomical location indicates carcass disarticulation by chopping (Layer D1-2, NISP = 2; Layer D3, NISP = 2). Similarly, the majority of the cut marks observed on these bones was caused by disarticulation (Layer D1-2: 70%,

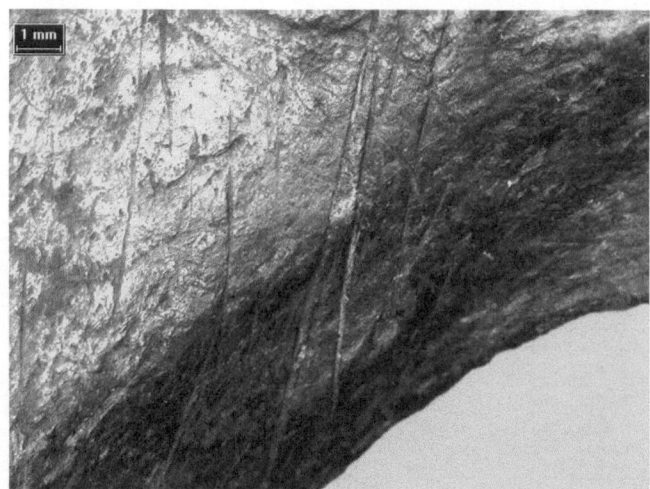

Fig. 5.32 Cut marks on a *Dama* sp. cervical vertebrae from Layer V-6 (no. 1,723)

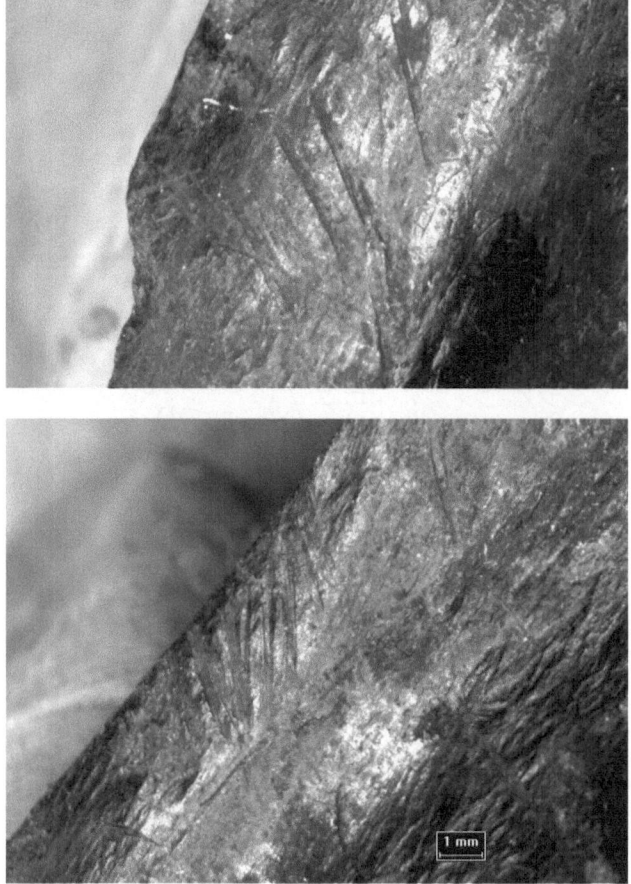

Fig. 5.33 Cut marks on a *Dama* sp. femur shaft from the JB (no. 12,668)

NISP = 23; Layer D3: 63%, NISP = 26). Cut marks resulting from filleting and skinning are also present (Rabinovich et al. 1997; Rabinovich 1998).

The consistent location of the cut marks on the *Dama* skeletal elements from all the D strata at Hayonim Cave

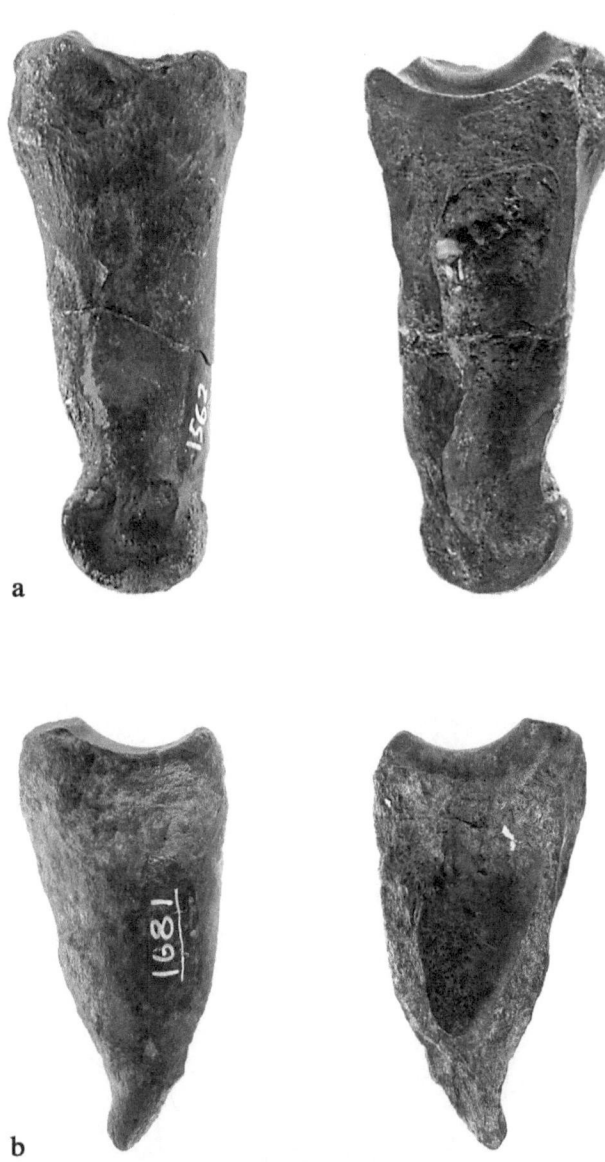

Fig. 5.35 Lateral and medial views of a *Dama* sp. split first phalange from **a** Layer V-6 (no. 1,562) and **b** Layer V-5 (no. 1,681)

(Fig. 5.36c, d) indicates that the occupants had an in-depth understanding of the species' anatomy. Rabinovich (1998) reconstructed an entire butchery sequence at Hayonim Cave D, starting from the skinning of the animal and ending with marrow extraction. This butchery pattern repeats itself in the Aurignacian strata at Hayonim Cave, evidence that in-situ exploitation of *Dama* carcasses obtained through hunting took place during the Aurignacian occupation of the cave (e.g., Stiner 2005).

The location of the butchery marks on the *Dama* bones from Hayonim Cave Layer D is similar to those of the GBY bones (Fig. 5.36a, b); we interpret this similarity as resulting from identical exploitation patterns. However,

Fig. 5.34 Percussion marks on a *Dama* sp. femur shaft fragment from Layer V-6 (no. 2,102)

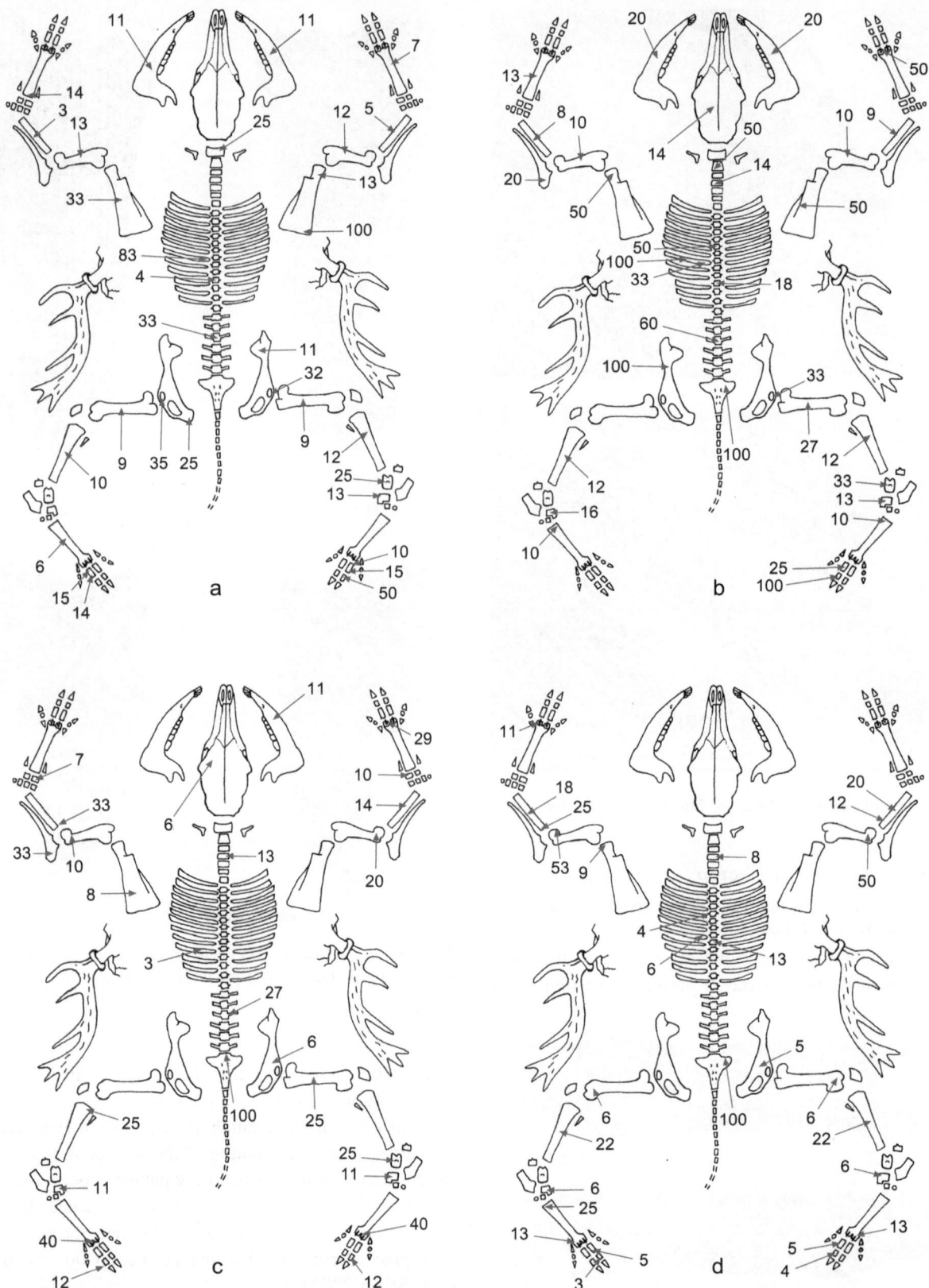

Fig. 5.36 Location and frequency of cut marks on *Dama* skeletal elements. Each number indicates the relative abundance (%) of cut marks on a particular skeletal element: **a** Layer V-6; **b** the JB; **c** Hayonim Cave, Layer D1–2; **d** Hayonim Cave, Layer D3. The sample size of cut marks from Layer V-5 is small and thus does not appear

Fig. 5.37 Frequency of animal species in Area C and the JB according to %NISP

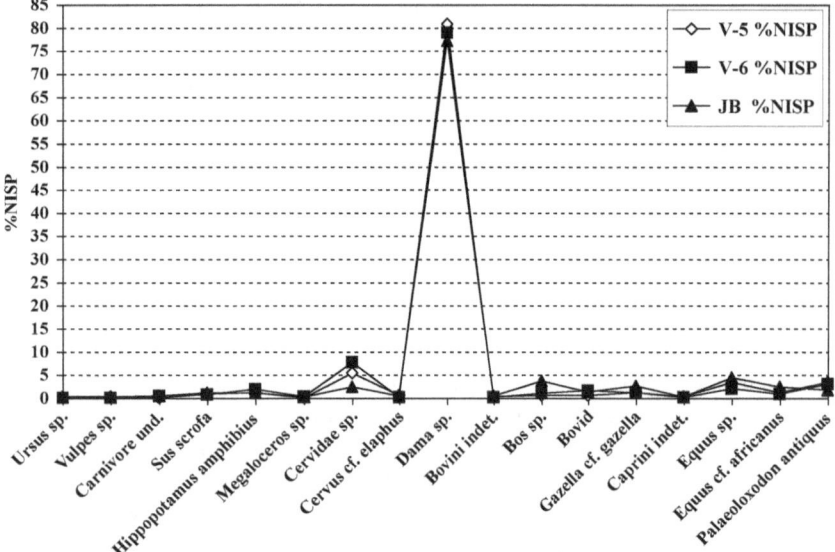

Fig. 5.38 Frequency of bone modifications from Area C and the JB (% of total number of recorded bones)

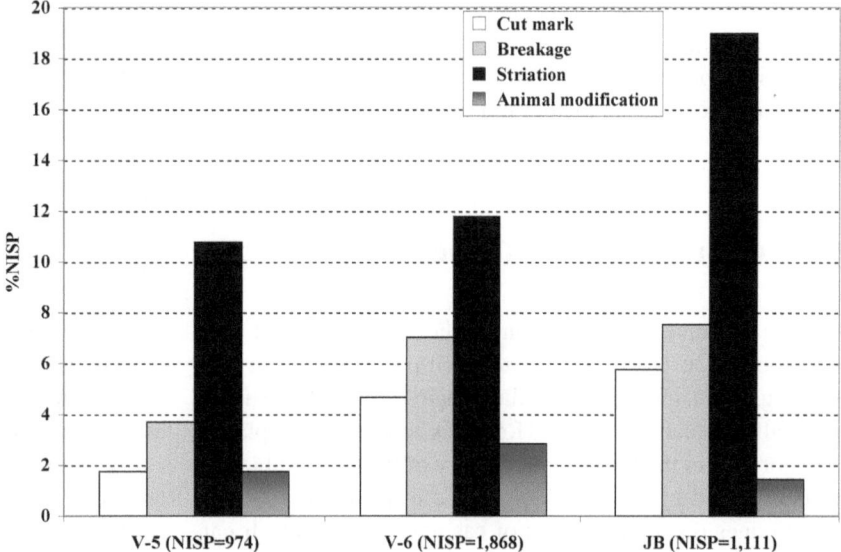

mark frequency on trunk bones at Hayonim Cave Layer D somewhat differ from those at GBY (Fig. 5.36c, d), a phenomenon that may be attributed to the dissimilar histories of the two sites and/or their differing nature (cave versus open-air site).

During our comparison of the two sites, we refrained from performing statistical tests because the only quantitative variable available in this context is the number of cut marks. This number can be useful in a qualitative assessment of butchering activity, but not as an indicator of the intensity or skill with which butchering was undertaken. An animal carcass can be butchered in a manner that leaves no marks, and a quantitative identification of marks can be mediated by a number of taphonomic factors; for example, the degree of fragmentation of a faunal assemblage or the raw material of the butchering tool (Dewbury and Russell 2007). In

conclusion, the evidence from Hayonim Cave corroborates the results of the analysis at GBY.

5.6 Animal- and Hominin-Induced Damage

At GBY, we observed several cases (NISP = 9) of hominin-induced cut marks and animal modifications on the same bone (Table 5.25). Three such instances display tooth scratches and tooth pits caused by rodents, two were gnawed and scratched by small carnivores, and two were scratched and pitted by large carnivores.

Of greatest interest are two bones featuring both hominin cut marks and large carnivore damage marks. The first is a young *Equus* sp. mandible from Layer V-6 (no. 2,033, r),

Table 5.25 Animal-induced modifications and cut marks on the same bone from Area C and the JB

Layer	Small-sized carnivore	Large-sized carnivore	Rodents	Unknown agent
V-5				
	PEL ISH *Dama* (no. 5,412, r)			FSH *Dama* (no. 1,753, r)
V-6				
	FSH *Dama* (no. 12,701)	MAND *Equus* (no. 2,033, r)	LBSH BSGE (no. 12,505)	MANF BSGC (no. 12,980)
			MANDF *Dama* (no. 1,478, l)	
JB				
		LBSH BSGD (no. 2,254)	RDS *Dama* (no. 12,670, r)	

which displays a large tooth mark as well as signs of tongue extraction, but due to the lack of superposition of the modification signs, we cannot suggest a particular scenario. The second bone is a long bone shaft from the JB (no. 2,254), seemingly porous (perhaps indicating a young age), exhibiting both cut marks and tooth scratches, but again with no clear superposition.

The rarity of cases, especially those of modified large carnivore bones, suggests minor carnivore activity. This proposal is supported by the overall low presence of carnivore traces when compared to hominin-induced modification.

5.7 Conclusions of the Taphonomic Analysis

The quantitative representation of the various animal taxa documented at GBY is consistent within the different layers (Fig. 5.37), indicating that the remains accumulated under generally similar conditions. Hominins appear to have played a major role in the taphonomic history of these assemblages, as is reflected by the clear dominance of *Dama* and *Dama*-sized bones displaying traces of carcass exploitation. It can thus be argued that hominins' faunal subsistence at the site focused primarily on *Dama*-procurement.

In addition to *Dama*, numerous species comprise the GBY faunal assemblage, although they have been documented in considerably lower numbers than *Dama* or *Dama*-sized taxa. A large variety of animal species of various body and weight classes have been identified. With NISP values of 15, Cervidae sp. and *Equus* sp. rank second in frequency after *Dama*. Given the low bone counts for the remaining taxa, it is quite striking that the numerous bones comprising seven taxa and the bones attributed to four body-size groups display traces of hominin influence (see Table 5.19). It should be noted that the entire butchering sequence, except for the initial carcass skinning, has been documented for these taxa as well. The large range of species of different body sizes and weight classes exploited by hominins for meat and marrow should be noted as it illustrates a high flexibility in prey choice by hominins at GBY.

5.7.1 Age and Sex Profiles and Occupation Seasons

Age and sex profiles of mammals can provide valuable insight into the exploitation modes, hunting methods, and occupation seasons. Animals are seasonal indicators. Thus, their appearance in the archaeological setting might reflect the season of hominin occupation. However, due to differences in individual variation, group/herd size and composition, seasonality, etc., separate assessments need to be made for each species.

Some of the recovered artiodactyls from the site are typically Mediterranean—the fallow deer, the mountain gazelle, and the wild boar. If we assume that a seasonality of distinct wet and dry seasons, similar to what presently occurs, prevailed during prehistoric occupation of the site, we may argue that birthing occurred during the spring months (March to May in the Mediterranean).

As noted earlier, the most common species in the GBY assemblages is *Dama*. A study of the age composition of *Dama* samples demonstrates the presence of juveniles and adults in each layer, as well as males and females. Three adults and one juvenile, both male and female, were found in Layer V-5; at least two juveniles and three adults, male and female, were found in Layer V-6; and at least two juveniles and two adults, reflecting at least one male and one female, were found in the JB.

Shed and unshed antlers of young specimens and juveniles of both sexes have been found in the JB, suggesting either a spring or winter presences at the site. This reconstruction is based on the fact that modern *Dama* birthing occurs in the Mediterranean region between March and May, whereas antlers are shed during February and March.

Mating of the modern female African ass (zebra) takes place in the spring, with a single foal born in early spring after an 11-month gestation period (Moehlman 2002). Thus, the presence in Layer V-6 of an *Equus* cf. *africanus* foal younger than 12 months may indicate its presence at the site from autumn to early winter. Only adults of this species type were found in the JB, and a single adult was found in Layers V-5 and V-6.

The gestation period of hippopotamus is approximately eight months. Since mating typically takes place during the dry season, most of the calves are born during the rainy season, when food is plentiful. Single births are by far the most common, although twins may be born on rare occasions. The calf stays close to its mother for its first two years (Eltringham 1999). The hippopotamus in Layer V-5 and the JB are less than three years old. The teeth of the specimen in Layer V-6 date it to approximately one year old, possibly indicating that its death occurred during the autumn or winter months.

The scarce *Sus scrofa* remains from the site include an animal aged approximately 12 months from Layer V-6, suggesting death occurred during winter to spring. Gazelle remains from in Layer V-6 belonged to an animal younger than two years old, and the JB yielded evidence of both a juvenile and an adult.

The remaining species provide very little information about occupation seasons, such as the red deer tooth from Layer V-6 that belonged to an animal less than 26 months old. Bones of *Bos* sp. from all layers are of adults. As for the largest species at the site, the elephant, young adults and adults are found in all three assemblages.

Based on the above data, we propose a mammalian autumn-winter-spring occupation at the site.

5.8 Summary and Conclusions

Results of the taphonomic analysis presented here enable the reconstruction of certain aspects of the site's taphonomic history. Above all, studies of hominin modification at the site have demonstrated that in-situ butchering of complete *Dama* carcasses occurred in all three layers at the site. Several additional mammalian species, though documented in considerably lower numbers, also display signs of butchering. Apparently they were brought to the site in the form of relatively complete carcasses, and it thus appears as though hominins were a major factor in the formation of the entire bone assemblage.

Striations are generally considered to be the result of direct interaction between sediment and bone surface due to trampling. Thus, the high amount of striated bones at GBY may account for heavy traffic at the site. A correlation between the occurrences of striations and the more heavily weathered bone surfaces, as could be expected, has not been established; rather, the opposite has been observed. Numerous striations occur on the better-preserved bone surfaces (i.e., *Dama* or *Dama*-sized bones), and a correlation was noted between the amount of striated bone surfaces and the frequency of hominin-induced butchering within a taxon.

Upon comparison of the number of bones displaying cut- and percussion marks with the number of striations per layer (Fig. 5.38), a correlation between the cut- and percussion marks frequency and the striation frequency can be established. When examining the number of modification occurrences out of the total number of recorded bones in a assemblage (e.g., bones identified to taxon, bones attributed to body-size groups, and unidentified mammal bones; see Table 4.3), we found that the percentages of bones with cut marks increased from 1.74% in Layer V-5, to 4.67% in Layer V-6, and to 5.77% in the JB. A similar increase is seen also in the percussion marks. In Layer V-5 and V-6, 3.69% and 7.03% of the bones, respectively, are characterized by percussion marks. A frequency of 7.54% was noted in the JB. Striated bones also increase accordingly: 10.78% in Layer V-5, 11.78% in Layer V-6, and 18.99% in the JB.

We further examined the modifications by general body-size group, (general body size group: the taxonomically identified and their respective body-size group) (Table 5.26a, b). The data is presented by a number of cases (Table 5.26a) and also as the relative frequency per each body size (%NISP BSG) (Table 5.26b). Bones with signs of honimin-induced damage are present in each body-size group, except for the elephant-sized group (BSGA). Among the remaining body-size group, the frequency of animal-induced damage is lower than the hominin-induced damage (see Table 5.26b).

Striation frequency in each body-size group fluctuates per layer. The most striated bones are those that belong to the general *Dama*-sized group (BSGD), followed by the larger (BSGC, giant deer, red deer, boar, bovine, 80–250 kg) and smaller (BSGE, gazelle, roe deer, 15–40 kg) body size (Fig. 5.39, Table 5.26a, b).

Positive significant correlation was found in Layer V-6 and the JB between the number of cut and striated bones grouped by body size (Layer V-6, r = 1.00, p < 0.001; JB: r = 1.00, p < 0.001). Positive significant correlation was found in all the assemblages between the number of bones grouped by body size displaying percussion marks and striations (Layer V-5; r = 0.943, p = 0.005; Layer V-6; r = 0.973, p < 0.001; JB: r = 0.868, p = 0.025). It therefore appears that the high amount of striations within an assemblage corresponds to the degree of hominin interference.

Since striations are usually regarded as the result of interaction with sediment, differences may be observed in the taphonomic histories of the bones from the studied assemblages. Against this background, the correlation between striated bone-surface frequency and hominin-induced butchering frequency within a taxon could be explained by arguing that the *Dama* assemblage represents the final butchering activities at the site before its abandonment and the probable quick burial of the bones. Consequently, remains of all other taxa must have been

Table 5.26 Summary of the bone-surface modifications by general body-size group (general body-size group). (**a**) Hominin-induced damage (Hom.); Striations (Str.); Animal-induced damage (Anm.)

	Hom. V-5	Str. V-5	Anm. V-5	NISP V-5	Hom. V-6	Str. V-6	Anm. V-6	NISP V-6	Hom. JB	Str. JB	Anm. JB	NISP JB
BSGA		1		22		2	1	36		1	1	17
BSGB	3	2		18	3	2	4	37	1	3		31
BSGC	5	11	10	92	21	23	33	177	22	32	3	153
BSGD	35	75	3	486	149	155	6	874	109	149	7	524
BSGE	8	12		111	30	24		220	8	23	1	133
BSGF	1			25	2	3		56				11
UNM	1	4	1	220	13	11	1	468	8	3		242
Total	53	105	14	974	218	220	45	1868	148	211	12	1111

(**b**) Relative frequency (%) of the bone-surface modifications by general body size group (general body-size group). Hominin-induced damage (Hom.); Striations (Str.); Animal-induced damage (Anm.)

	Hom. V-5	Str. V-5	Anm. V-5	Hom. V-6	Str. V-6	Anm. V-6	Hom. JB	Str. JB	Anm. JB
BSGA	0.00	4.55	0.00	0.00	5.56	2.78	0.00	5.88	5.88
BSGB	16.67	11.11	0.00	8.11	5.41	0.00	3.23	9.68	0.00
BSGC	5.43	11.96	0.00	11.86	12.99	2.26	14.38	20.92	1.96
BSGD	**7.20**	**15.43**	**2.06**	**17.05**	**17.73**	**3.78**	**20.80**	**28.44**	**1.34**
BSGE	7.21	10.81	2.70	13.64	10.91	2.73	6.02	17.29	0.75
BSGF	4.00	0.00	0.00	3.57	5.36	0.00	0.00	0.00	0.00
UNM	0.45	1.82	0.45	2.78	2.35	0.21	3.31	1.24	0.00
Total	5.44	10.78	1.44	11.67	11.78	2.41	13.32	18.99	1.08

Fig. 5.39 Frequency of striated bones from body-size groups, Area C, and the JB

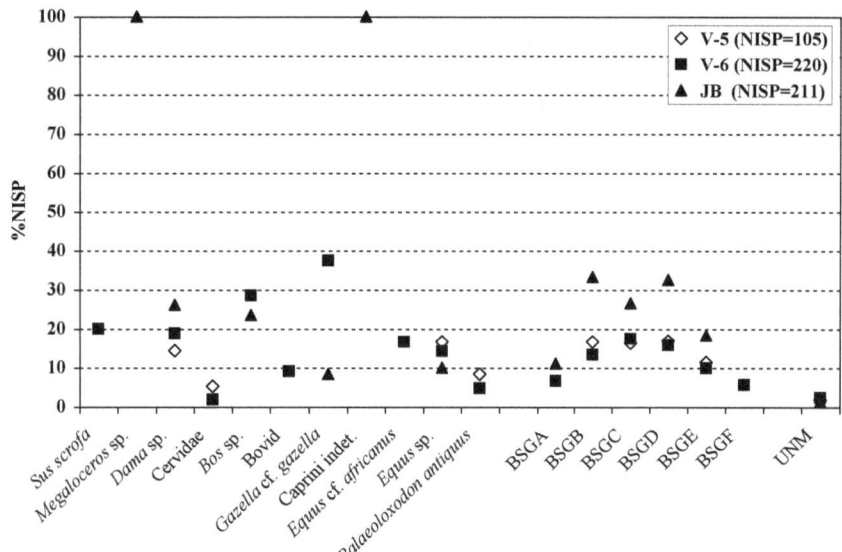

abandoned before these activities took place. The time delay would allow exposure to environmental conditions that induced weathering of the bones, which subsequently obliterated striations but not traces of hominin butchering. Following the evidence observed at GBY, one would assume that taxonomically and temporally sequenced butchering activities were repeated in every layer. Thus, this scenario, which suffers from methodological uncertainties, should be examined, since regular repetition of the proposed exploitation strategy requires uniform game availability over a longer period of time.

In order to better understand whether these are faunal assemblages with heterogeneous formation histories, a set of tumbling and trampling experiments was established to reproduce pre- and post-burial processes during early stages of bone deposition (see Chapter 6).

Chapter 6

Reconstructing Site-Formation Processes at GBY—The Experiments

Abstract A set of experiments were initiated to gain qualitative insight into the processes of bone modification and to assess the timing of the biostratonomic chronology at Gesher Benot Ya'aqov (GBY). Based on the results of the experiments, models for the internal operational sequence of an abrasional process due to water movement and trampling are presented. These models help to disentangle the taphonomic history at the site and have tremendous implications for future studies in bone taphonomy.

6.1 The Potential of Actualistic Studies in Taphonomic Research

Taphonomic analysis based on observations of modern processes is generally considered as fundamental to all zooarchaeological interpretations of deposits of bones with certain modifications (Cleghorn and Marean 2007). Since the late 1960s, the principles of bone taphonomy have become more and more influential for the interpretation of archaeological assemblages. As a consequence, the sequencing of the taphonomic chain and the recognition of specific taphonomic processes on faunal assemblages has become a major issue of research (Brain 1967a, b; Voorhies 1969; Behrensmeyer 1978; Hill 1978; Behrensmeyer and Hill 1980; Hill 1980; Binford 1981; Brain 1981; Shipman 1981; Behrensmeyer and Kidwell 1985). In particular, field studies in East and South Africa served to build a body of comparative information that was used to identify taphonomic processes and agents, as well as their effect on the qualitative and quantitative composition of fossil bone assemblages (Behrensmeyer 1975a, 1978; Brain 1981; Blumenschine 1986).

These actualistic studies are of crucial importance as they illustrate the potential complexity of the formation histories of archaeological sites. Many of these studies generalize their observations and classify taphonomic processes into characteristic substages and distinctive patterns (Voorhies 1969; Behrensmeyer 1978; Shipman 1981; Blumenschine 1986). More recent actualistic studies along the same lines pay more attention to the importance of concomitant circumstances for the progress of taphonomic processes (i.e., Andrews and Cook 1985; Coard and Dennell 1995; Coard 1999; Domínguez-Rodrigo and Barba 2006; Pickering and Egeland 2006). Studies undertaken by Voorhies (1969) and Coard and Dennell (1995) underline that the potential of bones for hydrodynamic sorting is highly dependent on carcass completeness and the degree of bone desiccation. Bone weathering, generally considered to be governed by an amalgam of climatically induced factors, was shown by Andrews and Cook (1985) mainly influenced by ultraviolet light. Pickering and Egeland (2006) emphasize that variability in the morphology of percussion marks can be induced by hammerstones made from different raw materials and percussion techniques.

These studies indicate the sensibility of taphonomic processes and reveal that they are governed by both the basic condition of the bone assemblage itself and a high susceptibility to environmental factors. They also illustrate the various consequences of these factors for the implementation and degree of taphonomic signals.

The current actualistic research is thus still in the process of reproducing taphonomic signals in order to isolate distinct patterns that would then be applied to the interpretation of archaeological sites. However, due to studies undertaken in the last 20 years it has become increasingly apparent that the enormous variability observed in taphonomic signals hampers their interpretation and the applicability of actualistic studies to fossil assemblages. Of course, single taphonomic processes are relatively easy to understand, but due to the fact that taphonomic processes are additive, reconstructing taphonomic histories is still a challenge. Thus, we tend to subdivide taphonomic processes into substages and distinctive patterns simply as a makeshift way to overcome these problems, to allow for their identification, and to ensure comparability (Voorhies 1969; Behrensmeyer 1975a, b; Blumenschine and Selvaggio 1988, 1991; Marean et al. 1992; Blumenschine 1995). As a result, we still lack a valid interpretative template that could be imposed on faunal assemblages.

Only in individual cases would it be possible to satisfactorily apply the results of actualistic studies to the interpretation of bone assemblages from archaeological sites. This is because actualistic studies usually describe only singular processes in the taphonomic chain in a very specific context. Actualistic studies must therefore pay attention to individual site-formation histories that can be recognized in archaeological sites. The advantage is that already known contexts can be then incorporated, linking the actualistic study to the taphonomic history of a given faunal assemblage. It is not the general recognition of taphonomic signal patterns which must be central to these studies, but the specific mode of operation of a taphonomic process.

6.2 Homogeneity or Non-homogeneity in the GBY Faunal Assemblage?

Results of the taphonomic analysis of the GBY bone assemblages suggest that a time delay may have occurred sometime between the butchering of the different taxa and the almost complete hominin exploitation of *Dama* carcasses before the site was abandoned. These taxonomically geared butchering sequences ending with *Dama* exploitation seem to have been repeated during the different occupation periods of GBY, which are represented by the assemblages discussed throughout the volume. It thus seems highly likely that such a scenario would suffer from severe methodological uncertainties. For instance, sequenced butchering oriented towards taxon during different hominin occupations would require uniform game availability over a long period of time, thus rendering this scenario quite unlikely.

In order to test the outlined interpretative scenario, the proposed homogeneity of the faunal assemblage has to be evaluated. This may be done by establishing the chronological sequencing of the known taphonomic processes that shaped the faunal assemblage, as revealed by the sedimentological and faunal analysis. The results of the analysis could have tremendous implications for the overall interpretation of the site, as they indicate that a larger time period was involved in the formation of the individual layers than has thus far been indicated by the available archaeological record.

In this context, bone-surface modifications and, in particular, striations are a unique feature of the GBY fauna and form key elements for disentangling the taphonomic chronology of the site.

Sedimentological analysis at GBY has revealed that the archaeological material was embedded in an homogeneous alluvial sediment, which was quickly deposited and suffered only minor secondary overprinting. Among other factors, this was indicated by the only minor fractioning of mollusk shells

and the in-situ preservation of such larger sediment particles as the body parts of crabs (Ashkenazi et al. 2005).

As a consequence, one should assume that only biostratonomic processes were responsible for bone-surface modification before the final burial of the bones. We witnessed the impact of biostratonomic processes on large mammalian fauna where several biotic and abiotic modifying agents were recognized. Biotic agents such as hominins and carnivores influenced the bone assemblage in an initial phase of the biotratonomic sequence. In contrast, abiotic processes responsible for bone preservation correlating with the body size of large mammals and striation formation point to a subsequent stage in the sequence.

In particular, the origin of striations is generally considered to result from interaction with sediment particles (e.g., Blasco et al. 2008; Domínguez-Rodrigo et al. 2009). In conclusion, we suggest that evident features of the biostratonomic processes that occurred both at the beginning of the biostratonomic sequence, following the death of the animals, and at the end of the sequence, with the burial of the bones, have thus far been recognized at GBY. A set of experiments simulating striation formation on bone surfaces against the background the various environmental conditions documented at the site was initiated to assess the timing of its biostratonomic chronology (see also Gaudzinski-Windheuser et al. 2010). These experiments were meant to mimic and evaluate the individual conditions at the site and are thus only of limited value for our general knowledge of bone taphonomy.

6.3 Materials

In order to evaluate the origin and character of the striations observed on the bone surfaces, as well as the differences in bone-surface preservation of the different taxa, a set of experiments was undertaken on both fresh and dry sheep and cow bones. These tumbling and trampling experiments were designed to mimic the effects of shoreline conditions at GBY. Finally, additional analyses were carried out in order to measure the effects of trampling and scratching, and of long-term burial under random conditions.

A selection of bone specimens from Layers V-5 and V-6, the JB, Area B and Trench IV were chosen for these analyses. This sample consisted of 125 bones, of which 87 displayed striations (see Chapter 5). The majority originated from *Dama* sp., as this was the most common mammalian species found at the site and its bone preservation was good. The study also includes specimens from the entire cervid skeleton. The long bones of large bovids differ from those of cervids in the rough surface structure of the bone compacta, as can be seen with the naked eye. The rough structure obscures the outline of the striations, thus rendering analysis all the more difficult.

A first set of experiments was undertaken using fresh bones. We will detail their characteristics and the treatment they underwent before describing the experiments themselves. The various treatments mimic the possible array of naturally and/or anthropogenically induced processes that occurred prior to deposition. All bones included in the experiments, as well as bone splinters resulting from fractures, were measured, weighed, and photographed.

1. Among other bones, our experiments utilized 15 fresh sheep metapodials. In order to expose the bone, the tissue was cut along the metapodial anterior and posterior faces parallel to their axis. The exposed anterior and posterior tendons were severed at the distal metaphysis. The metapodial was then pried out, leaving the distal epiphysis still in articulation and freeing the bone of the skin, meat and periosteum. Finally, the remaining scraps of tissue were removed with a surgical scalpel (Fig. 6.1). The process of defleshing and its resulting bone-surface modifications were meticulously documented in order to account for each modification produced on the bones in the early stages of the experiment *senso stricto* (prior to burial, tumbling, and trampling).

In about half the cases, metapodial processing resulted in fine vertical striations along the anterior face of the diaphysis. These modifications mimic the flat-based striations observed on bones from the GBY assemblages (see Fig. 5.4).

2. A fresh sheep hind leg was also included in the experiments, including pelvis, femur, tibia, astragalus, and the T4 (naviculocuboid) and attached calcaneum. The leg bones were previously defleshed by the butcher and further disarticulated with a metal knife. The periosteum was not removed, but tendons and scraps of meat attached to the proximal and distal ends of the long bones and the pelvis were removed. Each bone was individually weighed, with and without the attached meat and tendons. The maximum diameter and length of the long bones was measured, as was the maximum length and width of the pelvis, astragalus, calcaneum, and the T4 (naviculocuboid).

3. Two humeri and two radii/ulnae of domestic cow were also used. An articulated humerus and radius, with some meat remaining on the bones, were disarticulated and cleaned using a surgical scalpel. The ulnae were removed. The bones

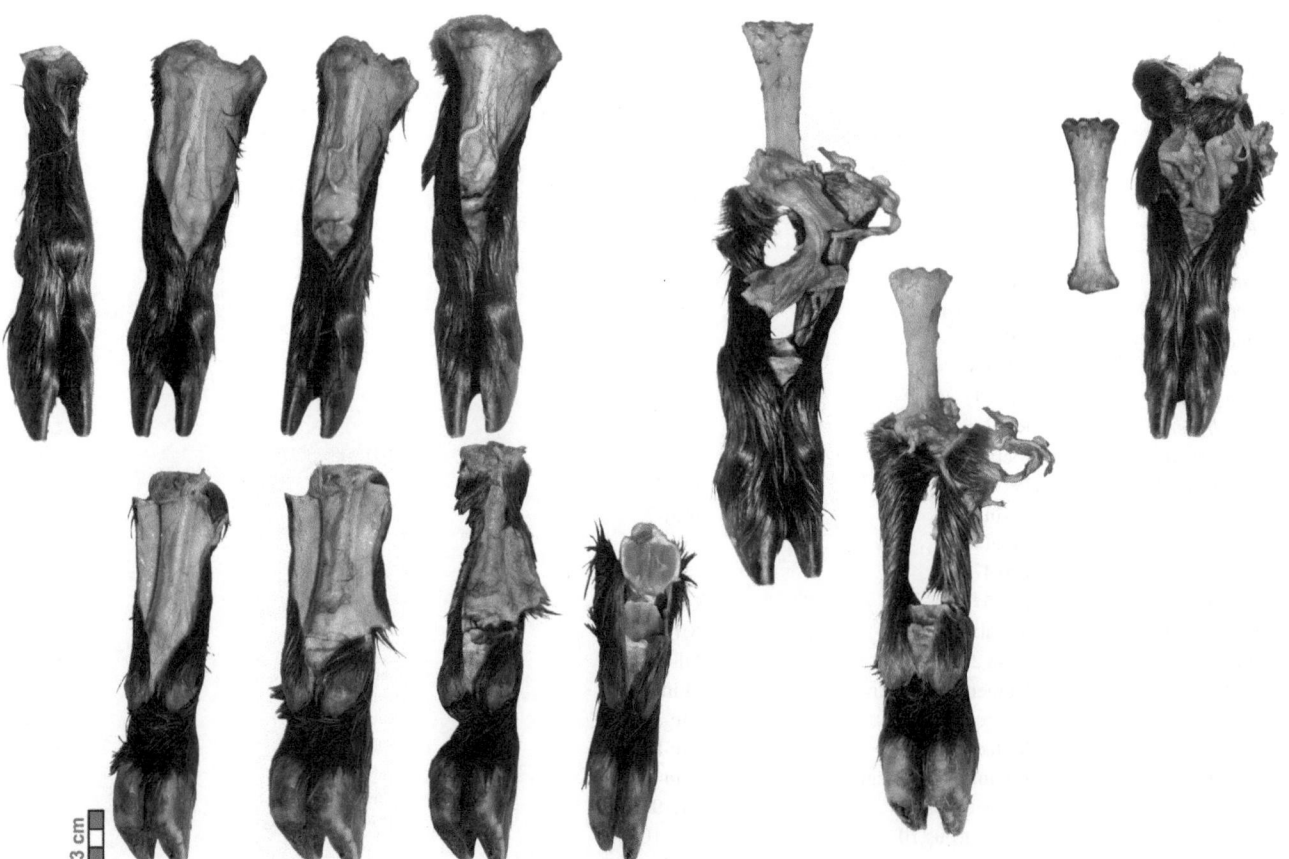

Fig. 6.1 Processing of sheep feet for the experiments. The *top row*, from left to right, shows feet from the anterior face in different processing stages. The *bottom row*, from left to right, shows feet in the posterior view in different processing stages

Table 6.1 The bones involved in the different experiments

Species	Bones	Abbreviation	Burial	Tumbling	Trampling
Ovis aries	Astragalus	Ast-1	1		
Ovis aries	Calcaneus	Calc-1	1		
Ovis aries	Femur 3	Fem-3	1		
Bos p. taurus	Femur 1, fragment 9	Fem-1-9	1		
Bos p. taurus	Femur 1, fragment 8	Fem-1-8	1		
Bos p. taurus	Femur 1, fragment 11	Fem-1-11	1		
Bos p. taurus	Femur 1, fragment 10	Fem-1-10	1	15	
Ovis aries	Pelvis 3	Pel-3	1		
Ovis aries	Naviculo-cuboid	T4-1	1		
Ovis aries	Tibia-6	Tib-6	1		
Bos p. taurus	Tibia 1, fragment 1	Tib-1-1	1		
Bos p. taurus	Tibia 1, fragment 2	Tib-1-2	1		
Ovis aries	Metacarpus 1	Mc-1		1	1
Ovis aries	Metatarsus 1	Mt-1		2	1
Ovis aries	Metacarpus 2	Mc-2		3	
Ovis aries	Metatarsus 2	Mt-2		4	
Bos p. taurus	Femur 2, fragment 1	Fem-2-1		5	
Bos p. taurus	Tibia 2, fragment 4	Tib-2-4		6	
Ovis aries	Metacarpus 3	Mc-3		7	
Ovis aries	Metatarsus 3	Mt-3		8	
Ovis aries	Metacarpus 4	Mc-4		9	
Ovis aries	Metatarsus 4	Mt-4		10	
Ovis aries	Metacarpus 5	Mc-5		11	
Ovis aries	Metatarsus 5	Mt-5		12	
Ovis aries	Metacarpus 6	Mc-6		13	
Ovis aries	Metatarsus 6	Mt-6		14	
Bos p. taurus	Radius 3, fragment 1	Rad-3-1		16	
Ovis aries	Metatarsus 7	Mt-7		17	
Bos p. taurus	Humerus 2, fragment 1	Hum-2-1		18	
Bos p. taurus	Femur 2, fragment 2	Fem-2-2			1
Bos p. taurus	Tibia 2, fragment 5	Tib-2-5			1
Bos p. taurus	Rib 1	Cost-1			2
Bos p. taurus	Radius 1	Rad-1			2
Hyaena hyaena	Rib 2	Cost-2			3
Hyaena hyaena	Metatarsus III 1	MtIII-1			3
Hyaena hyaena	Pelvis 1	Pel-1			3
Hyaena hyaena	Tibia 3	Tib-3			3
Hyaena hyaena	Ulna 1	Ul-1			3
Hyaena hyaena	Rib 3	Cost-3			4
Hyaena hyaena	Metatarsus IV 1	MtIV-1			4
Hyaena hyaena	Pelvis 2	Pel-2			4
Hyaena hyaena	Radius 2	Rad-2			4
Hyaena hyaena	Tibia-4	Tib-4			4
Hyaena hyaena	Rib 4	Cost-4			5
Hyaena hyaena	Rib 5	Cost-5			5
Hyaena hyaena	Metatarsus II 1	MtII-1			5
Hyaena hyaena	Ulna 2	Ul-2			5
Bos p. taurus	Humerus 1, fragment 1	Hum-1-1			6
Ovis aries	Metacarpus 7	Mc-7			6
Ovis aries	Metacarpus 8	Mc-8			6
Bos p. taurus	Radius 3, fragment 2	Rad-3-2			6
Ovis aries	Metacarpus 9	Mc-9			7
Ovis aries	Metacarpus 10	Mc-10			7
Bos p. taurus	Radius 4, fragment 1	Rad-4-1			7
Ovis aries	Tibia-5	Tib-5			7

were dried for three days and afterwards fragmented with a blunt quartzite pebble.

One of the humeri (Humerus 1) was broken by means of blows to the anterior face, causing it to break into two pieces. Additional blows were struck to the lateral and medial faces, resulting in three long bone shaft fragments. One fragment featured percussion marks on the postrior/lateral and posterior/medial face. Humerus 2 was fractured into two pieces by a blow to the medial diaphysis. A further blow was struck to the anterior face, resulting in an additional long bone shaft fragment. A conical percussion mark was produced during this process.

Radius 1 was struck by several blows to the posterior diaphysis and broken into two pieces. In order to obtain additional shaft fragments, several blows were struck to the proximal cranial face, resulting in a longitudinal break along the proximal epiphysis. No conical impact fractures were produced during this bone percussion. Radius 2 was smashed by blows to the posterior diaphysis. The first blow produced a bone flake still attached to the point of impact. A second blow to the same spot resulted in a conical impact fracture that broke the bone into three pieces. In addition, another bone flake fracture was produced as the second blow reverberated through the bone to the opposite side. Further blows to the lateral face of the bone resulted in an additional long bone shaft fragment.

4. We also experimented with part of the hind leg of a domestic cow (Femur 1 and Tibia 1). The bones were still in articulation, but a local butcher removed all the meat and then separated the bones from each other with a metal knife. As this experiment was designed to smash the bones and mimic bone-marrow extraction, it was necessary to thoroughly clean the bone surfaces. Any remaining meat and large sections of the periosteum were removed using a metal knife.

The cow tibia was laid on an anvil and smashed with a blunt-tipped quartzite pebble. The bone shaft fractured into two major fragments displaying conical impact marks. The anvil consisted of a large calcareous block used to steady the bone during smashing. A considerable effort was required to successfully break the bone on the anvil. We discovered that removing the anvil actually facilitated the bone-breaking process. We also noted that the bones broke more easily if the meat and periosteum were thoroughly removed.

The same procedure was used to smash the cow's femur. Meat and periosteum still attached to the bone were partially removed with a metal knife before the bone was smashed, this time without the use of an anvil. Several blows with a blunt-tipped quartzite pebble were needed to open the marrow cavity. Additional blows were required to produce the six resulting bone fragments.

5. Another femur of a domestic cow (Femur 2) was included in the experiments. Its periosteum was partly removed and the bone-marrow cavity exposed by a blunt-tipped pebble. The bone was fractured into two pieces by blows to the posterior/lateral bone face. Further blows to the anterior face produced several additional bone fragments. The bone was left to dry for eight days, at which point the remaining periosteum was then removed, where possible, with a surgical scalpel. During periosteum removal, longitudinal and horizontal cut marks were intentionally produced.

6. To end, we fragmented another tibia (Tibia 2) from a domestic cow. To obtain the fragments, the meat and flesh were first removed with a surgical scalpel and a kitchen knife. Most of the periosteum remained attached to the bone. Numerous blows with a blunt stone were struck to the posterior face of the tibia. These blows produced fractures, but the periosteum held the bone intact. Several strikes along the fracture lines were required to complete the bone breakage. The bone was left to dry for eight days, after which the periosteum remained tightly attached to the bone and could not be removed without serious damage to the bone surface.

A second set of experiments was undertaken with bones from the comparative collections of the National Natural History Collections, at the Hebrew University. The complete bones used are as follows: (a) *Bos taurus*: radius, second rib; and (b) juvenile *Hyaena hyaena*: one radius, two ulnae, MT II, MT III, MT IV, two tibiae, two pelvises, and four ribs. The organic content of the hyena and cow bones had been considerably reduced after having been buried for several years.

All the bones involved in the different experiments are listed in Table 6.1.

6.4 Description of the Experiments

6.4.1 Scratching Experiment

The pointed helix of mollusks and fragmented mollusks are capable of scratching bone surfaces. In order to document the effects of mollusk scratches on bone surfaces, fresh bones—a dried fresh domestic cow radius shaft and a fresh sheep metacarpus—were scratched and striated in a variety of ways by a female weighing 50 kg handling a large *Viviparus* and a pointed *Melanopsis* helix with different pressure.

Considerable force was necessary in order to produce wide, blunt, and irregular striations of differing depth. Striation depth correlated with the degree of pressure that was applied to the bones. The results of the scratching experiment are illustrated in Fig. 6.2.

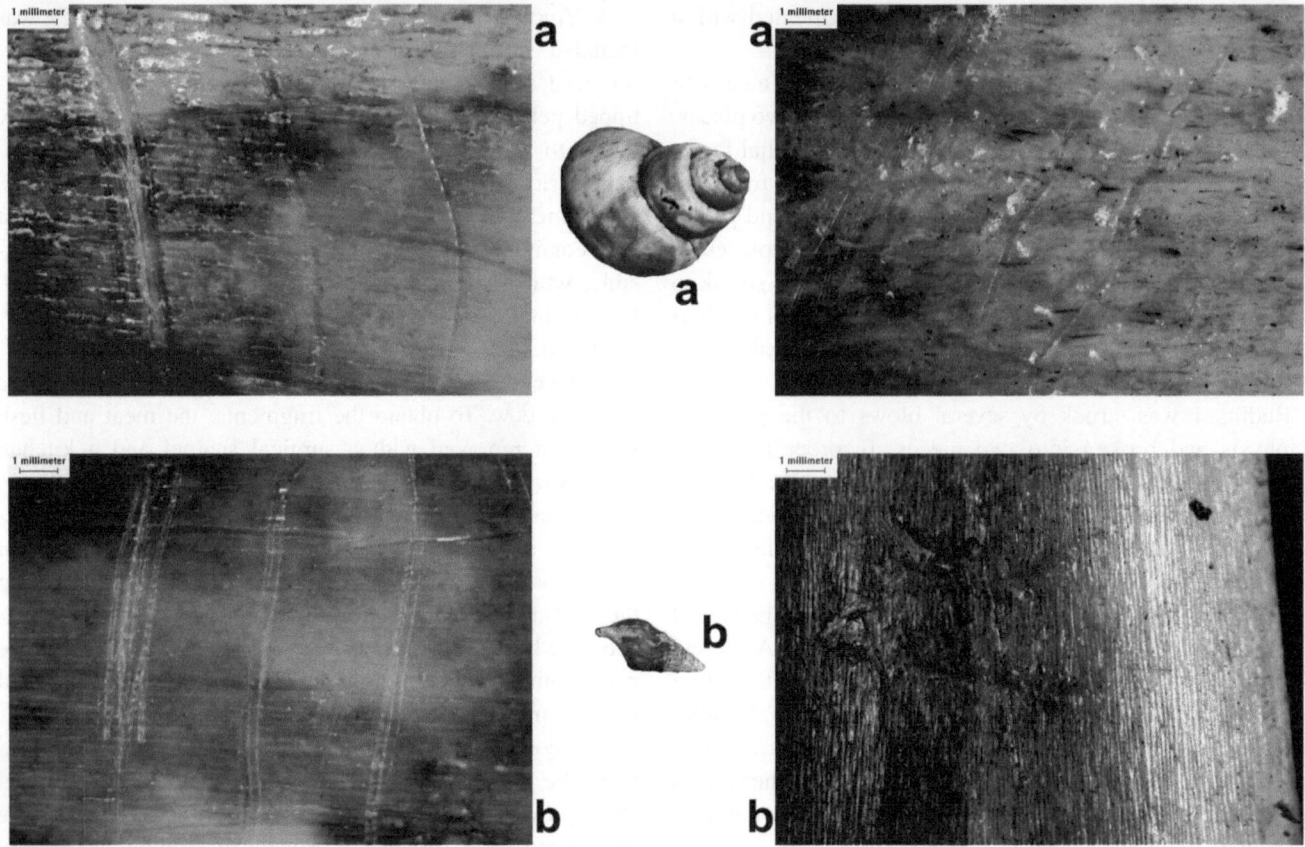

Fig. 6.2 Scratching experiment: striations produced by **a** a large *Viviparus* and **b** the pointed helix of a *Melanopsis*. On the *left* is a dried fresh domestic cow radius shaft, and on the *right* is a fresh sheep metacarpus

6.4.2 Burial Experiment

We buried the bones near Khirbet Sa'adim in the Judean Hills in order to observe the effects of burial on bone-surface conditions. This location was selected because of the constant traffic of picnickers and hikers, who constantly step on and pack the earth in which the bones were shallowly deposited. We assumed that the unpredictable activities in the picnic area would produce distinct, yet random, modification patterns. The bones comprised part of the hind leg of a sheep, a tibia (Tibia 1) and femur fragments (Femur 1) of a domestic cow, as well as fresh *Bos* shaft fragments (Fig. 6.3) because of their greater susceptibility to weathering and drying than complete bones.

The burial matrix was created from 5.3 kg of washed coquina from Layers V-5 and V-6 (Sediment sample: GBY 14.09.97; Layer V-6; square and sub-square coordinates: 183/130c; elevation 60.11/60.07–59.86/59.80 m above msl) mixed with 1 kg of gravely sand, and wrapped together with the bones in a piece of mesh fabric. A pit measuring 1.0 × 0.50 m and 15 cm deep was dug next to a picnic table and the mesh containing the bones and sediment was placed

in it and covered by some 10 cm of gravely sand. We marked and photographed the exact location of the pit.

The bones were buried on August 5, 2004, and re-excavated nearly one year later, on July 21, 2005. Excavation of the mesh was undertaken with blunt-tipped picks. The mesh enclosing the bones and matrix was still covered with approximately 10 cm of the original sediment; however, it became clear that the upper half of the mesh was open and only some of the bones were still covered with sediment. Three bovid long bone fragments (Femur 1, Fragment 10; Tibia 1, Fragment 2; Femur 1, Fragment 11) and a sheep astragalus survived. The surviving bones were washed, photographed, and analyzed with the image analyzer.

The bones displayed different types of surface modifications. Femur 1 (Fragment 10) displayed slightly rounded broken edges. The bone surface was abraded, resulting in the removal of the outermost bone surface, which, in turn, revealed the underlying porous bone structure. Slight exfoliation and singular striations were visible on the central part of the bone fragment and its edge (Fig. 6.4).

Femur 1 (Fragment 11) was not significantly rounded, and only tiny sections of the bone surface were exfoliated or

Fig. 6.3 Smashing of a *Bos* bone for the burial experiment. On the *top left* is the positioning of the bone on a large calcareous block before smashing with a blunt-tipped quartzite pebble. On the *top right* is a detail of the shaft of a *Bos* tibia after smashing. The periosteum of the bone prevented the disintegration of the bone. On the *bottom* is a *Bos* femur after smashing

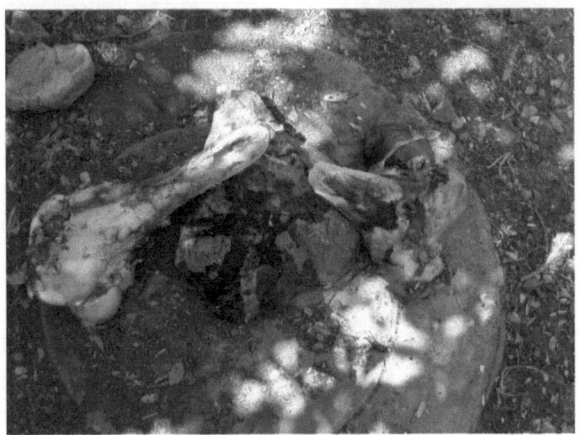

abraded. As in the case of Femur 1 (Fragment 10), Tibia 1 (Fragment 2) was also slightly rounded. Much of its surface was exfoliated and displayed dry fractures, and it had been gnawed by a rodent (Fig. 6.5).

The sheep astragalus appeared almost unaffected by burial. Dry meat was still attached to the bone, whose surface displayed no signs of modification.

The burial experiment failed to produce the expected large number of surface striations. Striations were observed only in two cases, both connected with exfoliation due to removal of the periosteum before burial took place. However, the clearest and most interesting result of this experiment was the abrasion of long bone shaft fragments, suggesting that considerable interaction or movement between bones and sediment must have taken place. Therefore, the lack of striations does not indicate a lack of bone movement within the matrix. The lack of striations is surprising, given the fact that the sediment contained a high amount of sharp shell fragments.

We tentatively conclude that striations were not necessarily the result of interaction between the bones and coquina. But since striations occurred at GBY, it can be assumed that they were the result of specific environmental and/or sedimentological conditions at the site. In order to substantiate this hypothesis, detailed tumbling and trampling experiments were undertaken in order to mimic the conditions at GBY.

6.4.3 Tumbling Experiments

As mentioned earlier, our tumbling experiments were designed to mimic the constant movement of sediment and bone in a low-velocity stream- or lake-shore setting. A machine that provides constant and calm movement is a cement mixer, whose chamber rotates 25 times per minute. The amount of water added to the sediment and the duration of rotation varied for each experiment. The sediment mixture was replaced for each experiment.

In the first experiments (Tumbling Experiments 1–8), we placed a bone together with mollusks and sediment into a water-permeable cloth bag and affixed the bag to the water-filled rotation chamber of the cement mixer. This set of experiments was designed to mimic constant unidirectional shoreline movement. In subsequent experiments (Tumbling Experiments 9–18), bone, sediment, and water were placed in a plastic container and affixed to the mixer's rotation chamber (Fig. 6.6), and continuously tumbled.

In contrast to the unidirectional Experiments 1–8, here the water movement was multidirectional and random.

6.4.3.1 Tumbling Experiment 1

A fresh sheep metacarpus was used for the first tumbling experiment. Deep, longitudinal striations resulted from

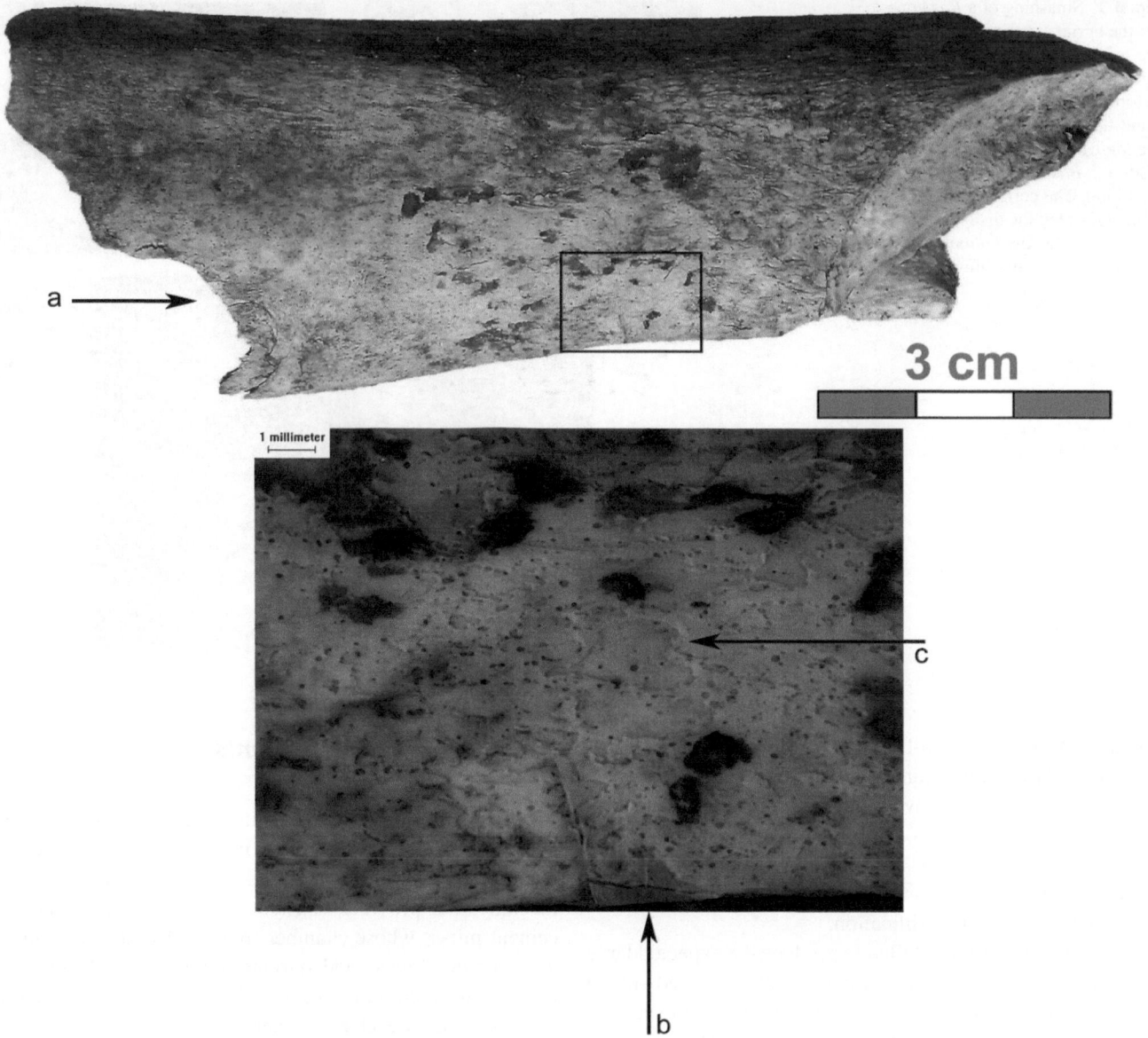

Fig. 6.4 Burial experiment. Modifications on a domestic cow femur fragment (Fem-1-10) after burial: **a** porous bone structure, edge abrasion and rounding; **b** singular striation; **c** exfoliation of the bone surface

removal of the periosteum on the posterior face of the diaphysis, adjacent to the sulcus. Periosteum remained attached to the proximal and distal metaphyses.

This experiment utilized a mixture of 1.5 kg of mollusks (GBY 15/07/97; Layer V-6; 184/129 a+b, 60.17/60.09–59.22/59.86) and 1 kg of clay. Bone and sediment were placed in a cloth bag and soaked in ten liters of water for ten minutes. The wet weight of the bag and its contents was 3.75 kg. The bag was affixed to the rotation chamber, which was then filled with 100 liters of water, and tumbled for two hours. After tumbling, the sediment sorted into the small-, medium-, and large-sized fractions.

Apart from a few areas of periosteum removal, the bone surface of the proximal metaphysis remained intact. Deepening of the sulcus was observed on the posterior face of the diaphysis. The striation next to the sulcus remained a narrow incision (Fig. 6.7). The central section of the diaphysis' lateral face was slightly abraded.

6.4.3.2 Tumbling Experiment 2

A fresh sheep metatarsus was featured in the second tumble experiment, after removal of its skin, meat, and periosteum.

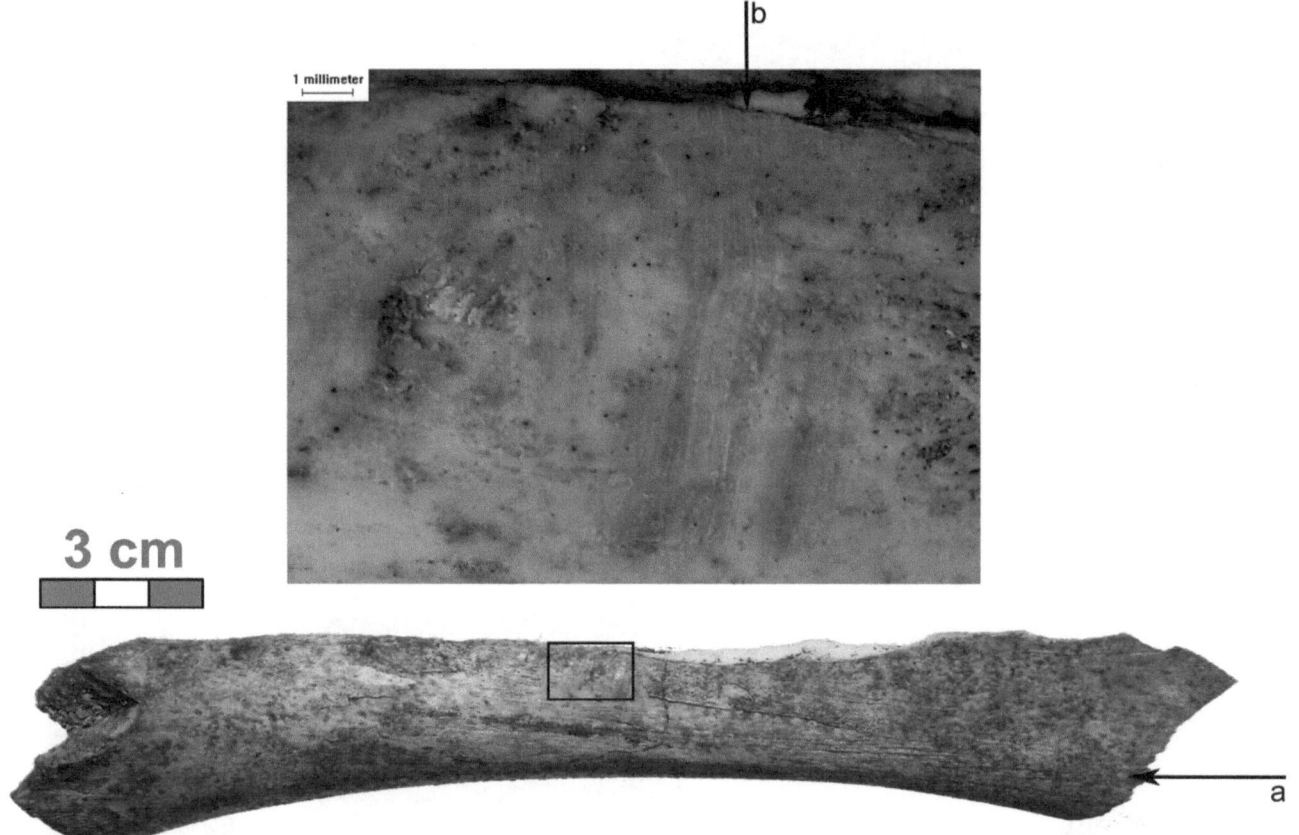

Fig. 6.5 Burial experiment. Modifications on a domestic cow tibia fragment (Tib-1-2) after burial. The fragment shows **a** dry fractures and **b** gnawing damage by a rodent

Removal of the latter resulted in deep, longitudinal striations on the central portion of the anterior face of the diaphysis. The periosteum remained attached to the proximal and distal metaphyses.

This experiment utilized a mixture of 1 kg of mollusks (GBY 15/07/97; Layer V-6; 184/129 a+b, 60.17/60.09–59.22/59.86) and 0.5 kg of clay. Bone and sediment were placed in a cloth bag and soaked in ten liters of water for ten minutes. When wet, the bag and its contents weighed 2.75 kg. The bag was affixed to the rotation chamber, filled with 100 liters of water, and tumbled for three hours. As in Tumbling Experiment 1, after tumbling the sediment sorted into small-, medium-, and large-sized fractions.

The longitudinal striations on the anterior face of the diaphysis and of the sulcus appeared more pronounced than before tumbling (Fig. 6.8). Tumbling resulted in the slight abrasion of the outer bone surface.

6.4.3.3 Tumbling Experiment 3

A fresh sheep metacarpus was used for the third tumble experiment. After removal of the skin and meat, no visible surface modifications were noted. Periosteum remained attached to large parts of the proximal and distal metaphyses.

This experiment utilized a mixture of 1 kg of mollusks (GBY 16/09/97; Layer V-6; 185/129 a+b, 60.41/60.31–59.94/59.68) and 0.5 kg of clay. Bone and sediment were placed in a cloth bag and soaked in ten of liters water for ten minutes. When wet, the bag and its contents weighed 2.5 kg. The bag was affixed to the rotation chamber, which was then filled with 100 liters of water, and tumbled for three hours. After the tumbling, the sediment sorted into small-, medium-, and large-sized fractions.

After tumbling, prominent parts of the bone were abraded. This was well illustrated by changes in the porous structure of the distal metaphysis. In contrast, the shape of the foramina on the posterior face remained unaffected by abrasion (Fig. 6.9).

6.4.3.4 Tumbling Experiment 4

A fresh sheep metacarpus was used for the fourth tumbling experiment. Skin, meat, and the periosteum were removed, with the latter resulting in two deep longitudinal striations on the posterior face of the diaphysis. Prior to tumbling,

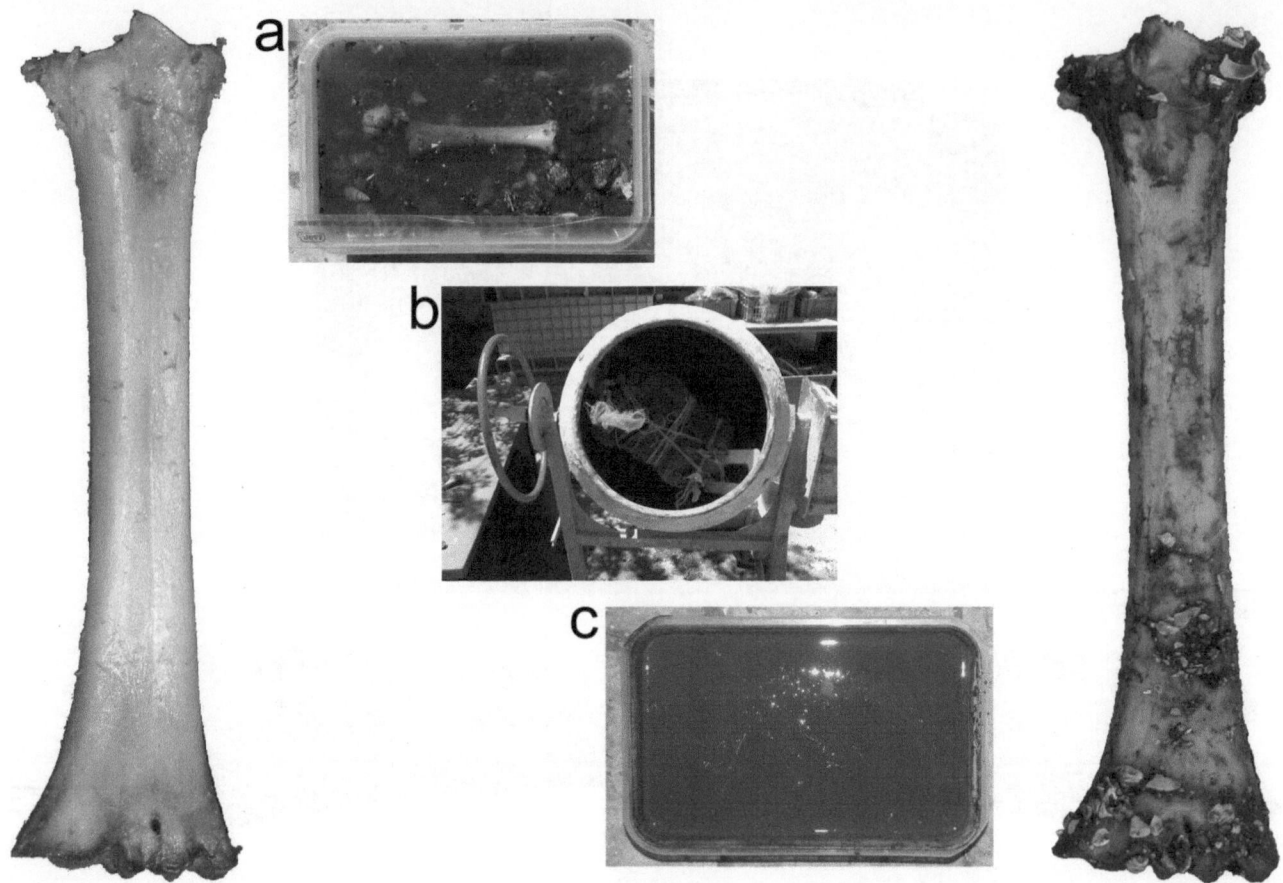

Fig. 6.6 Preparations for a tumbling experiment: **a** the bone was placed in a plastic container with sediment and water; **b** the container was affixed to the rotation chamber for tumbling; **c** the contents of the plastic container after tumbling. A sheep metapodial is seen prior to (*left*) and after tumbling (*right*)

periosteum covered large parts of the proximal and distal metaphyses.

The bone was placed in a cloth bag with a mixture of 0.5 kg of mollusks and clay (GBY 15/07/97; Layer V-6; 184/129 a+b, 60.17/60.09–59.22/59.86), affixed to the rotation chamber, and tumbled for three hours. After tumbling, bone and sediment remained unsorted.

Following tumbling, the longitudinal striations on the anterior face of the diaphysis were more pronounced, being smoother and wider, as is well-illustrated by the changed morphology of the striations' start- and endpoints (Fig. 6.10). The sulcus on the diaphysis also changed, becoming shallower and less narrow. The new morphology seen on the surface of the distal metaphysis highlights the higher degree of abrasion that occurs when water is involved (Fig. 6.11).

6.4.3.5 Tumbling Experiment 5

A domestic cow femur was used for the fifth tumbling experiment. After removal of the skin and meat, a number of

shallow, well-defined longitudinal and vertical cut marks were visible. Large chunks of periosteum remained attached to the bone. We also noted a slight exfoliation of the periosteum.

This experiment utilized a mixture of 0.75 kg of mollusks (GBY 05/09/97; Layer V-6; 181/130d, 60.20/60.13–59.84/59.76) and 0.25 kg of clay. Bone and sediment were placed in a cloth bag and soaked in ten liters of water for ten minutes. When wet, the bag and its contents weighed 3.75 kg. The bag was affixed to the rotation chamber of the mixer, which was filled with 70 liters of water and tumbled for three hours. After tumbling, the sediment sorted into small-, medium-, and large-sized fractions.

Following tumbling, the bone was soaked in water. The edges of the bone fragment were rounded, and its outermost surface had disappeared due to abrasion. While the longitudinal cut marks had almost vanished, the vertical cuts had become shallower but remained visible (Fig. 6.12). After tumbling longitudinal cracking along the breaking edge occurred accompanied by levelling of breakage morphology (Fig. 6.13).

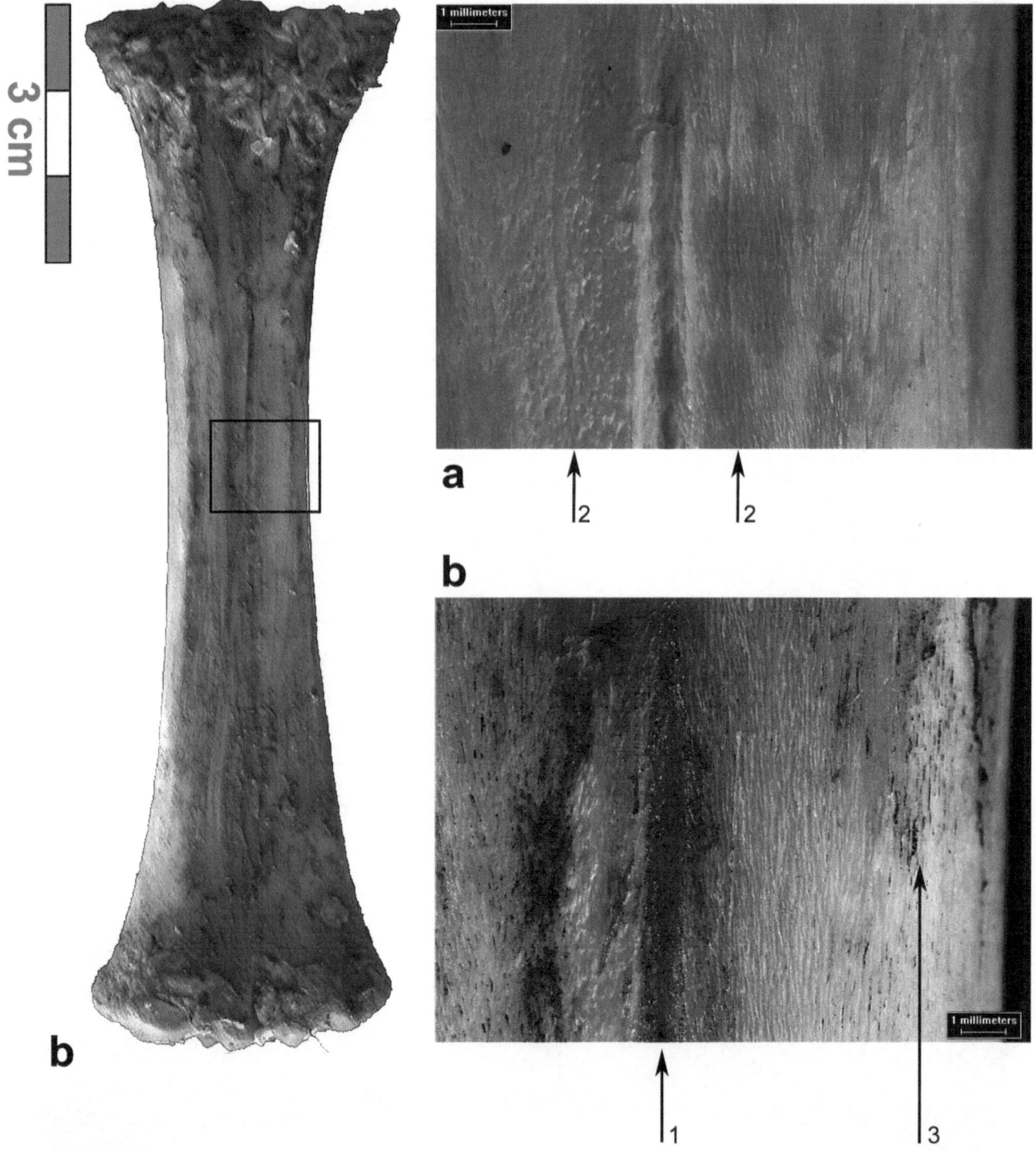

Fig. 6.7 Tumbling Experiment 1, sheep metacarpus. The posterior face of the diaphysis **a** before and **b** after tumbling. Deepening of the sulcus (1), striation (2), and abrasion (3)

6.4.3.6 Tumbling Experiment 6

A fresh tibia fragment of a domestic cow was used for the sixth tumbling experiment. Meat and skin were removed with a surgical scalpel and a kitchen knife, producing small, well-defined, deep longitudinal cuts on the distal end of the bone, and horizontal cuts on its cranio-medial edge. The deep cut marks became shallower towards each end. We produced additional cut marks by scraping with a kitchen knife.

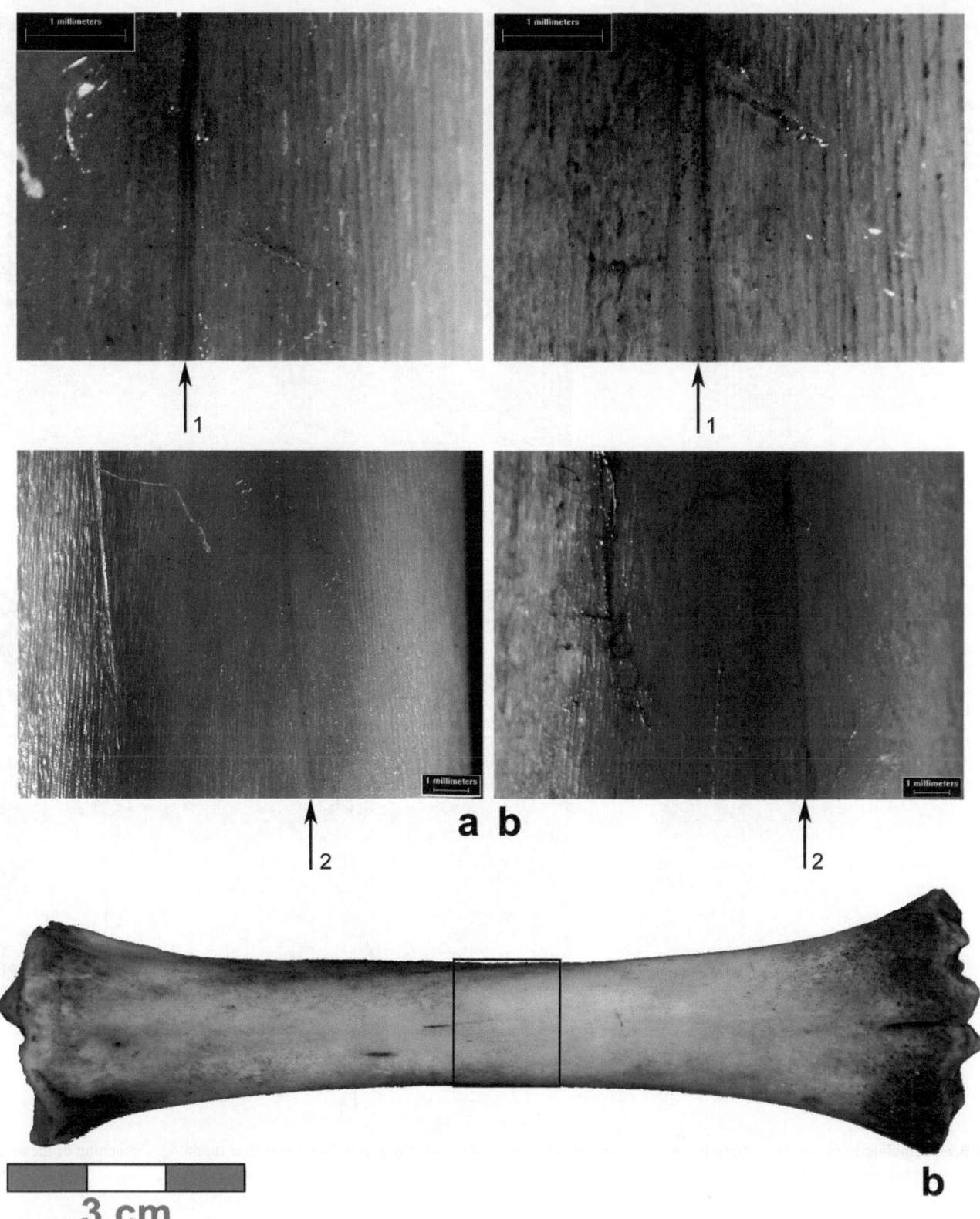

Fig. 6.8 Tumbling Experiment 2, sheep metacarpus. The anterior face of the diaphysis **a** before and **b** after tumbling. *Top row*: pronunciation of the sulcus (1); *bottom row*: pronunciation of the longitudinal striation (2). Both rows show an identical detail of the sheep metacarpus at different degrees of magnification

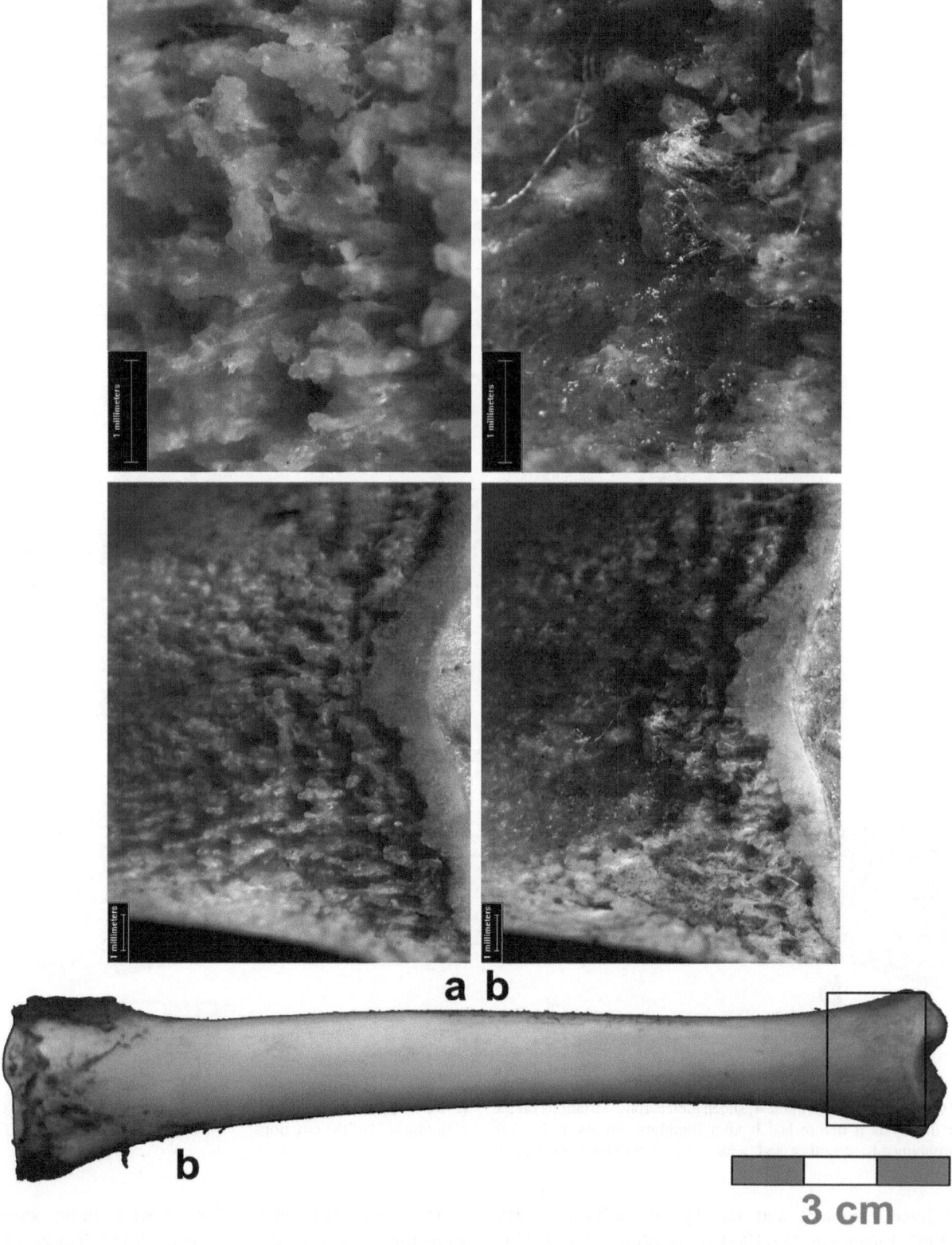

Fig. 6.9 Tumbling Experiment 3, sheep metacarpus. The distal metaphysis **a** before and **b** after tumbling. Abrasion of the bone caused by tumbling led to changes in the porous bone structure. The *top* and *bottom rows* show an identical detail of the sheep metacarpus at different degrees of magnification

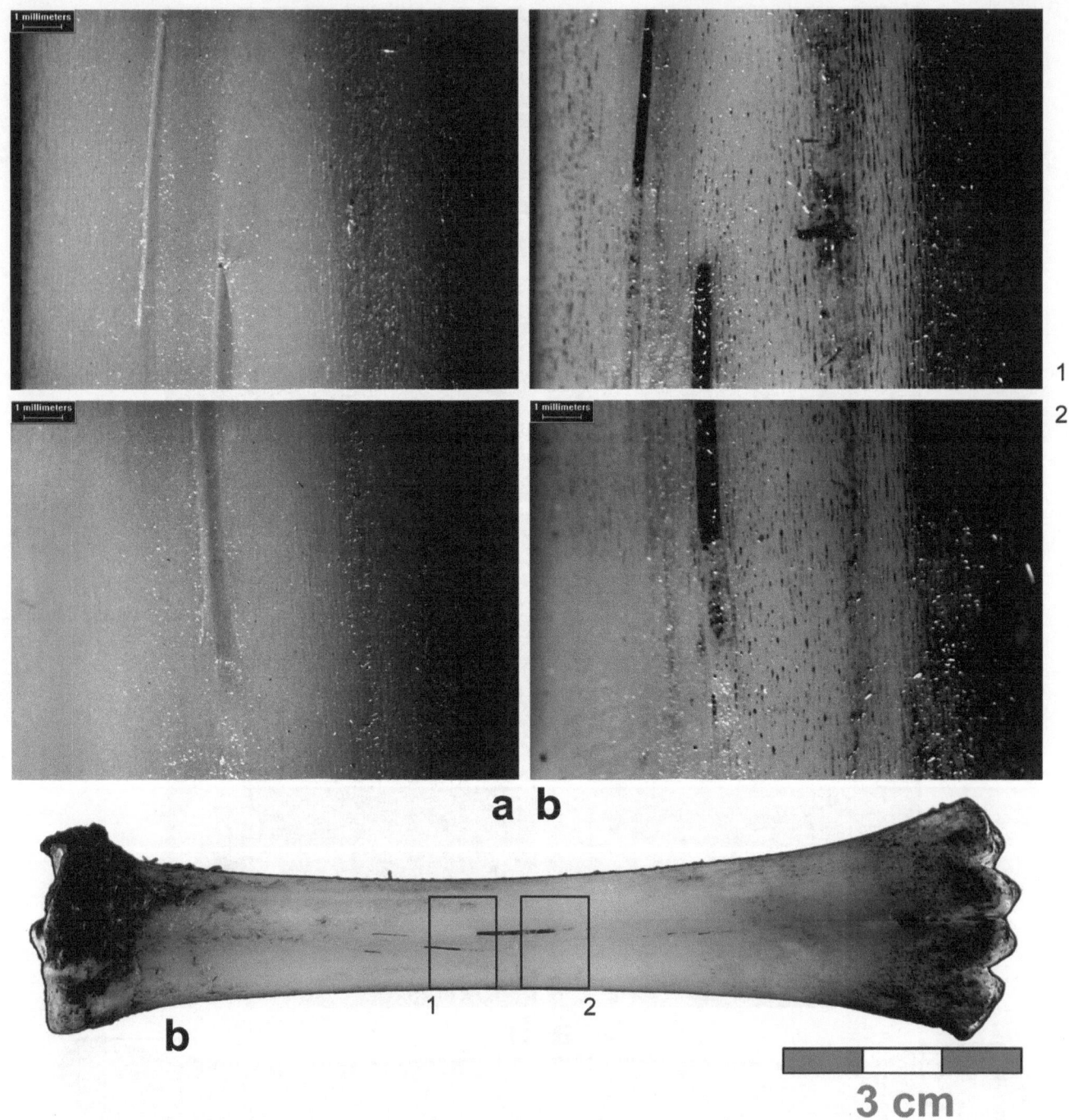

Fig. 6.10 Tumbling Experiment 4, sheep metacarpus. Striations on the anterior diaphysis **a** before and **b** after tumbling. Striations became more pronounced, smoother and wider after tumbling occurred, as is well illustrated by the changed morphology of the start- (1) and endpoints (2) of the striations

The bone fragment and 0.5 kg of mollusks (GBY 15/09/1997; Layer V-6; 181/130d, 60.20/60.13–59.84/59.76) were placed in a cloth bag, whose overall weight was 0.75 kg. The bag was affixed to the rotation chamber, which was then filled with 70 liters of water and tumbled for 2.5 hours. After tumbling, sediment remained unsorted.

In the course of tumbling, large portions of the periosteum were removed from the bone (Fig. 6.14). Traces of abrasion on the bone surface are only seen in those spots that had already been cleaned of periosteum prior to tumbling. The broken edges of the tibia fragment were slightly rounded (Fig. 6.15).

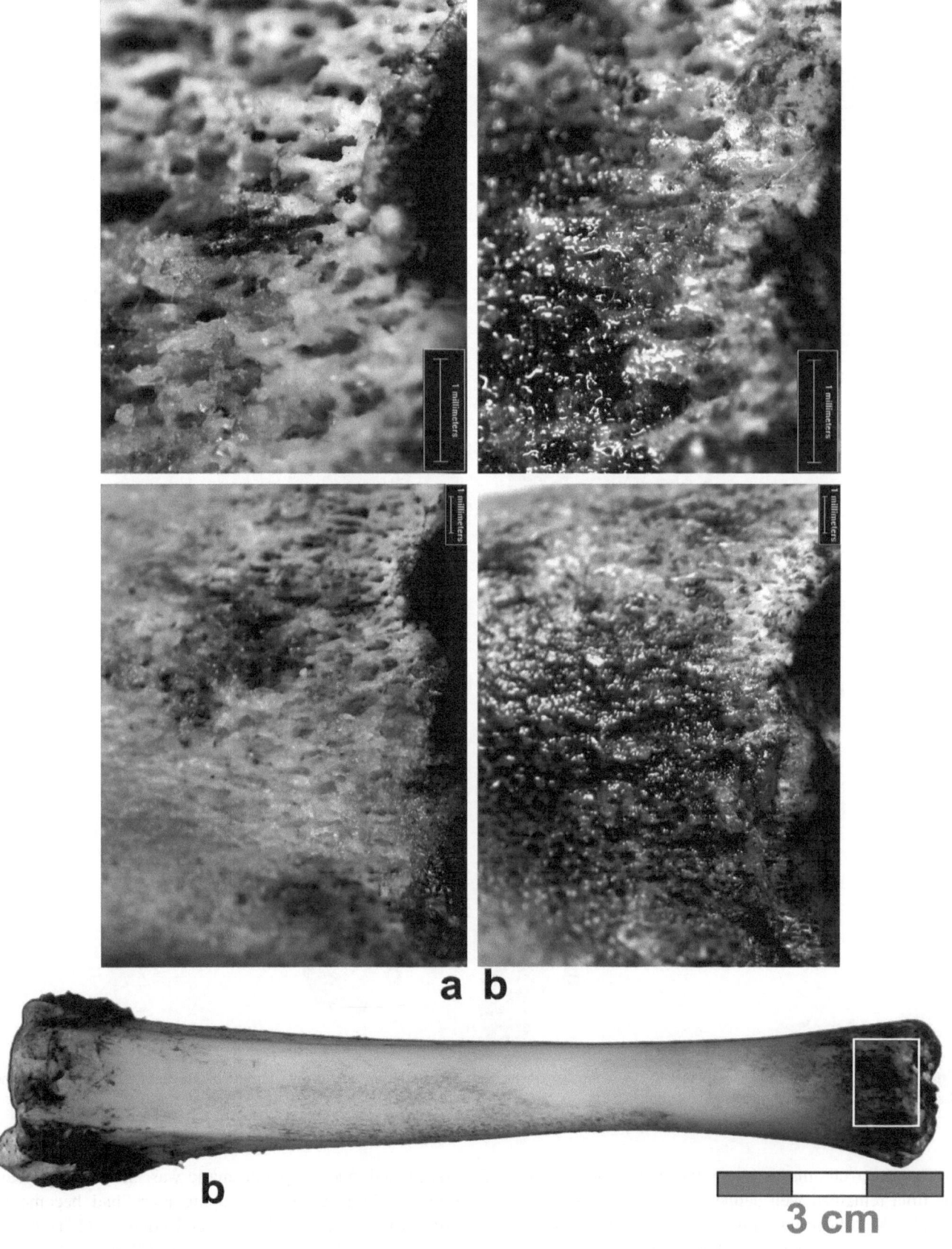

Fig. 6.11 Tumbling Experiment 4, sheep metacarpus. The distal metaphysis **a** before and **b** after tumbling. Morphological changes due to abrasion were created by tumbling. Both columns show an identical detail of the sheep metacarpus at different degrees of magnification

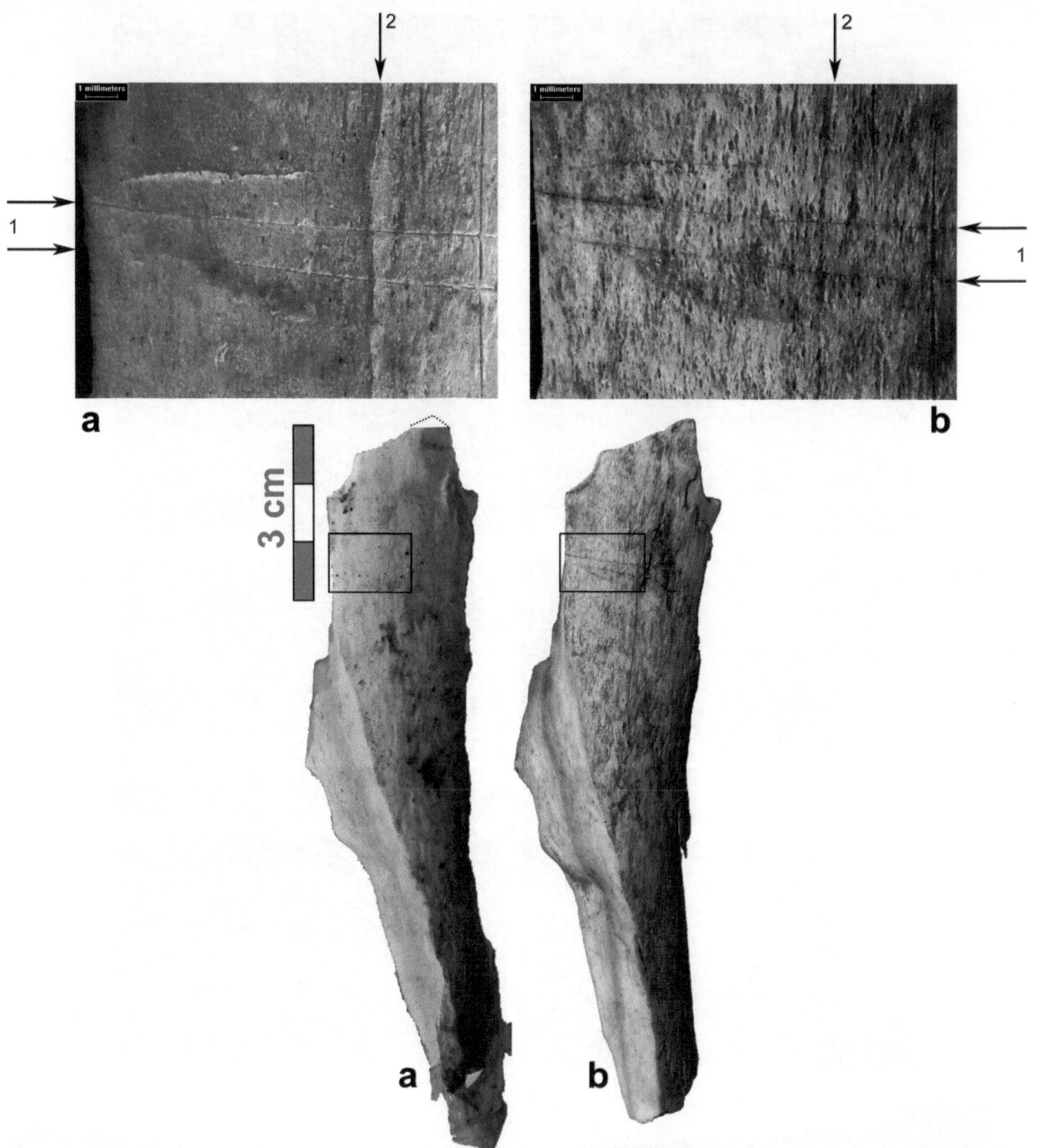

Fig. 6.12 Tumbling Experiment 5, domestic cow femur fragment. Cut marks **a** before and **b** after tumbling. Vertical cut marks became more shallow (1); longitudinal cut marks almost vanished (2)

The deep cut marks along the cranio-medial edge of the tibia remained well defined, but became smoother and rounder. Modifications on the distal end of the anterior side changed in appearance. One fine cut mark was completely erased and one deep cut mark was flattened, its edges rounded. In addition, a scraping mark had become significantly smoother, with rounded edges, and its internal structure was almost entirely destroyed (Fig. 6.16).

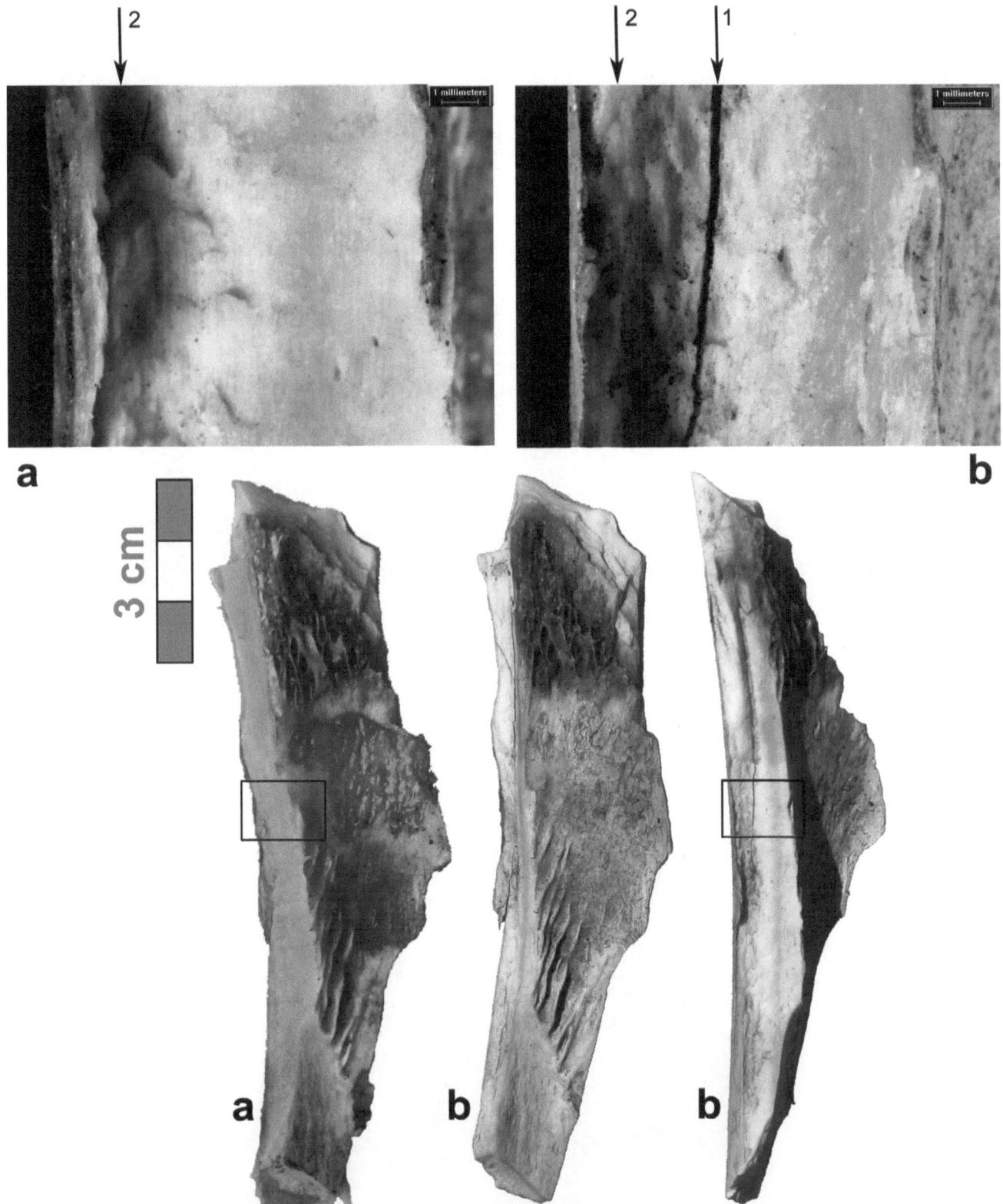

Fig. 6.13 Tumbling Experiment 5, domestic cow femur fragment. Bone-breakage patterns **a** before and **b** after tumbling. Longitudinal cracking along the breaking edge (1) was caused by tumbling and accompanied by leveling of breakage morphology (2)

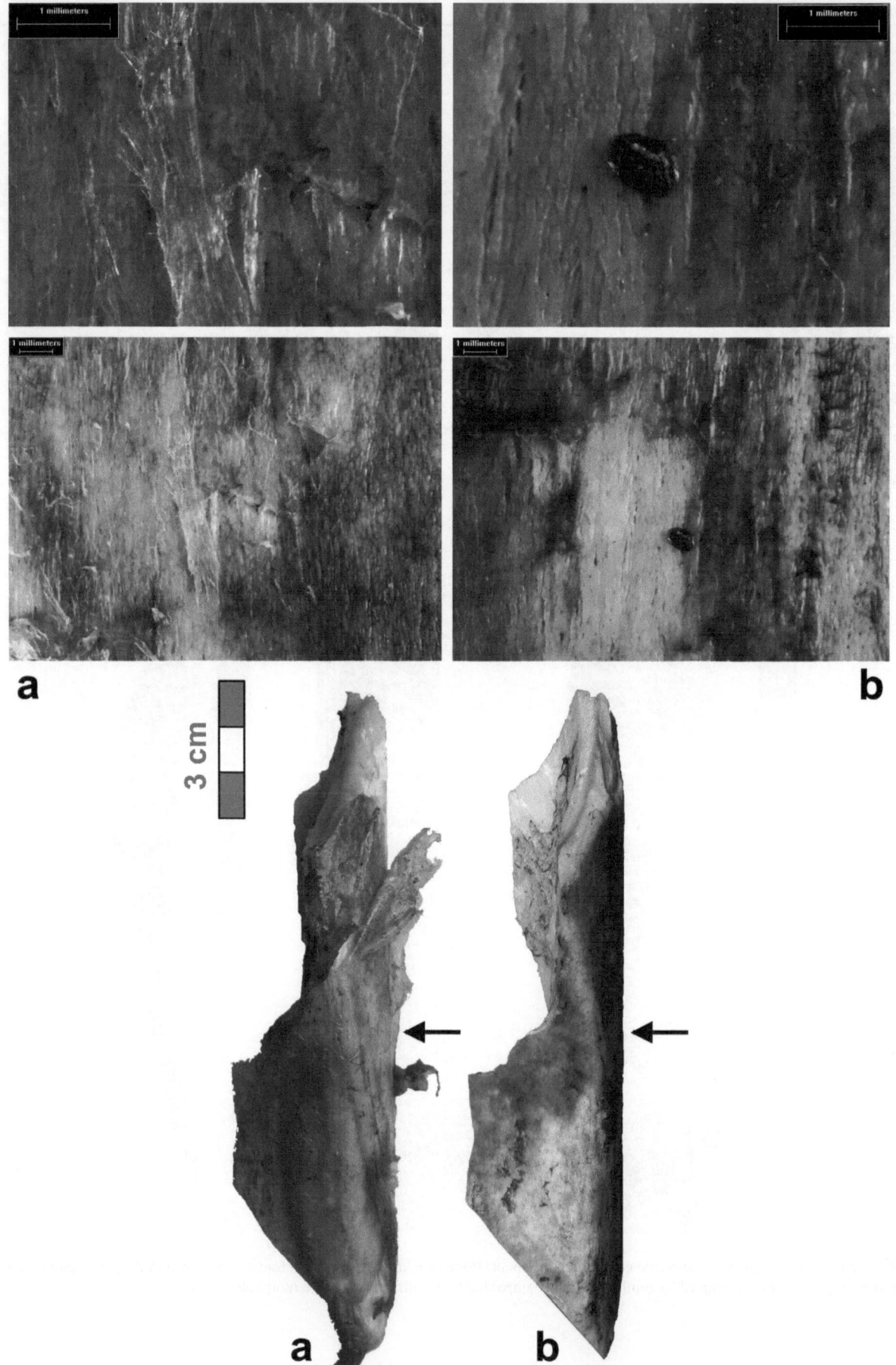

Fig. 6.14 Tumbling Experiment 6, domestic cow tibia fragment. Periosteum preservation **a** before and **b** after tumbling. *Arrows* indicate the area of the bone that was magnified. Both columns show an identical detail of the bone at different degrees of magnification

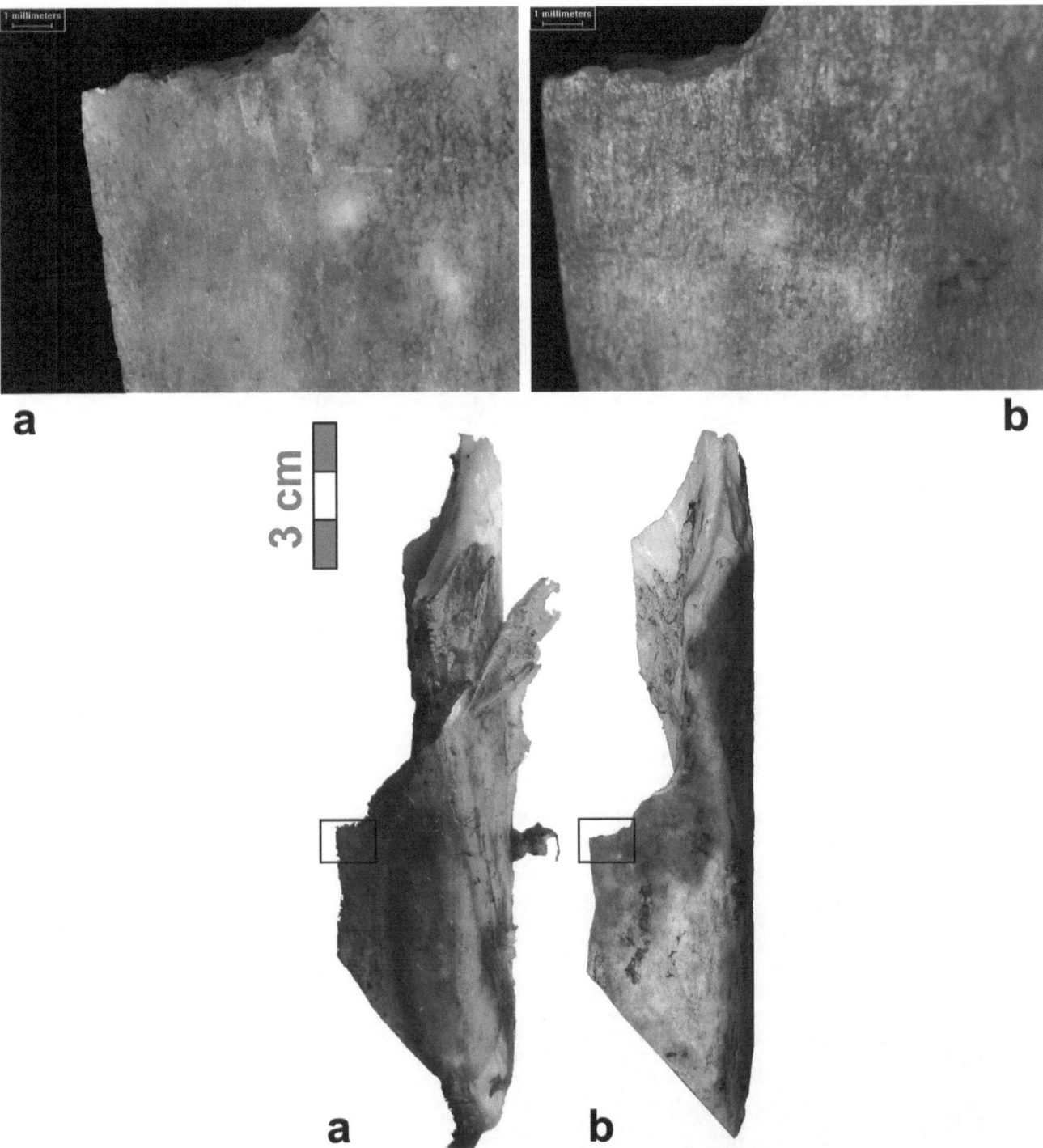

Fig. 6.15 Tumbling Experiment 6, domestic cow tibia fragment. Bone-breakage patterns **a** before and **b** after tumbling. Rounding of bone edges was caused by tumbling

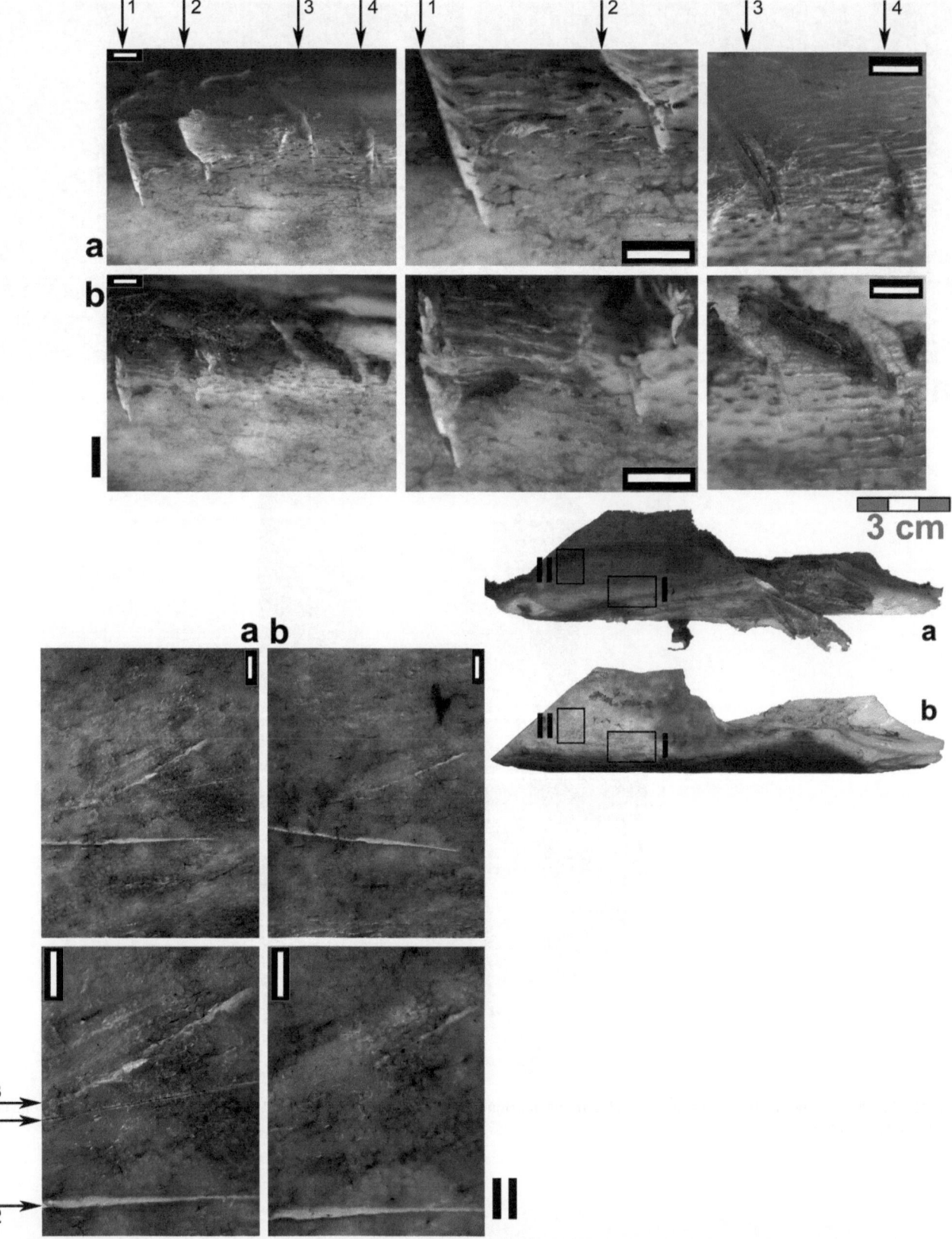

Fig. 6.16 Tumbling Experiment 6, domestic cow tibia fragment. Anthropogenically induced cut- and scraping marks **a** before and **b** after tumbling. (I) Smoothing and rounding of the cut marks were caused by tumbling. Both rows show an identical detail of the bone at different degrees of magnification. The morphological characteristics of the cut marks are indicated by *arrows* (1–4). (II) Tumbling erased the fine cut mark (1). Tumbling flattened and rounded the morphology of the deep cut mark (2) and smoothed the morphology of the scraping mark (3). Both columns show an identical detail of the bone at different degrees of magnification. Site scale bar of the microscope images is 1 mm

6.4.3.7 Tumbling Experiment 7

A fresh sheep metacarpus was used for seventh tumbling experiment. After removal of the skin and meat, periosteum remained attached to large parts of the proximal and distal metaphyses, and a horizontal scratch was visible on the diaphysis, next to the posterior foramina. Prior to tumbling, the bone was dried for eight days.

This experiment utilized a mixture of 1 kg of mollusks (GBY 15/09/97; Layer V-6, 181/130d, 60.20/60.13–59.84/59.76) and 0.5 kg of clay. The mixture contained only a small number of large mollusks. It was placed together with the bone in a cloth bag and soaked in ten liters of water for ten minutes. When wet, the bag and its contents weighed 2.5 kg. It was affixed to the rotation chamber, which was then filled with 70 liters of water, and tumbled for three hours. After tumbling, bone and sediment remained unsorted.

Following tumbling, the bone diaphysis was slightly abraded (Fig. 6.17). On the posterior face of the diaphysis, the foramina and the fine horizontal scratch appear broader, with broadening also noted on the anterior sulcus. Here, a vertical striation with uneven edges and a rough morphology is observed (Fig. 6.18).

6.4.3.8 Tumbling Experiment 8

A fresh sheep metatarsus was used for the eighth tumbling experiment. After removal of the skin and meat, periosteum remained attached to the proximal and distal metaphyses. Periosteum removal resulted in deep longitudinal striations along the anterior face of the diaphysis, on both sides of the sulcus. Low-magnification analysis of the bone surface revealed horizontal scraping marks accompanied by fine

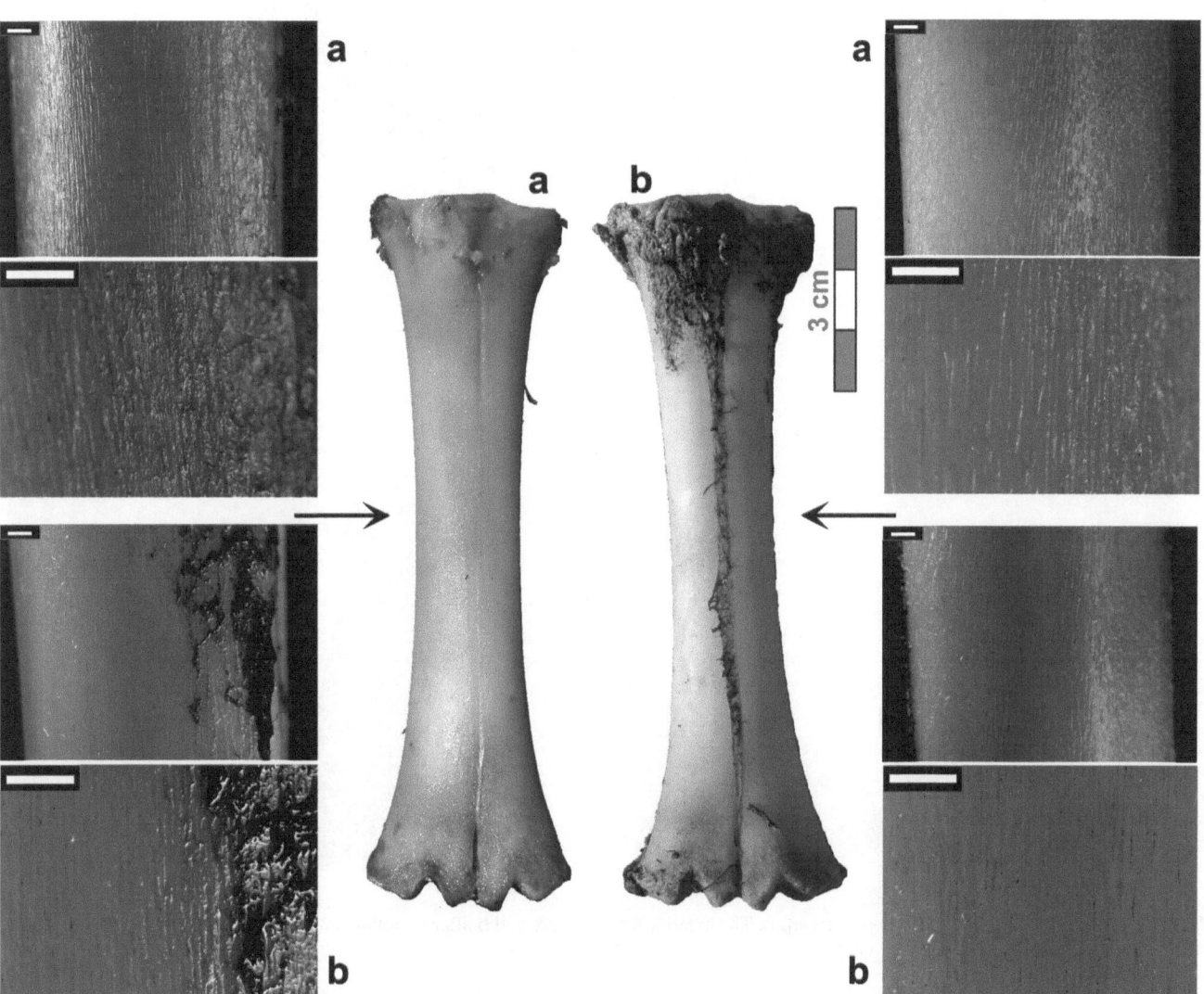

Fig. 6.17 Tumbling Experiment 7, sheep metacarpus. **a** unabraded diaphysis and **b** abraded diaphysis after tumbling. On the *left* are details of the lateral face of the metacarpus at different degrees of magnification.

On the *right* are details of the medial face of the metacarpus at different degrees of magnification. *Arrows* indicate the areas of the bone that were magnified. The scale bar of the microscope images is 1 mm

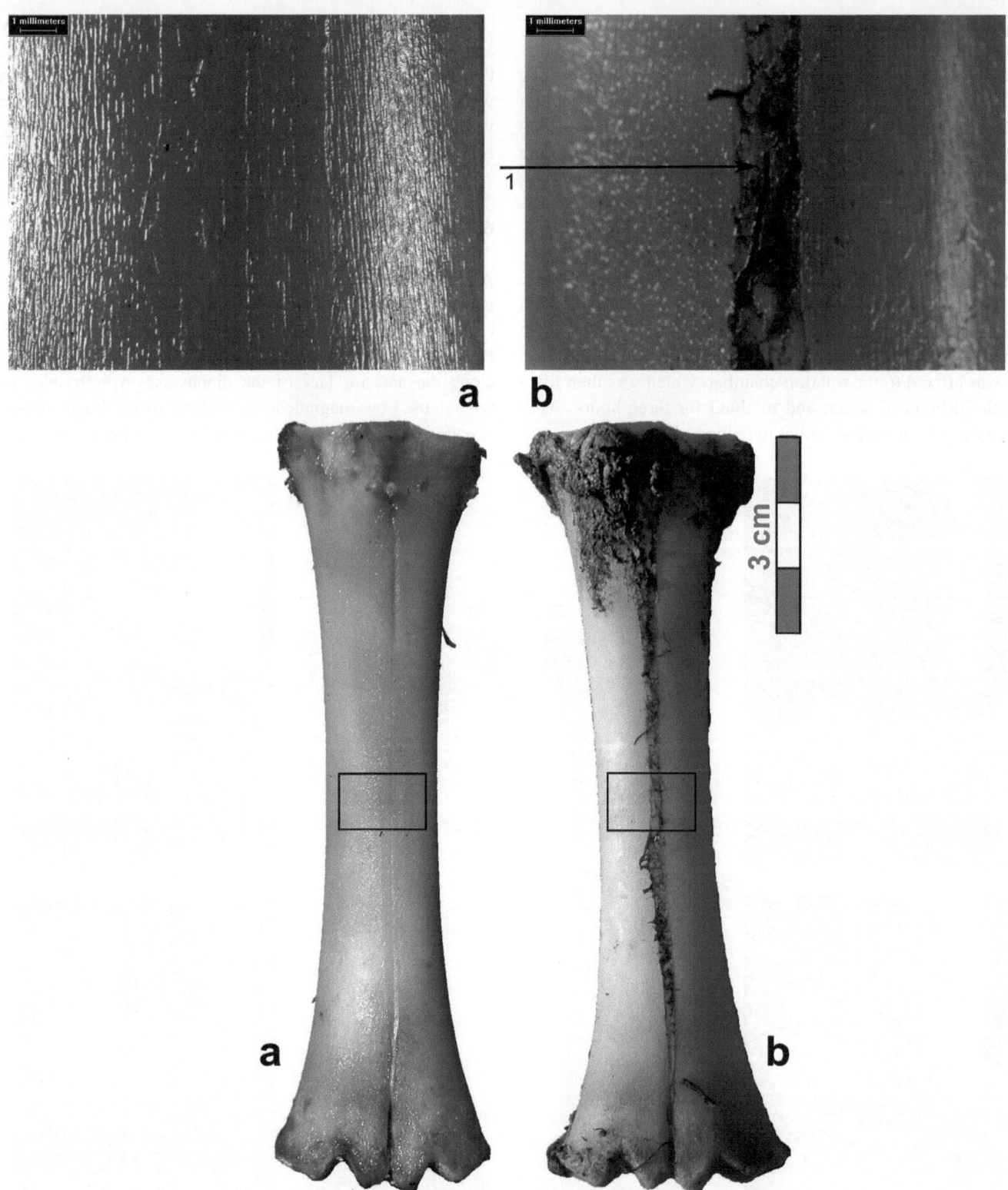

Fig. 6.18 Tumbling Experiment 7, sheep metacarpus. The anterior face **a** before and **b** after tumbling. A vertical striation (1) with uneven edges and a rough morphology was created by tumbling

parallel vertical striations on the lateral face of the diaphysis. In addition, fine horizontal and diagonal striations were noted on the anterior face along the proximal and distal metaphysis. Prior to tumbling, the bone was dried for eight days.

This experiment utilized a mixture of 0.75 kg of mollusks (GBY 15/09/97; Layer V-6; 181/120d, 60.20/60.13–59.84/59.76) and 0.2 kg of clay. Bone and sediment were placed in a cloth bag and soaked in ten liters of water for five minutes. The bag and its contents weighed 1.75 kg. The bag was affixed to the rotation chamber, which was then filled with 70 liters of water, and tumbled for six hours. After tumbling, bone and sediment remained unsorted.

Following tumbling, meat and periosteum remained attached to large parts of the proximal metaphysis. Consequently, the fine marks on the posterior face were only slightly modified (Fig. 6.19). The vertical marks produced by periosteum removal from the anterior face of the bone were broadened (Fig. 6.20). Along the lateral face of the bone, the vertical fine striations produced during scraping were more pronounced, as were the horizontal scraping marks (Fig. 6.21). The posterior face of the distal metaphysis exhibited fine, oval and round punctures, while the traces observed before tumbling had vanished and were replaced by fine striations (Fig. 6.22).

6.4.3.9   Tumbling Experiment 9

A fresh sheep metacarpus was used in the ninth tumbling experiment. After removal of the skin and meat, periosteum remained attached to the proximal and distal metaphyses. During removal of the periosteum, a fine vertical striation was produced along the cranio-medial edge of the bone. In addition, fine diagonal striations were produced on the proximal anterior/lateral part of the metaphysis, as well as on the posterior diaphysis. Fine horizontal marks, resulting from the scraping off of the periosteum, are seen on the medial edge of the bone's posterior face. Prior to tumbling, the bone was dried for ten days.

This experiment utilized a mixture of 1 kg of mollusks (GBY 16/09/97; Layer V-6; 185/129a+b, 60.41/60.03–59.94/59.68) and 0.25 kg of clay. Bone and sediment were placed in a plastic container, to which one liter of water was added. The overall weight of the mixture was 2.25 kg. The container was affixed to the middle of the rotation chamber and tumbled for three hours. Following tumbling, the clay was covered by a layer of mollusks and bone, which, in turn, were covered by the water.

After tumbling, the fine vertical striation along the cranio-medial edge of the bone disappeared, as did the diagonal marks on the proximal metaphysis of the lateral face. In addition, the horizontal scraping marks on the medial edge

of the posterior face vanished, and were replaced by fine vertical striations (Fig. 6.23). The diagonal marks on the anterior metaphysis and the posterior diaphysis were more pronounced, and the anterior metaphysis was covered by irregular oval and round punctures (Fig. 6.24).

6.4.3.10   Tumbling Experiment 10

A fresh sheep metatarsus was used for the tenth tumbling experiment. After removal of the skin and meat, periosteum remained attached to large parts of the proximal and distal metaphyses. In the course of periosteum removal, longitudinal striations were produced on the anterior face of the diaphysis. The posterior face displayed fine diagonal marks along the lateral edge. Prior to tumbling, the bone was dried for 12 days.

This experiment utilized a mixture of 1.25 kg of mollusks (GBY 14/09/97; Layer V-6; 182/130c, 60.12/60.02–59.85/59.80) and 0.5 kg of clay. Bone and sediment were placed in a plastic container, to which two liters of water were added. The mixture weighed 4 kg. The container was affixed to the middle of the rotation chamber and tumbled for three hours. Similar to Tumbling Experiment 9, the tumbling action sorted the container contents so that the basal clay was covered by a layer of mollusks and bone, which formed a thick amalgamation of sediment with water on top.

Following tumbling, the vertical striations on the bone's anterior face were smoothed and broadened (Fig. 6.25). The few diagonal marks on the posterior face that survived after tumbling (Fig. 6.26) were also smoothened. In addition, irregular oval and round punctures were observed on the bone surface. The lateral diaphysis was similarly modified (Fig. 6.27).

6.4.3.11   Tumbling Experiment 11 and 11.1

A fresh sheep metacarpus was used for the eleventh tumbling experiment. After removal of the skin and meat, periosteum remained attached to the proximal and distal metaphyses. Periosteum removal produced fine vertical striations on the anterior face of the diaphysis. Prior to tumbling, the bone was dried for 12 days.

This experiment utilized a mixture of 1 kg of mollusks (GBY 16/09/97; Layer V-6; 185/129a+b, 60.41/60.31–59.94/59.68) and 0.25 kg of clay. Bone and sediment were placed in a plastic container filled with 1 liter of water. The mixture weighed 2.5 kg. The container was affixed to the rotation chamber and tumbled for four hours. Again, the tumbling action sorted the container contents so that the clay was covered by a layer of mollusks and bone, which, in turn, was covered by the water.

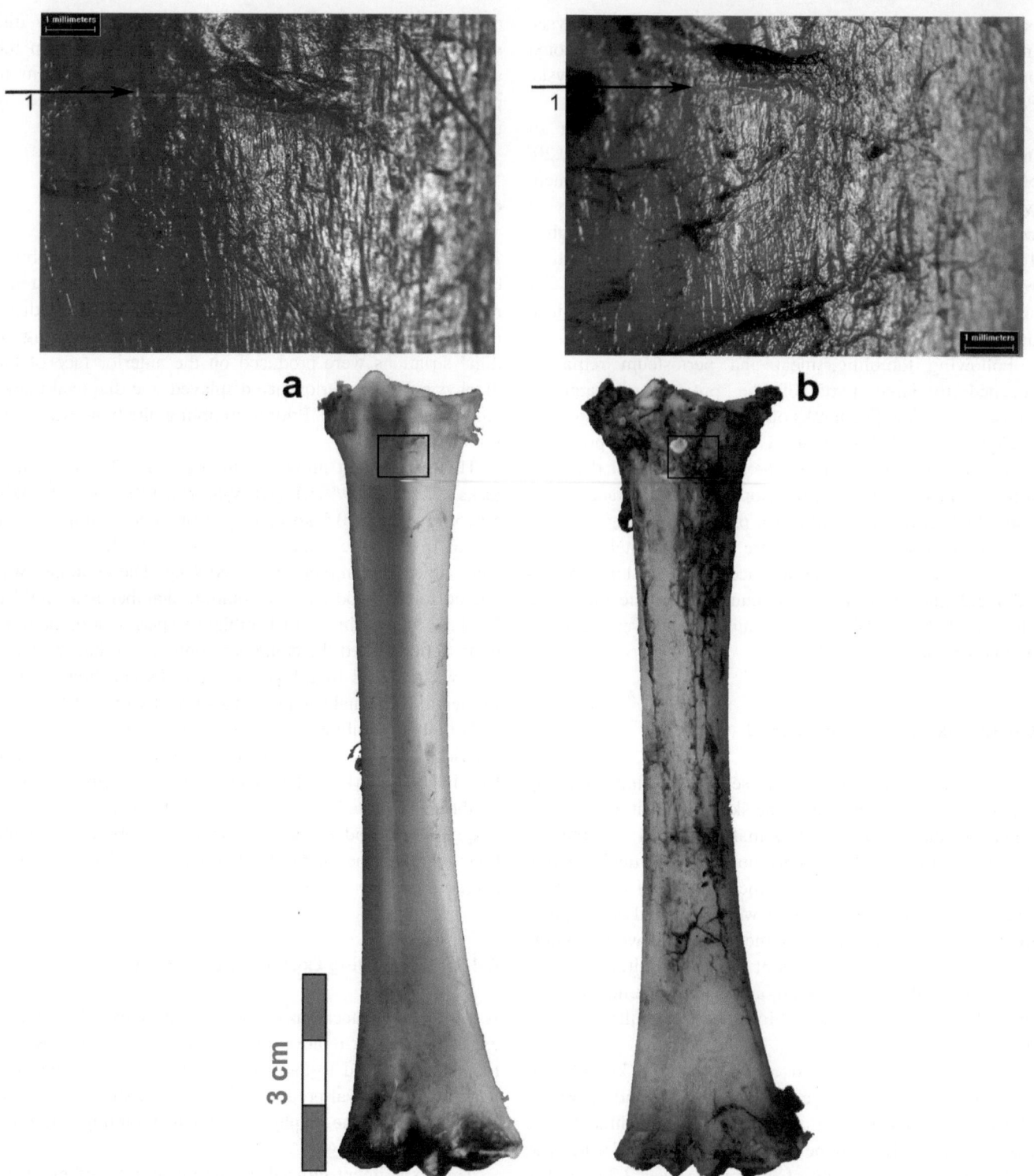

Fig. 6.19 Tumbling Experiment 8, sheep metatarsus. Cut marks on the posterior bone surface **a** before and **b** after tumbling. Only slight modifications of the cut mark (1) were created by tumbling

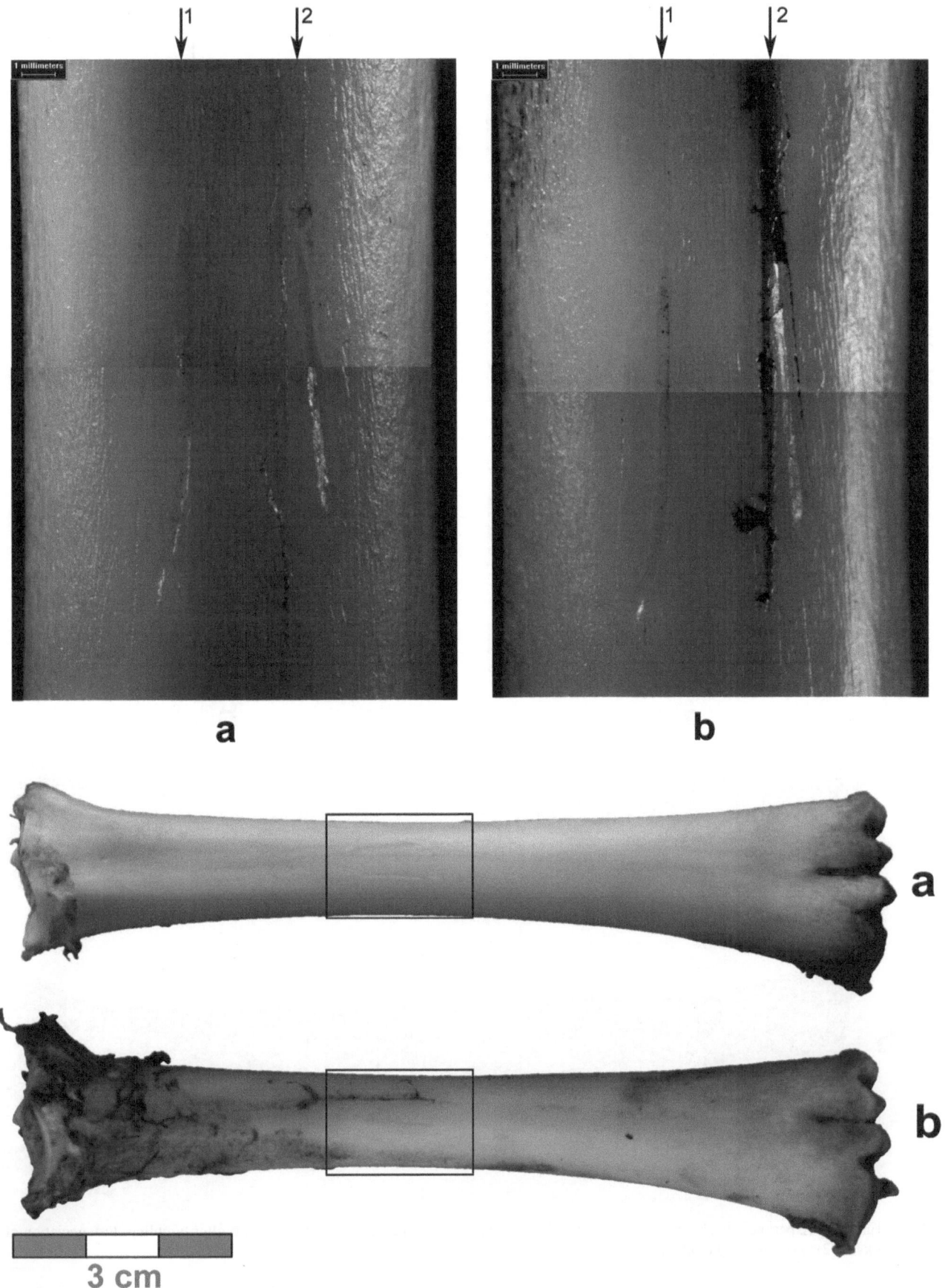

Fig. 6.20 Tumbling Experiment 8, sheep metatarsus. Traces of periosteum removal (1, 2) on the anterior bone surface **a** before and **b** after tumbling

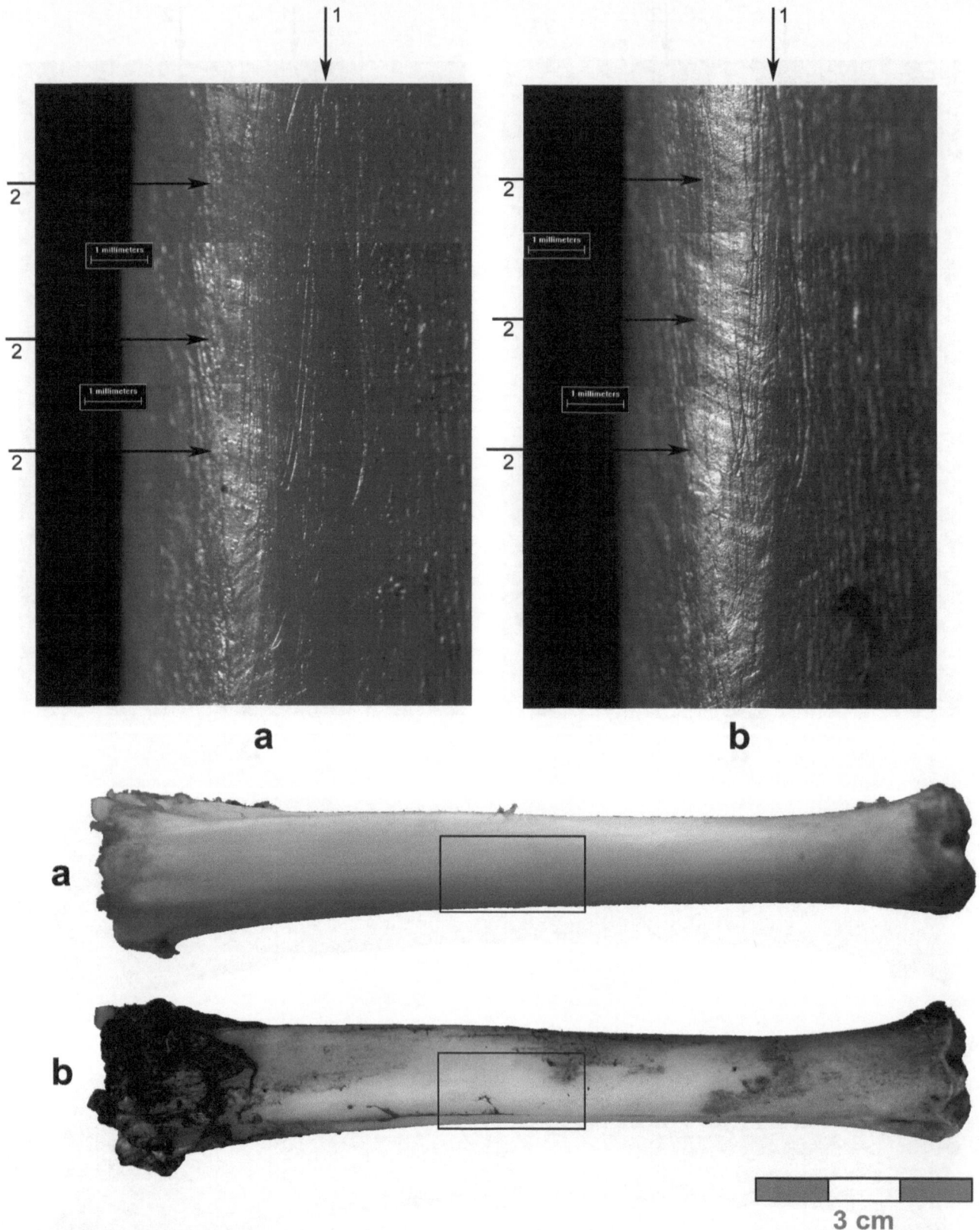

Fig. 6.21 Tumbling Experiment 8, sheep metatarsus. Striations on the lateral bone surface **a** before and **b** after tumbling. Pronunciation of the scraping marks (1, 2) was caused by tumbling

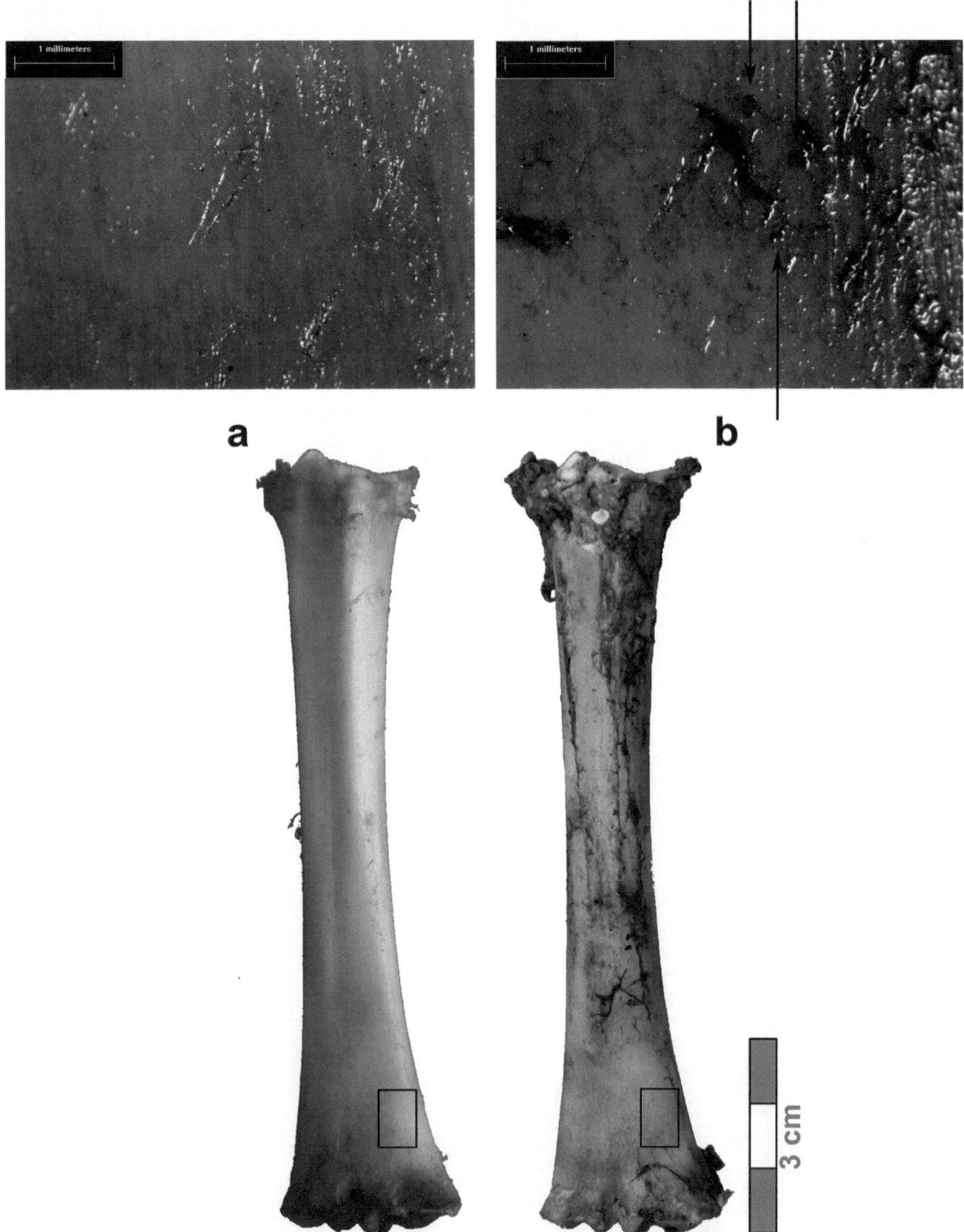

Fig. 6.22 Tumbling Experiment 8, sheep metatarsus. The posterior face **a** before and **b** after tumbling. Fine oval- and round-shaped punctures, indicated by *arrows*, were created by tumbling

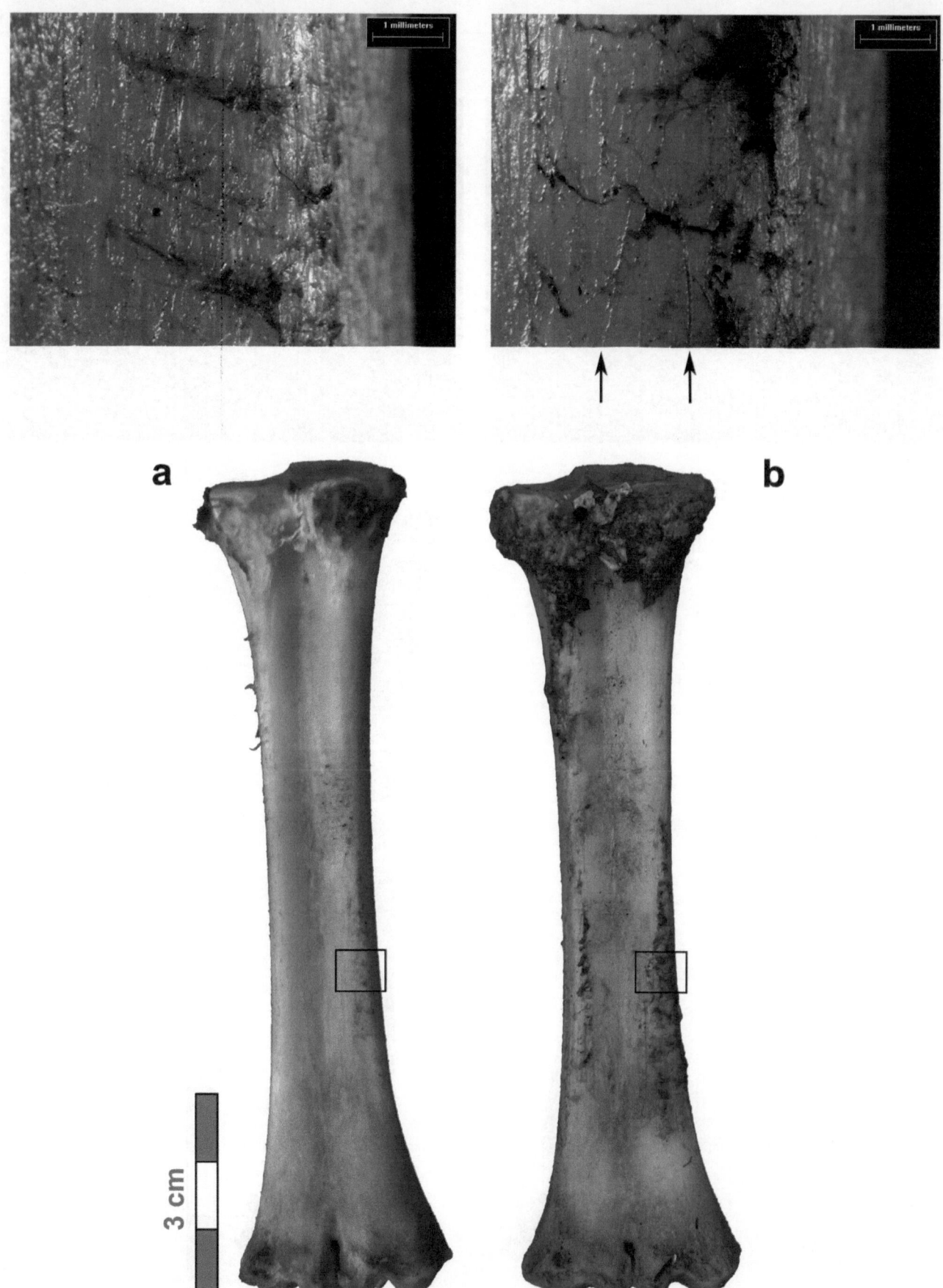

Fig. 6.23 Tumbling Experiment 9, sheep metacarpus. The medial edge of the posterior face **a** before and **b** after tumbling. Removal of the scraping marks and the appearance of fine vertical striations, indicated by *arrows*, were caused by tumbling

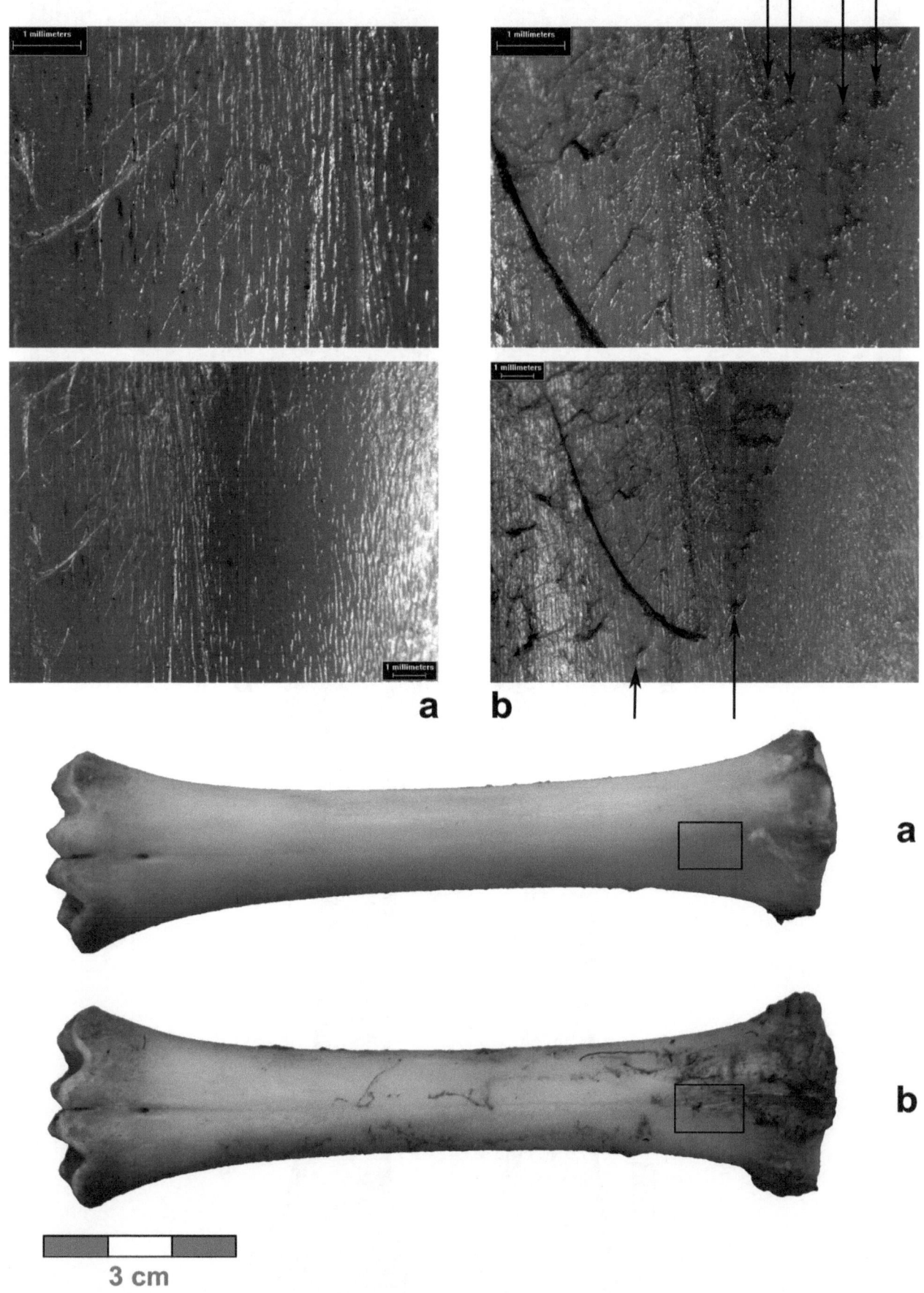

Fig. 6.24 Tumbling Experiment 9, sheep metacarpus. The anterior metaphysis **a** before and **b** after tumbling. Irregular-, oval-, and round-shaped punctures, indicated by *arrows*, were created by tumbling. Both columns show an identical detail of the bone at different degrees of magnification

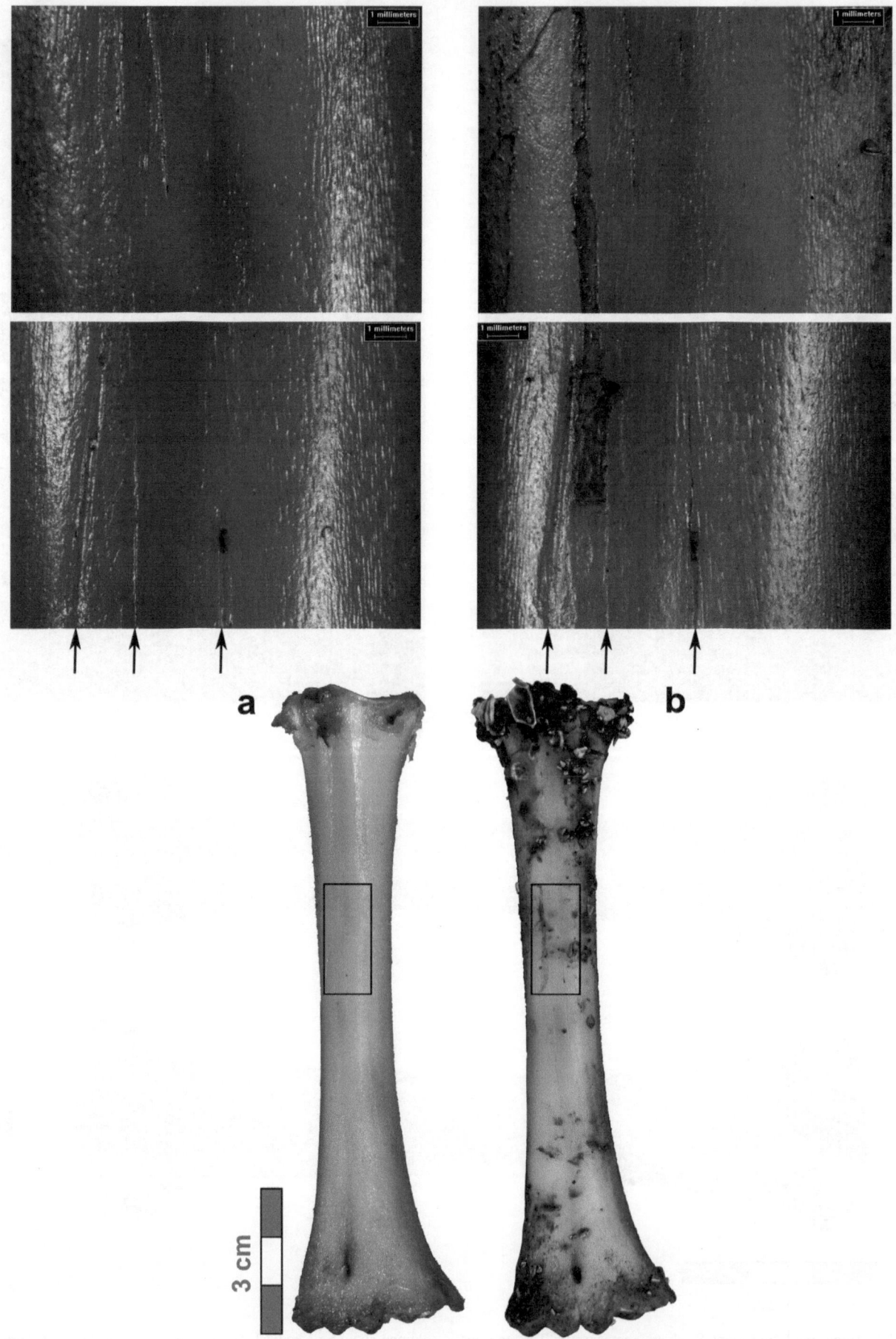

Fig. 6.25 Tumbling Experiment 10, sheep metatarsus. The anterior face **a** before and **b** after tumbling. Striations, indicated by *arrows*, were smoothed and broadened by tumbling. Both columns show an identical detail of the bone at different degrees of magnification

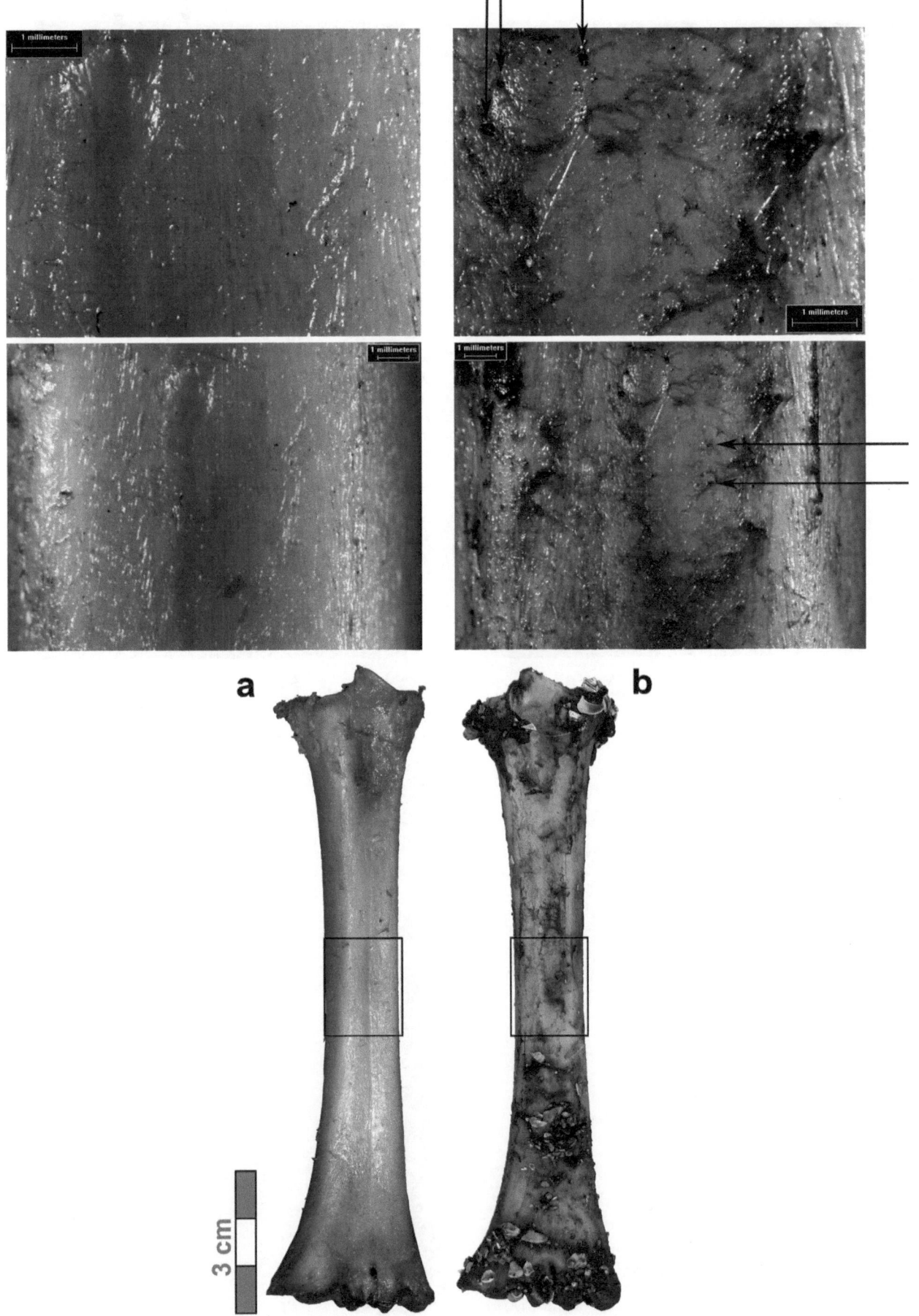

Fig. 6.26 Tumbling Experiment 10, sheep metatarsus. The posterior face **a** before and **b** after tumbling. Diagonal marks were smoothed by tumbling. Irregular-shaped punctures, indicated by *arrows*, were created by tumbling. Both columns show an identical detail of the bone at different degrees of magnification

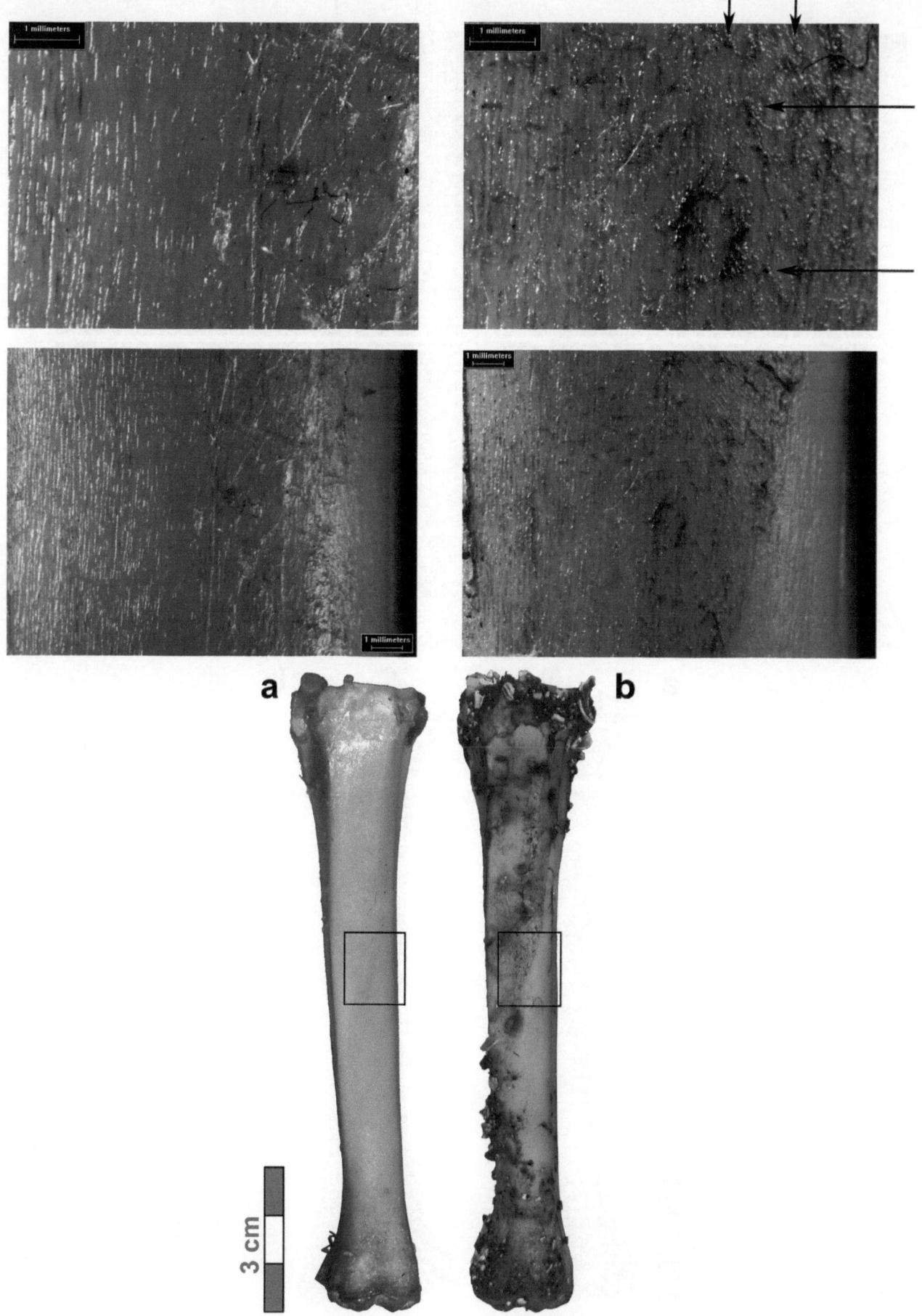

Fig. 6.27 Tumbling Experiment 10, sheep metatarsus. The lateral diaphysis **a** before and **b** after tumbling. Irregularly-shaped punctures, indicated by *arrows*, were created by tumbling. Both columns show an identical detail of the bone at different degrees of magnification

Following tumbling, slight abrasion was observed on each bone face, especially the distal metaphysis. The proximal metaphysis featured irregular oval and round punctures (Fig. 6.28). The sulcus of the anterior face appeared deeper and broader in shape. The fine striations along the anterior diaphysis had almost disappeared.

The same sheep metacarpus used in Tumbling Experiment 11 was used again for a follow-up experiment, Tumbling Experiment 11.1. This experiment utilized a mixture of 1.5 kg of mollusks (GBY 15/09/97; Layer V-6; 181/130d, 60.31/60.20–59.94/59.76) and 0.5 kg of clay. Bone and sediment were placed in a plastic container filled with 0.75 liters of water. The mixture weighed 3 kg. The container was affixed to the middle of the rotation chamber and tumbled for six hours. After tumbling, sediment remained unsorted.

Following tumbling, the entire bone was rounded and abraded, particularly at the distal metaphysis (Fig. 6.29) and the edges of the anterior sulcus' incision. During Tumbling Experiment 11, the fine vertical striations on the anterior diaphysis almost disappeared. This process continued during Experiment 11.1, and only one narrow and flat, but well-defined, striation remained visible (Fig. 6.30). The medial face of the proximal metaphysis exhibited a deep diagonal striation (Fig. 6.31).

6.4.3.12 Tumbling Experiment 12

A fresh sheep metatarsus was used for the twelfth tumbling experiment. After removal of the skin and meat, periosteum remained attached to the proximal and distal metaphyses. Periosteum removal resulted in fine vertical striations along the anterior diaphysis. On the anterior face of the proximal metaphysis, fine diagonal traces run into the sulcus. Prior to tumbling, the bone was dried for 15 days.

This experiment utilized 1.5 kg of mollusks (GBY 16/09/97; Layer V-6; 184/130a, 60.35/60.19–60.10/60.00) mixed with 0.5 kg of clay. Bone and sediment were placed in a plastic container filled with 1.5 liters of water. The mixture weighed 3.7 kg. The container was affixed to the middle of the rotation chamber and tumbled for four hours. Again, the tumbling action sorted the container contents so that the clay was covered by a layer of mollusks and bone, which, in turn, was covered by the water.

Following tumbling, the entire bone surface was slightly abraded, especially on the distal metaphysis. The striations along the anterior face deepened and became more pronounced (Fig. 6.32). In contrast to the smooth and rounded diagonal marks, the sulcus on the anterior proximal metaphysis was also more pronounced (Fig. 6.33), as was the sulcus on the distal metaphysis. Tumbling resulted in a broad, flat diagonal u-shaped striation on the posterior face of the proximal metaphysis (Fig. 6.34). The medial part of the

proximal metaphysis displayed a deep narrow striation and fine irregular punctures (Fig. 6.35).

6.4.3.13 Tumbling Experiment 13

A fresh sheep metacarpus was used for the thirteenth tumbling experiment. After removal of the skin and meat, periosteum remained attached to the proximal and distal metaphyses. No bone surface modifications were observed. Prior to tumbling, the bone was dried for 16 days.

This experiment utilized a mixture of 1.1 kg of mollusks (GBY 16/09/97; Layer V-6; 185/129a+b, 60.41/60.31–59.94/59.68) and 0.25 kg of clay. Bone and sediment were placed in a plastic container filled with 0.75 liters of water. The mixture weighed 2.4 kg. The container was affixed to the rotation chamber and tumbled for three hours. The tumbling action sorted the container contents so that the basal clay was overlain by a layer of mollusks and bone, and then the water.

Following tumbling, a vertical trace identical to that produced by periosteum removal was observed on the anterior face of the diaphysis. Due to its fineness, this striation was not noted before tumbling (Fig. 6.36). Slight abrasion affected the entire bone, especially the distal metaphysis.

6.4.3.14 Tumbling Experiment 14

A fresh sheep metatarsus was used for the fourteenth tumbling experiment. After removal of the skin and meat, periosteum remained attached to the proximal and distal metaphyses. Periosteum removal resulted in vertical striations along the anterior face of the diaphyses. Prior to tumbling, the bone was dried for 16 days.

This experiment utilized a mixture of 1.5 kg of mollusks (GBY 15/09/97; Layer V-6; 181/130d, 60.20/60.31–59.94/59.76) and 0.5 kg of clay. Bone and sediment were placed in a plastic container, but no water was added. The mixture weighed 2.5 kg. The container was affixed to the rotation chamber and tumbled for five hours. After tumbling, bone and sediment remained unsorted.

Following tumbling, typical abrasion features were seen on the entire bone. Evidence of abrasion was much more pronounced in this experiment than in the other tumbling experiments. The vertical striations and the sulci on the anterior diaphysis were deepened and broadened. The edges of the striations were not significantly rounded (Figs. 6.37, 6.38, and 6.39).

6.4.3.15 Tumbling Experiment 15

The domestic cow femur (Fragment 10) used in the burial experiment was used again in the fifteenth tumbling

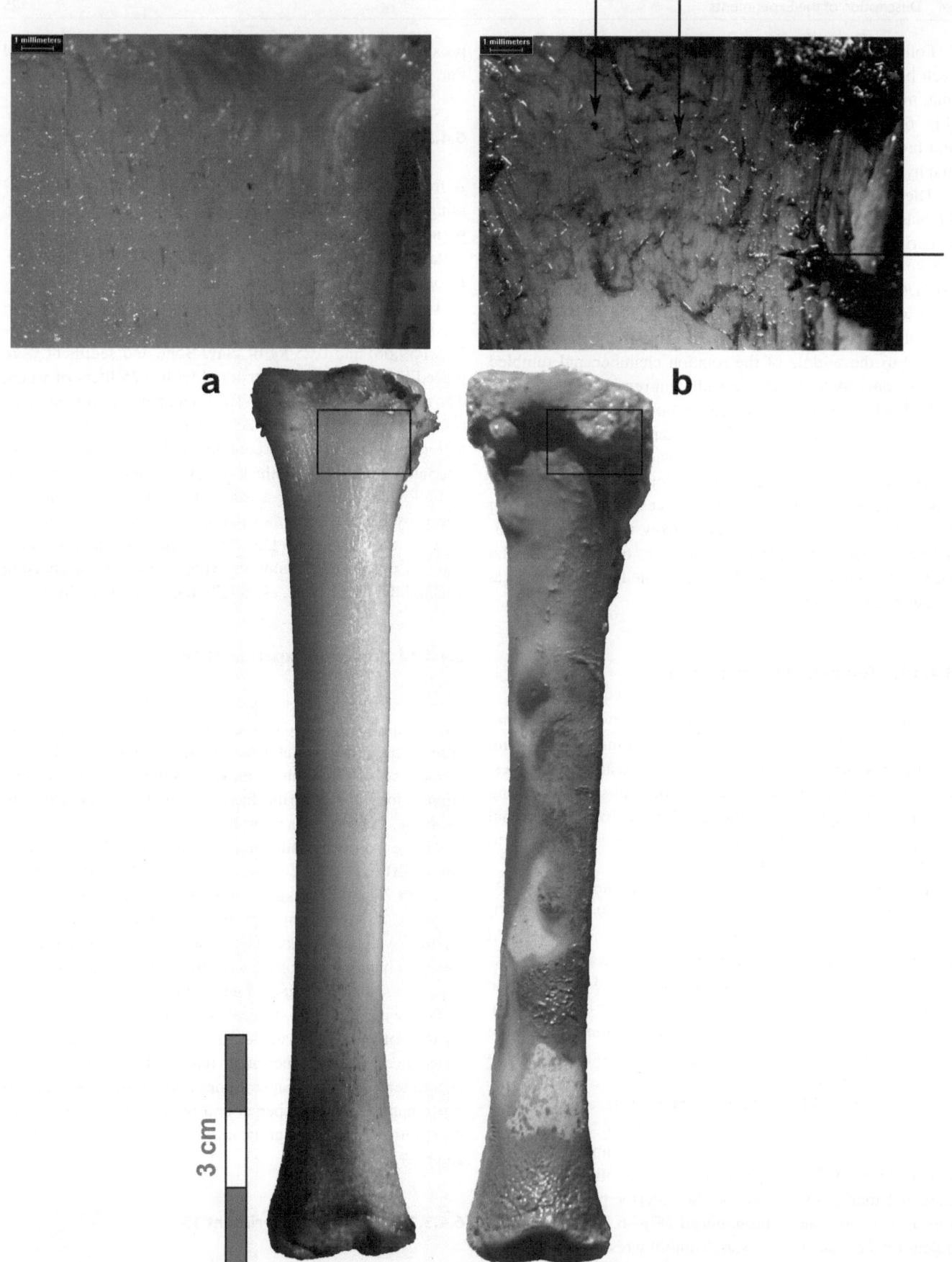

Fig. 6.28 Tumbling Experiment 11, sheep metacarpus. The proximal metaphysis **a** before and **b** after tumbling. Irregular-, oval-, and round-shaped punctures on the proximal metaphysis, indicated by *arrows*, were created by tumbling

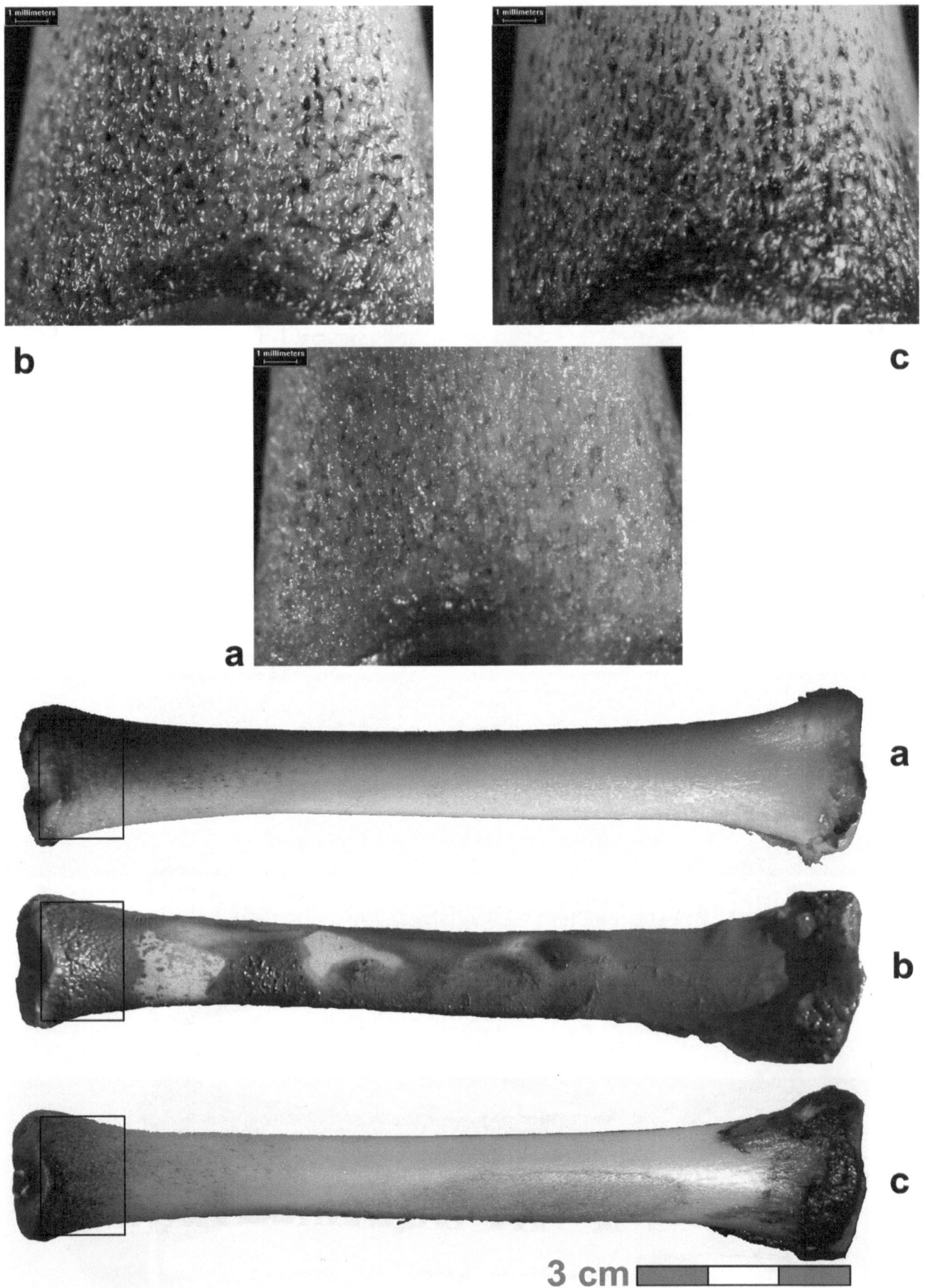

Fig. 6.29 Tumbling Experiment 11.1, sheep metacarpus. The distal metaphysis **a** before and **b** after four hours of tumbling, and **c** after ten hours of tumbling. Abrasion and rounding of the distal metaphysis increased with the duration of tumbling

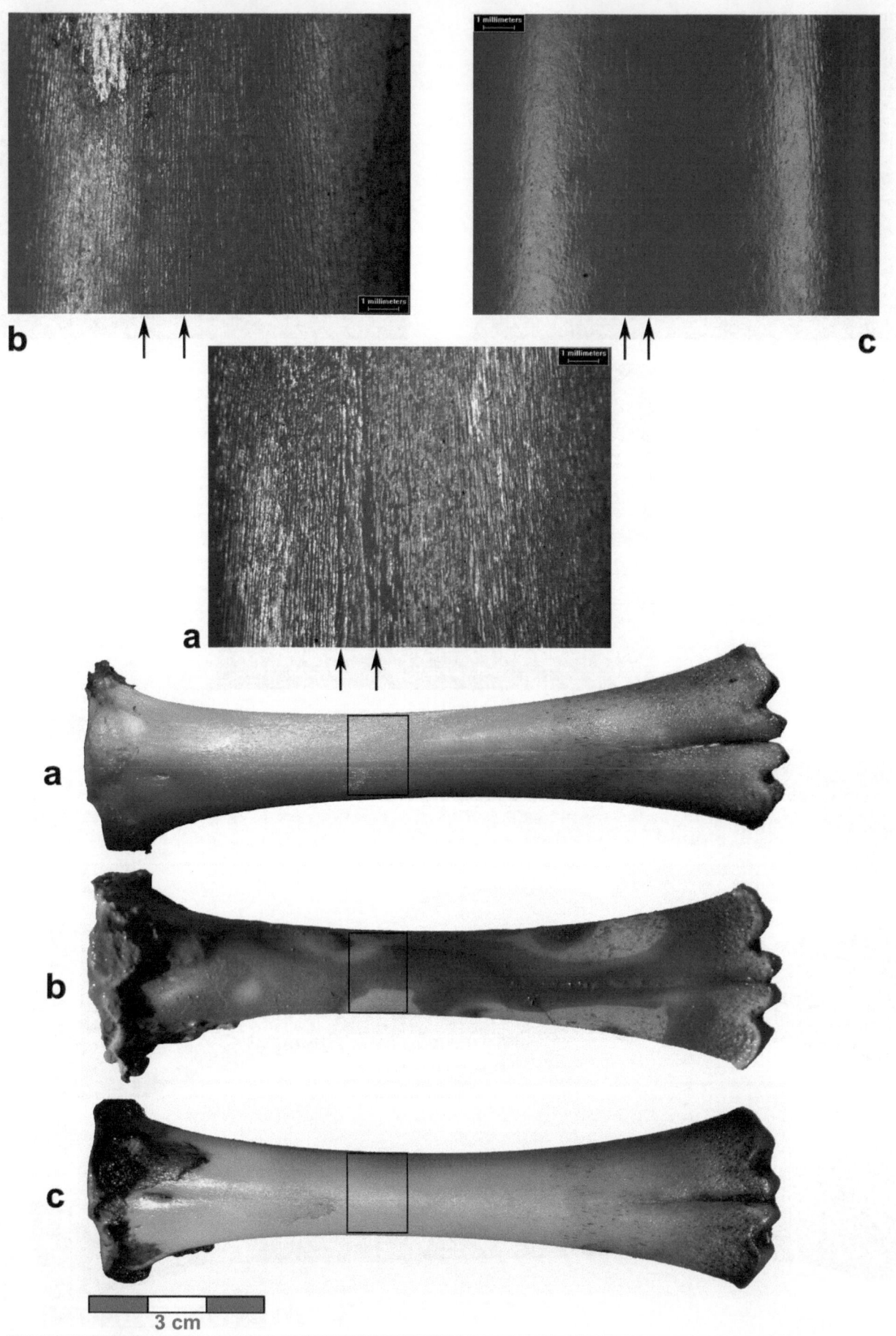

Fig. 6.30 Tumbling experiment 11.1, sheep metacarpus. The anterior diaphysis **a** before and **b** after four hours of tumbling, and **c** after ten hours of tumbling. Removal of the vertical striations, indicated by *arrows*, was caused by tumbling

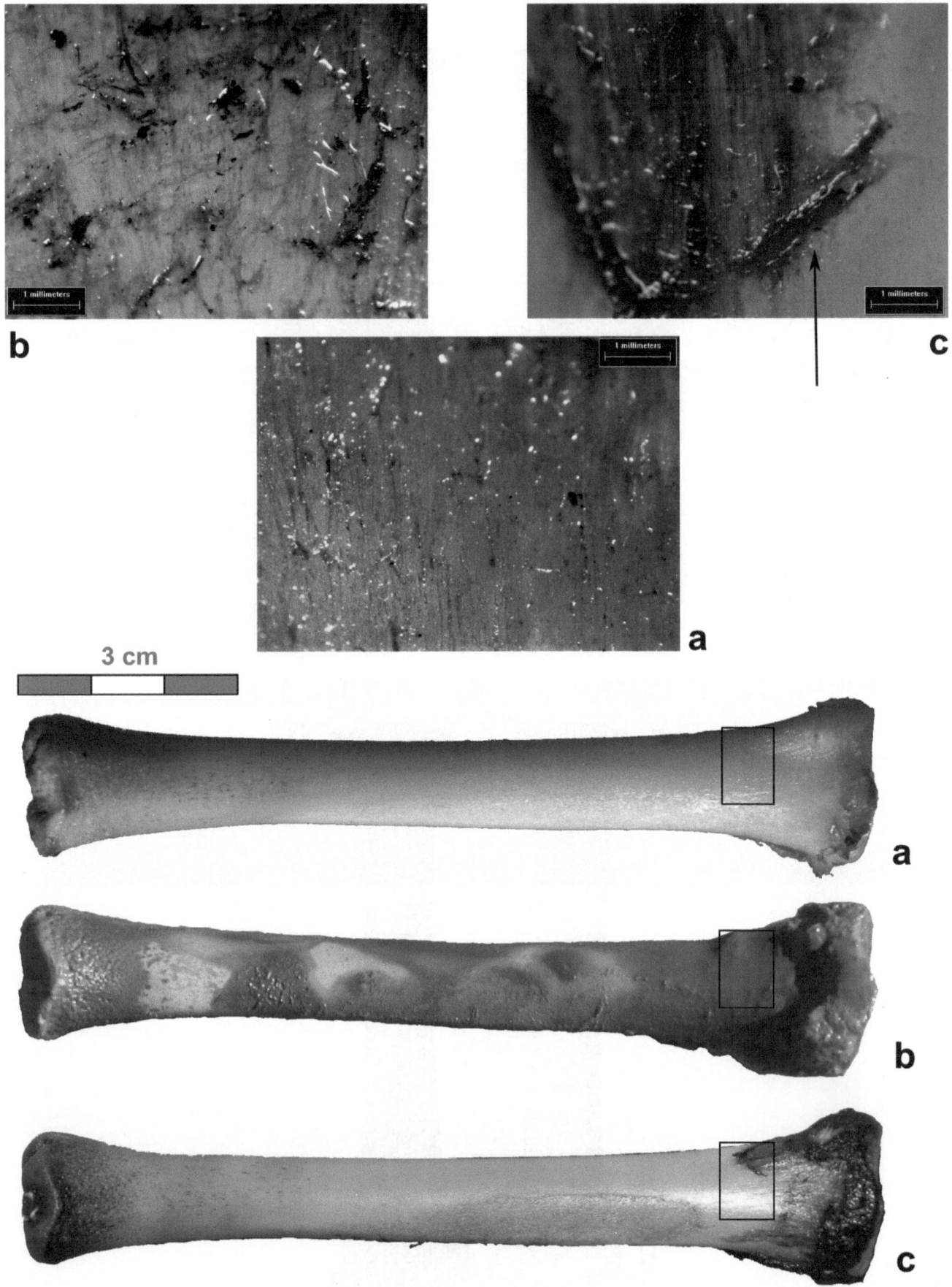

Fig. 6.31 Tumbling Experiment 11.1, sheep metacarpus. The proximal metaphysis of the medial face **a** before and **b** after four hours of tumbling, and **c** after ten hours of tumbling. A deep vertical striation, indicated by an *arrow*, was created by tumbling

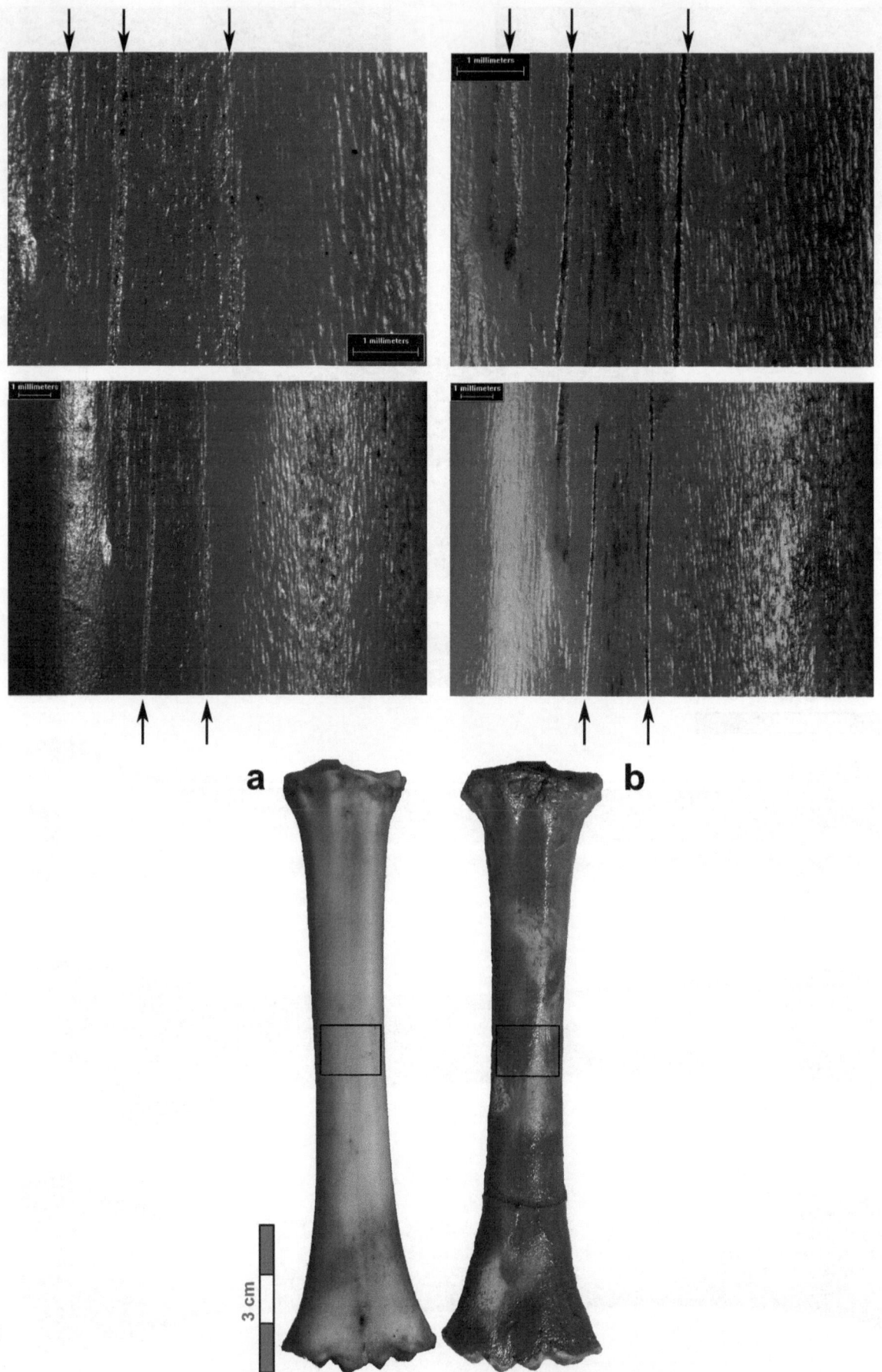

Fig. 6.32 Tumbling Experiment 12, sheep metatarsus. The anterior face **a** before and **b** after tumbling. The striations, indicated by *arrows*, deepened and became more pronounced after tumbling. Both columns show an identical detail of the bone at different degrees of magnification

Fig. 6.33 Tumbling Experiment 12, sheep metatarsus. The anterior proximal metaphysis **a** before and **b** after tumbling. Pronunciation of the morphology of the sulcus, indicated by an *arrow*, was caused by tumbling. Both columns show an identical detail of the bone at different degrees of magnification

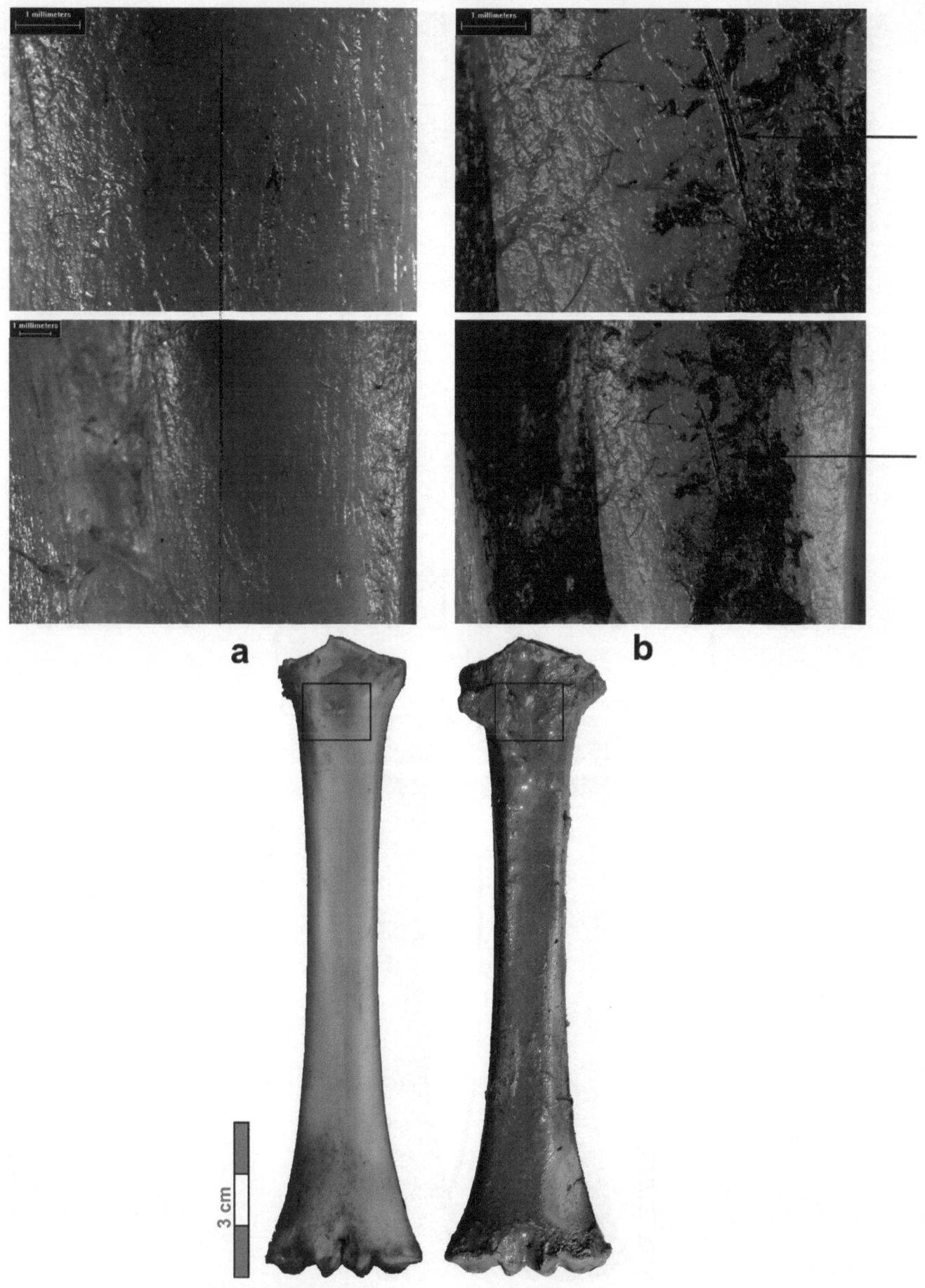

Fig. 6.34 Tumbling Experiment 12, sheep metatarsus. The posterior face of the proximal metaphysis **a** before and **b** after tumbling. U-shaped striations, indicated by an *arrow*, were created by tumbling.

Both columns show an identical detail of the bone at different degrees of magnification

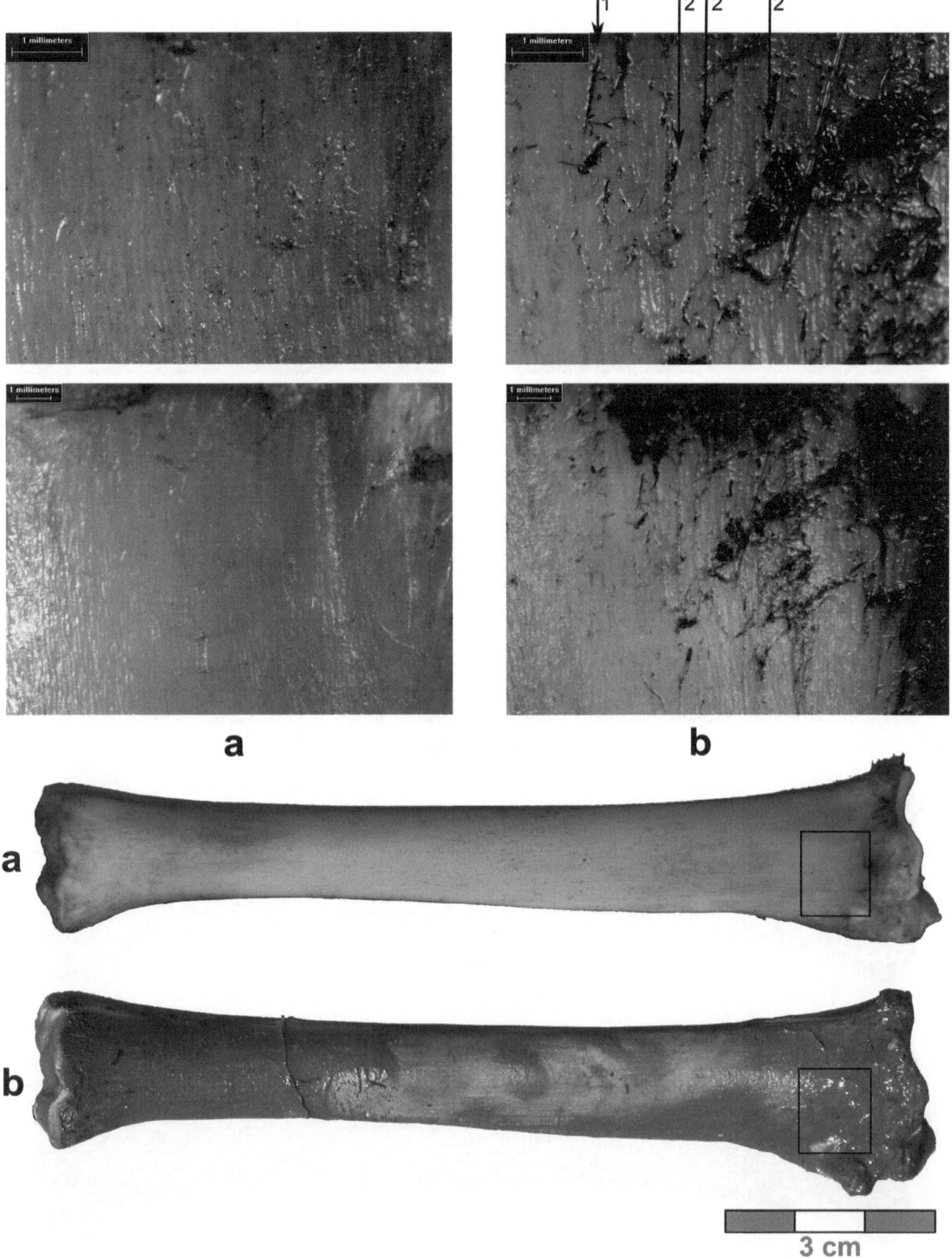

Fig. 6.35 Tumbling Experiment 12, sheep metatarsus. The medial face of the proximal metaphysis **a** before and **b** after tumbling. A deep narrow striation (1) and fine irregular-shaped punctures (2) were created by tumbling. Both columns show an identical detail of the bone at different degrees of magnification

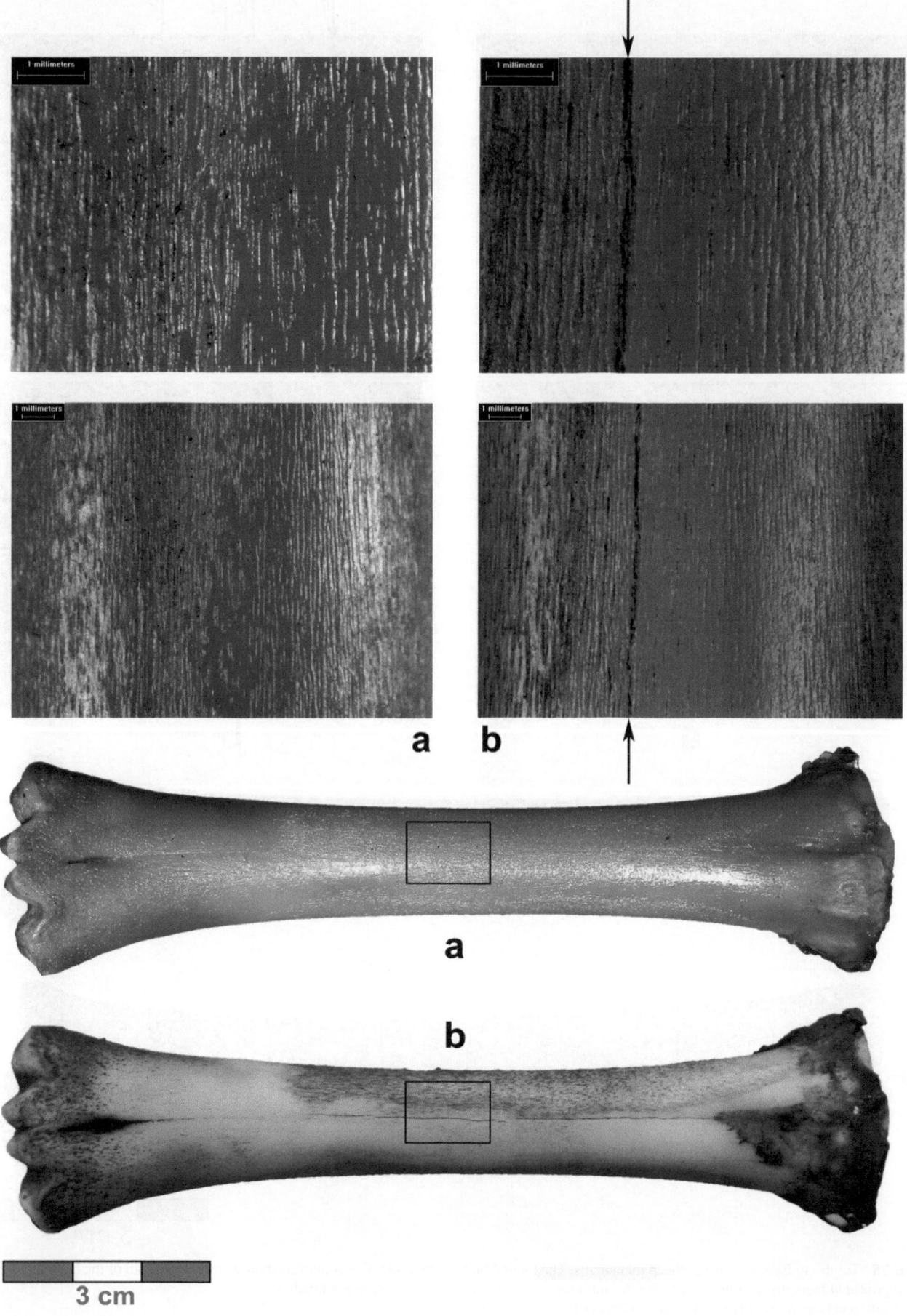

Fig. 6.36 Tumbling Experiment 13, sheep metacarpus. The anterior face of the diaphysis **a** before and **b** after tumbling. A striation, indicated by an *arrow*, was created by tumbling. Both columns show an identical detail of the bone at different degrees of magnification

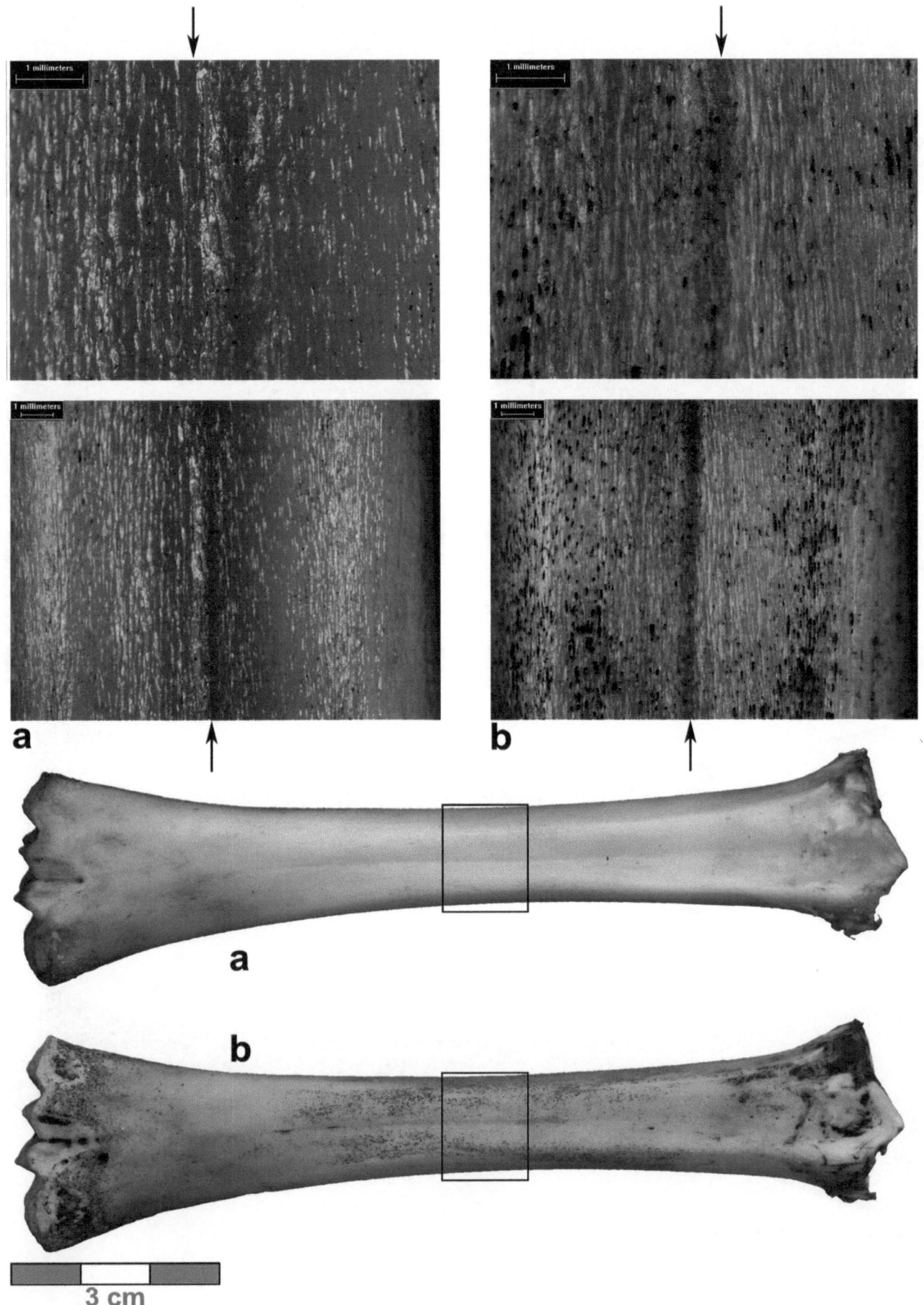

Fig. 6.37 Tumbling Experiment 14, sheep metatarsus. The posterior face of the diaphysis **a** before and **b** after tumbling. Pronunciation of the sulcus, indicated by *arrows*, was caused by tumbling. Both columns show an identical detail of the bone at different degrees of magnification

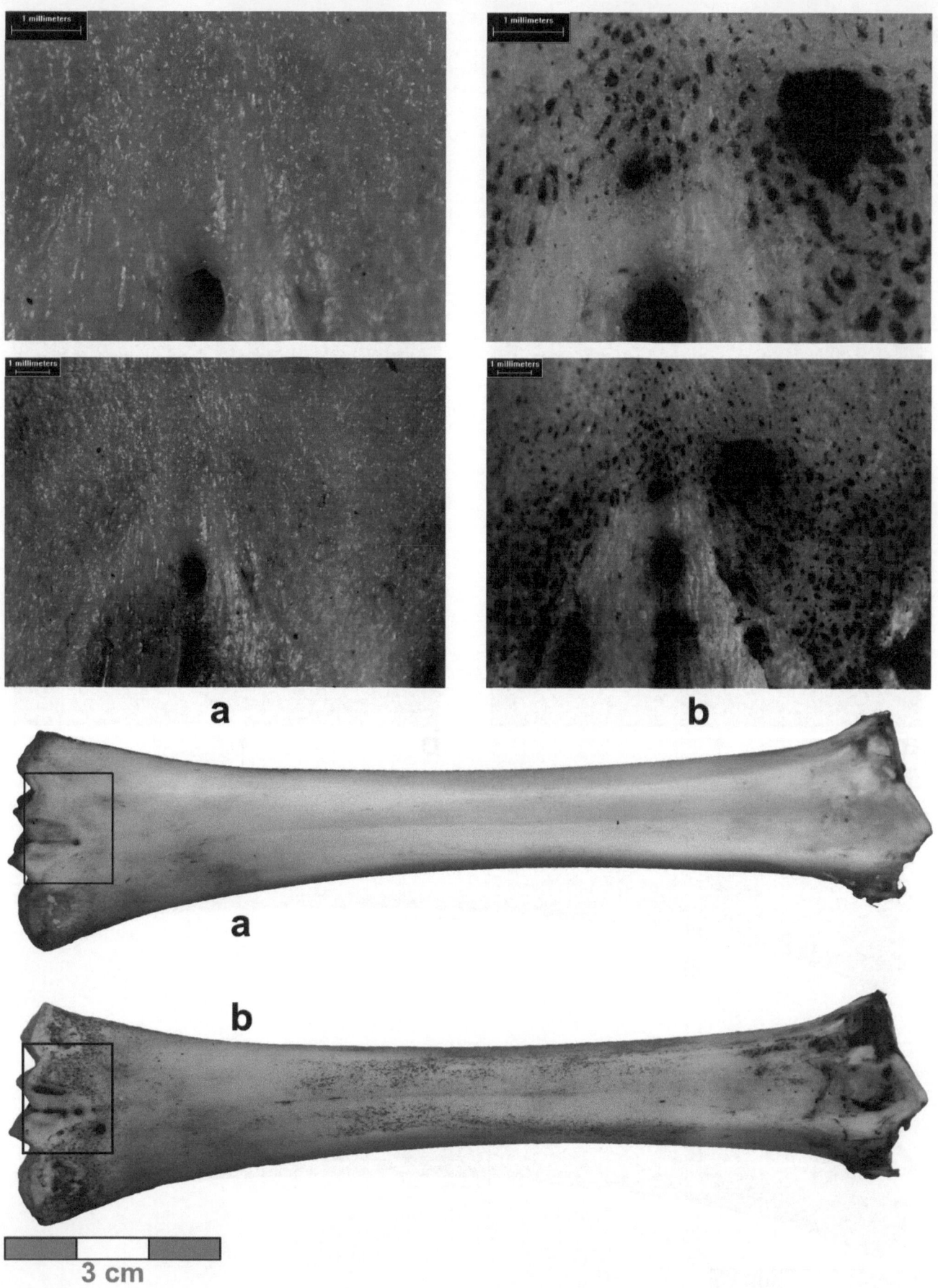

Fig. 6.38 Tumbling Experiment 14, sheep metatarsus. The posterior face of the distal metaphysis **a** before and **b** after tumbling. Abrasion on the distal metaphysis was caused by tumbling. Both columns show an identical detail of the bone at different degrees of magnification

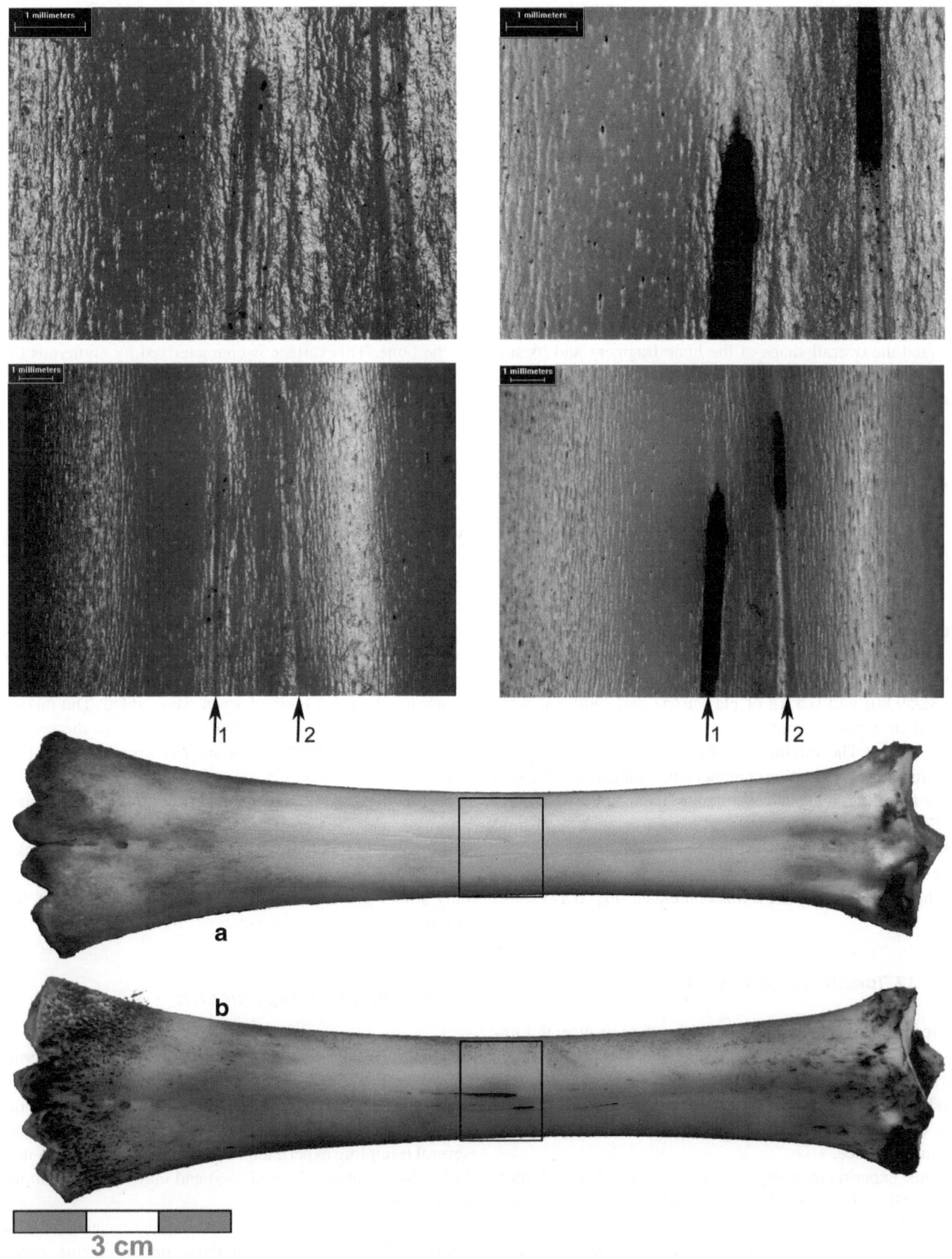

Fig. 6.39 Tumbling Experiment 14, sheep metatarsus. The anterior diaphysis **a** before and **b** after tumbling. The deepening and broadening of the striations (1, 2) were caused by tumbling. Both columns show an identical detail of the bone at different degrees of magnification

experiment. After burial, the broken edges of the bone were slightly rounded. The bone surface was abraded in such a way that the outermost portion had been removed and the porous structure underneath had become visible. Slight exfoliation occurred along with singular striations on the central part of the fragment and on its edge.

This experiment utilized a mixture of 1.5 kg of mollusks (GBY 14/09/97; Layer V-6; 182/130c, 60.12/60.42–59.85/59.80) and 0.5 kg of clay. Bone and sediment were placed in a plastic container, to which 1.5 liters of water were added. The mixture weighed 3.75 kg. The container was affixed to the rotation chamber and tumbled for 16 hours. After tumbling, bone and sediment remained unsorted.

Following tumbling, heavy abrasion and polishing affected the overall shape of the bone fragment and its surface structure. It should be noted that no striations were produced during tumbling (Fig. 6.40).

6.4.3.16 Tumbling Experiment 16

A fresh radius fragment (Radius 3, Fragment 1) of a domestic cow was used for the sixteenth tumbling experiment. Periosteum and meat remained partly attached to the bone. A tiny cut mark was produced with a surgical scalpel on the lateral edge of the fragment.

This experiment utilized a mixture of 1.5 kg of mollusks (GBY 14/09/97; Layer V-6; 182/130c, 60.12/60.42–59.85/59.80) and 0.5 kg of clay. Bone and sediment were placed in a plastic container, to which 1.5 liters of water were added. The mixture weighed 3.75 kg. The container was affixed to the rotation chamber and tumbled for 16 hours. After tumbling, bone and sediment remained unsorted.

Following tumbling, heavy abrasion and polishing affected the overall shape of the bone fragment and its surface structure (Fig. 6.41). No striations were produced during tumbling, and the cut mark was obliterated (Fig. 6.42).

6.4.3.17 Tumbling Experiment 17

A fresh sheep metatarsus was used for the seventeenth tumbling experiment. After removal of the skin and meat, a vertical striation along the anterior face of the diaphyses was produced in the course of periosteum removal. In addition, a fine cut mark was produced on the proximal metaphysis of the anterior face.

This experiment utilized a mixture of 1.5 kg of mollusks (GBY 14/09/97; Layer V-6; 182/130c, 60.12/60.42–59.85/59.80) and 0.5 kg of clay. Bone and sediment were placed in a plastic container, to which 1.5 liters of water were added. The mixture weighed 3.75 kg. The container

was affixed to the rotation chamber and tumbled for 16 hours. After tumbling, bone and sediment remained unsorted.

Abrasion produced during tumbling reduced the overall size of the proximal epiphysis and, in particular, the distal metaphysis. The diaphysis displayed considerable rounding along its lateral and medial edges (Fig. 6.43). The most striking damage occurred on the proximal epiphysis, where two large perforations formed between the articulated surfaces (Fig. 6.44). Large parts of the cancellous interior vanished and the marrow cavity emptied. Only the abraded remains of the anterior sulcus survived.

The vertical striation on the sulcus was smoothened (Fig. 6.45), and the cut mark vanished (Fig. 6.46). Periosteum survived only in the form of scraps on the posterior face of the bone. This surface is characterized by numerous tiny round punctures accompanied by diagonal irregularly shaped striations (Fig. 6.47).

6.4.3.18 Tumbling Experiment 18

A humerus fragment (Humerus 2.1) of a domestic cow was used for the final tumbling experiment. The periosteum remained attached to the surface, and the bone displayed a single vertical cut mark.

This experiment utilized only 2 kg of clay and no mollusks in order to identify the modifying variables within the sediments. Bone and sediment were placed in a plastic container, to which 1.5 liters of water were added. The mixture weighed 3.75 kg. The container was affixed to the rotation chamber and tumbled for 16 hours. After tumbling, the bone was still embedded within the sediment.

Following tumbling, the periosteum was partially removed and the fragment edges were slightly rounded (Fig. 6.48). We also observed exfoliation along the broken edges. The porous structure of the bone surface was revealed only where periosteum had been removed (Fig. 6.49). The cut mark was altered (Fig. 6.50).

6.4.4 Trampling Experiments

Trampling, especially by hoofed mammals and pachyderms, causes both sediment movement and compaction. We propose that these actions produce striations on bones bedded within sediment. In order to test this hypothesis, we designed several trampling experiments, all utilizing a mixture of mollusks (from Layers V-5 and V-6) and clay, and all trampled by the same person.

Trampling in the first experiment was repeated for short periods of time over several days, under varying conditions of water saturation. This experiment was designed to

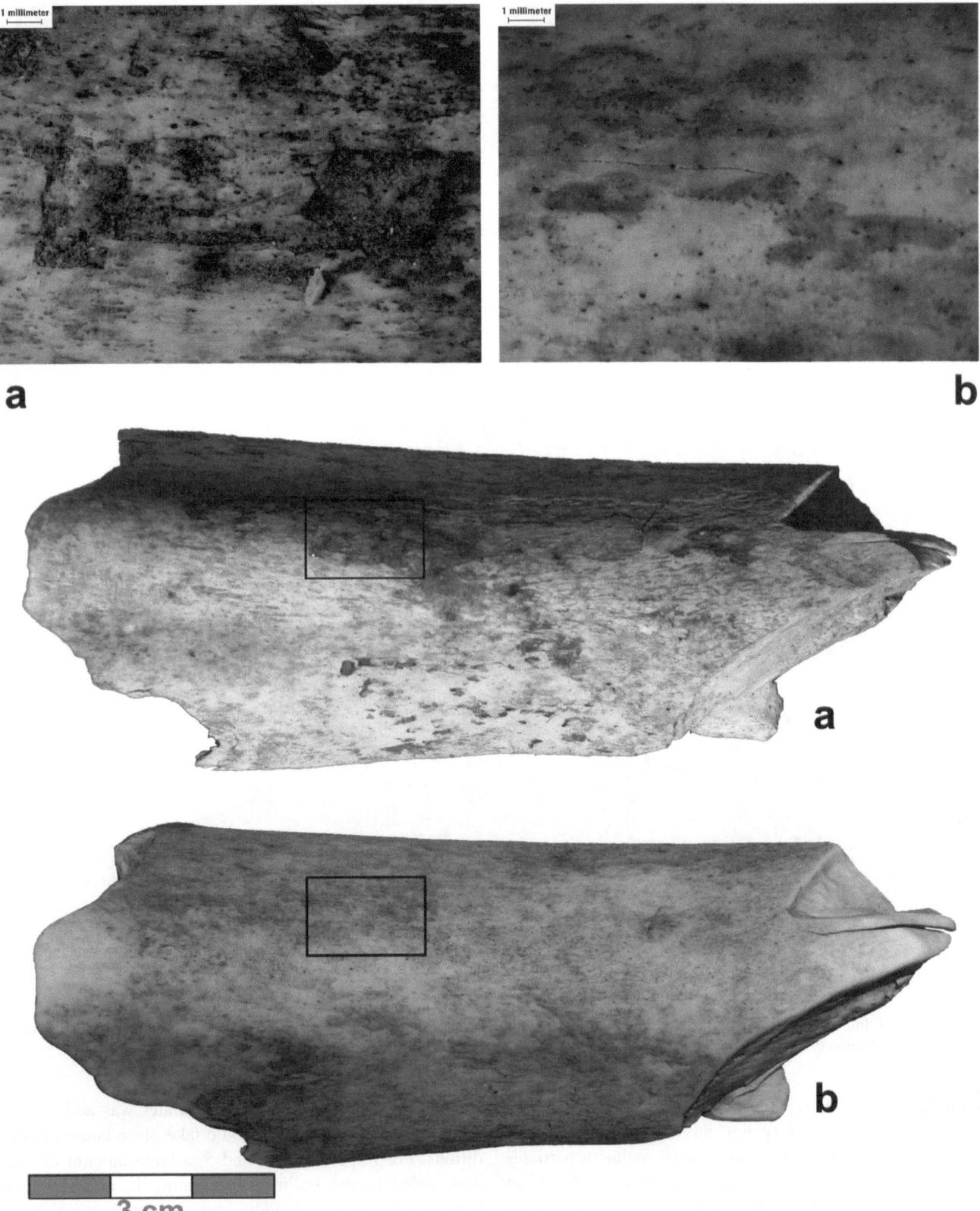

Fig. 6.40 Tumbling Experiment 15, domestic cow femur fragment. Surface modifications **a** before and **b** after tumbling. Polishing of the abraded bone surface was caused by tumbling

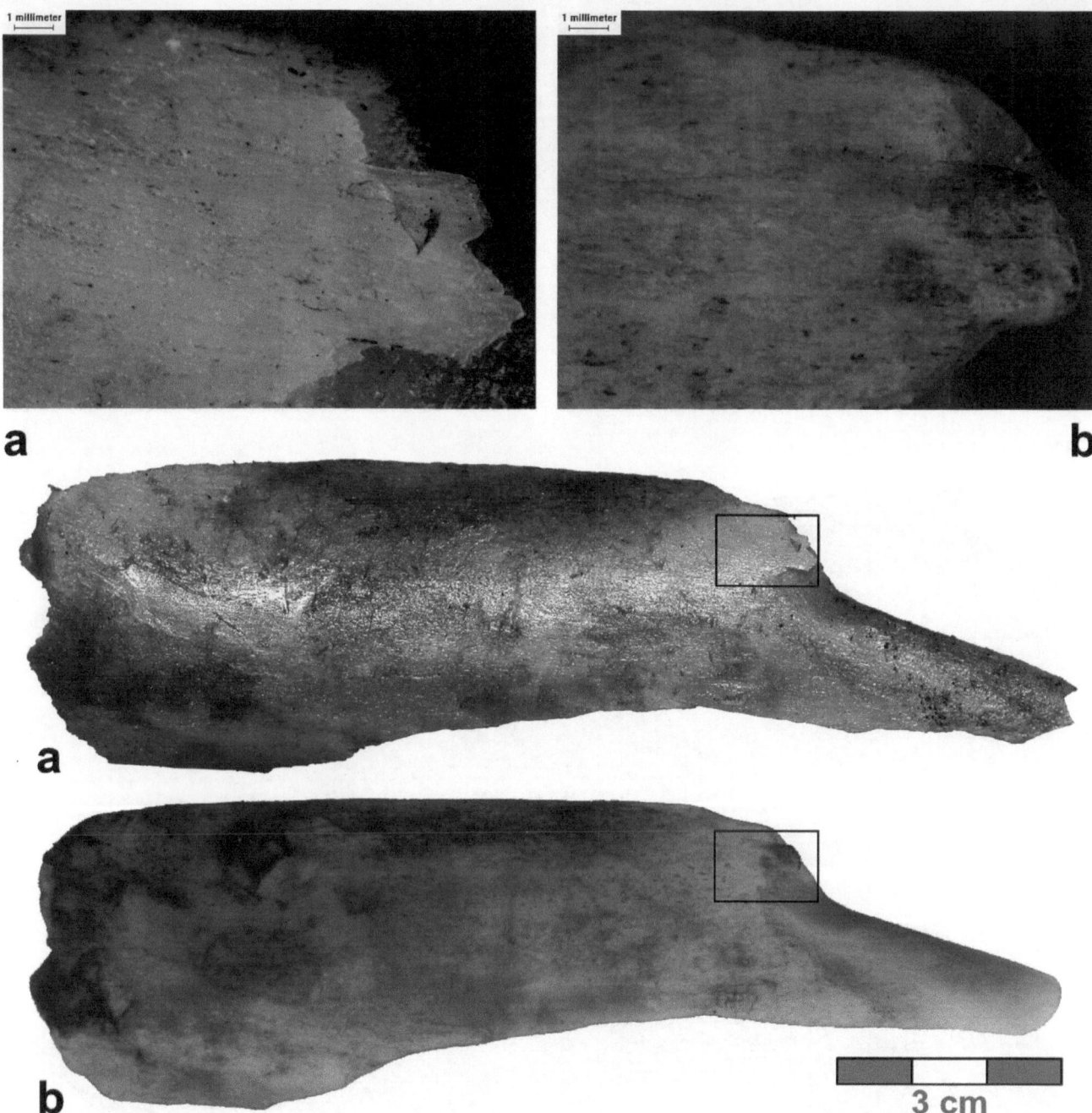

Fig. 6.41 Tumbling Experiment 16, domestic cow radius fragment. Surface modifications **a** before and **b** after tumbling. Heavy abrasion, rounding, and polishing of the fragment were caused by tumbling

mimic the effects of repetitive, but infrequent, trampling. We observed that trampling not only caused serious bone damage, but also altered the condition of sediment particles (i.e., increased mollusk fragmentation). Since the degree of bone-surface damage was comparable to that observed in the containing sediment (e.g., fragmentation, surface alteration of components), changing the amount of water added to the sediment allowed us to test the effects of water saturation in the trampling process.

In Trampling Experiment 3, no water was added to the sediment so that trampling would take place under dry conditions. For Experiments 2 and 5, a large amount of water was added to the sediment, saturating it, and simulating very wet conditions. Muddy conditions were mimicked in Experiments 2.2, 4, 6, and 7. Before and after each experiment, a 500-ml sample was separated from the remaining sedimentary matrix with a 5-mm sieve and complete mollusks and mollusk fragments were counted (Table 6.2).

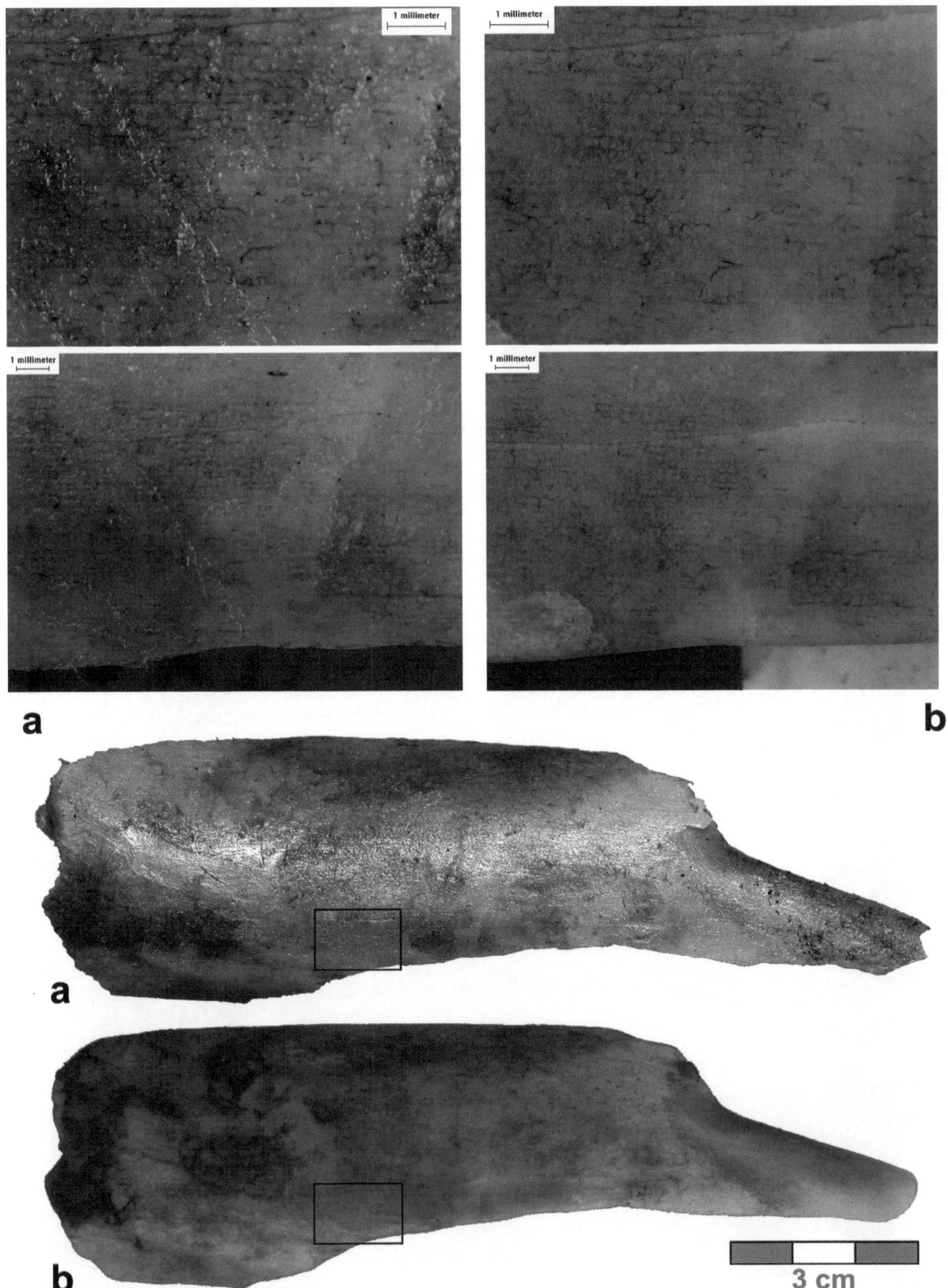

Fig. 6.42 Tumbling Experiment 16, domestic cow radius fragment. The lateral face **a** before and **b** after tumbling. Obliteration of the cut mark was caused by tumbling. Both columns show an identical detail of the bone at different degrees of magnification

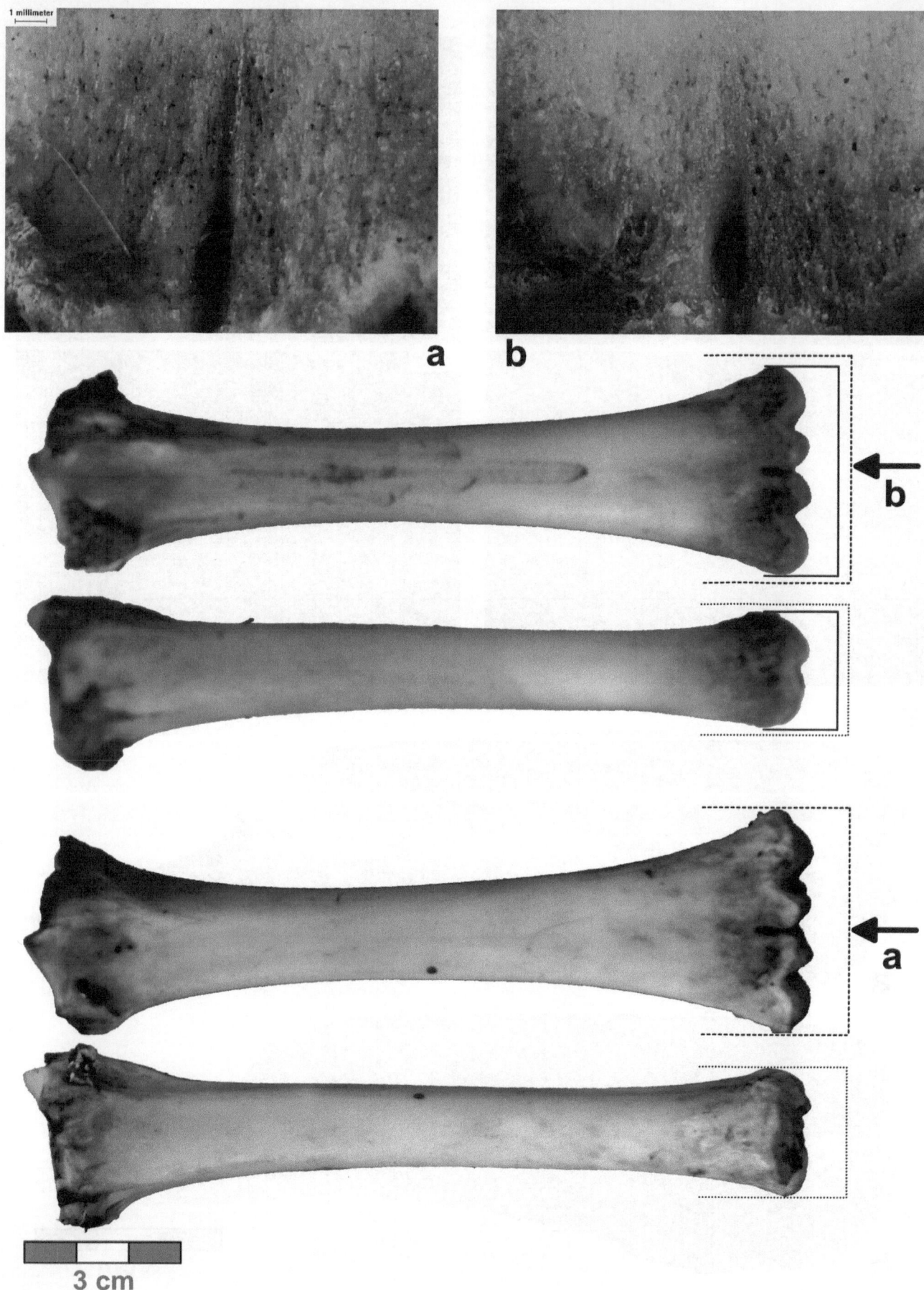

Fig. 6.43

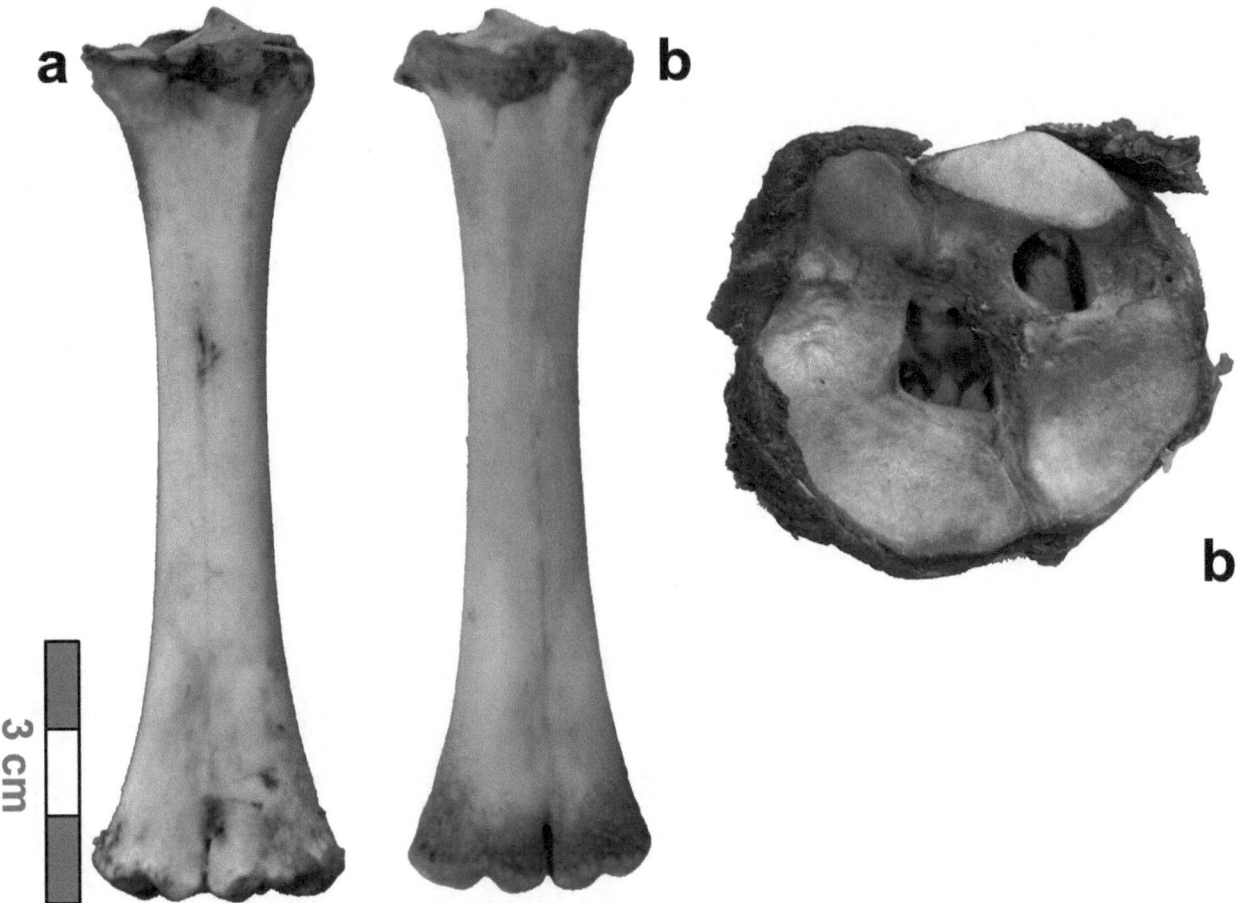

Fig. 6.44 Tumbling Experiment 17, sheep metatarsus. **a** before and **b** after tumbling. Damage to the proximal articulation that occurred after tumbling exposed the marrow cavity

The degree of water saturation in the experiments had an obvious effect on shell preservation; trampling under dry conditions had no effect on shell condition. Nevertheless, a higher number of shells were recorded in the 500-ml sample after trampling, indicating that sorting according to grain size must have occurred.

As mentioned above, trampling under wet conditions resulted in increased shell fragmentation. However, the number of medium-sized, dense complete mollusks, such as *Melanopsis*, did not change, suggesting only minor fragmentation of large complete shells, such as *Viviparus apamaea.*

Trampling under muddy conditions led to a significant decrease in the number of complete shells. However, three experiments under muddy conditions resulted in only a minor increase in shell fragmentation. Aside from these examples,

trampling in muddy sediment had the most destructive effect on shell preservation.

6.4.4.1 Trampling Experiment 1

A complete sheep metacarpus and metatarsus, and a domestic cow femur and tibia shaft fragments were used in the first trampling experiment. Prior to trampling, the two sheep bones were also used in Tumbling Experiments 1 and 2. The two long bone fragments were broken during marrow extraction, and dried for 12 days before trampling.

As a result of tumbling, the sheep metapodials bore slight traces of abrasion as well as fine vertical striations on the anterior face produced in the course of periosteum removal.

Fig. 6.43 Tumbling Experiment 17, sheep metatarsus **a** before and **b** after tumbling. Considerable rounding of the bone was caused by tumbling. The *dotted lines* indicate the distal width and depth of the metaphysis prior to tumbling; the *black lines* indicate the distal width

and depth of the metaphysis after tumbling. *Arrows* indicate the area of the bone that was highly magnified, illustrating the morphological change created by tumbling

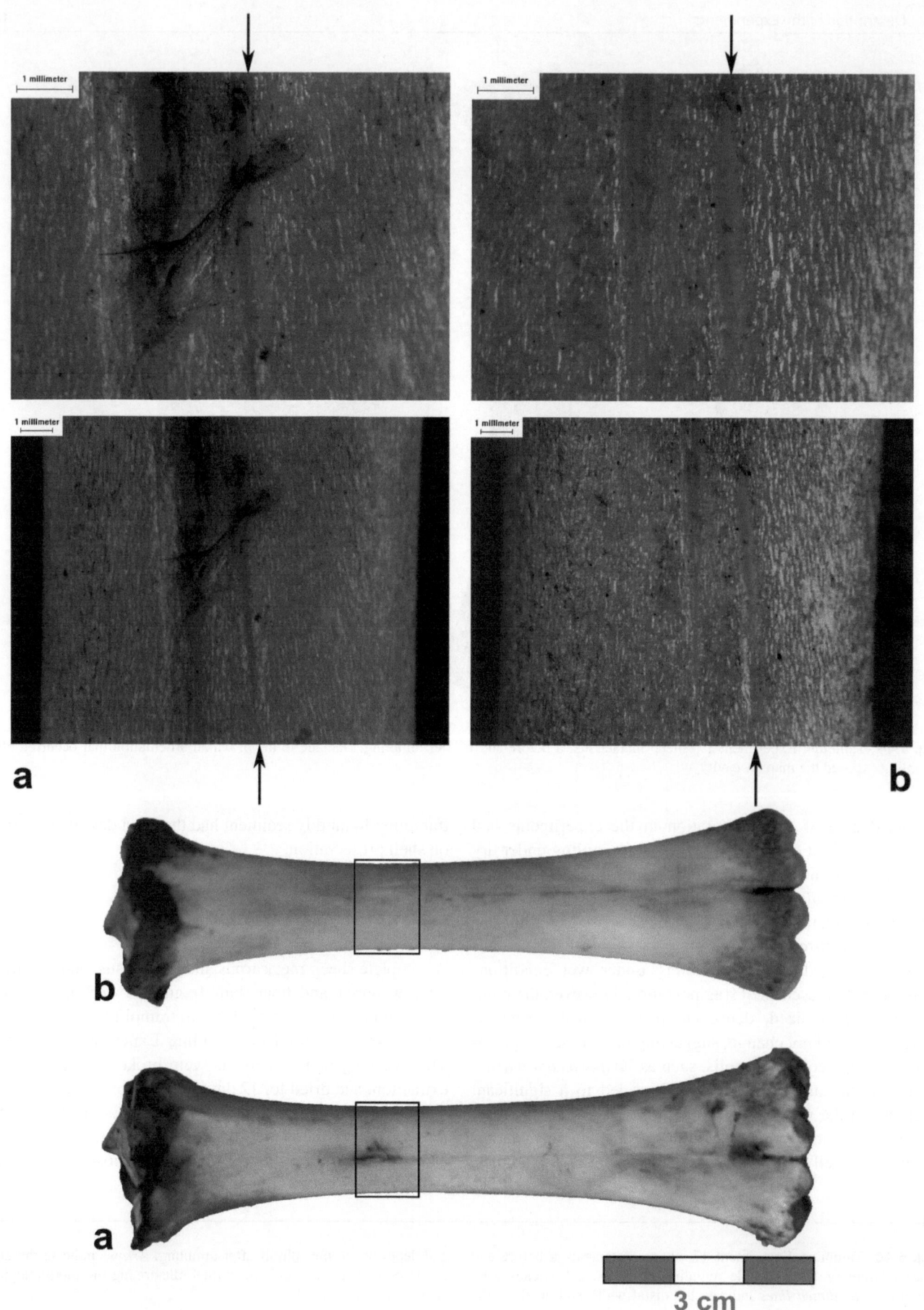

Fig. 6.45 Tumbling Experiment 17, sheep metatarsus. Bone-surface modification on the anterior face of the diaphysis **a** before and **b** after tumbling. Smoothing of the vertical striation, indicated by an *arrow*, was caused by tumbling. Both columns show an identical detail of the bone at different degrees of magnification

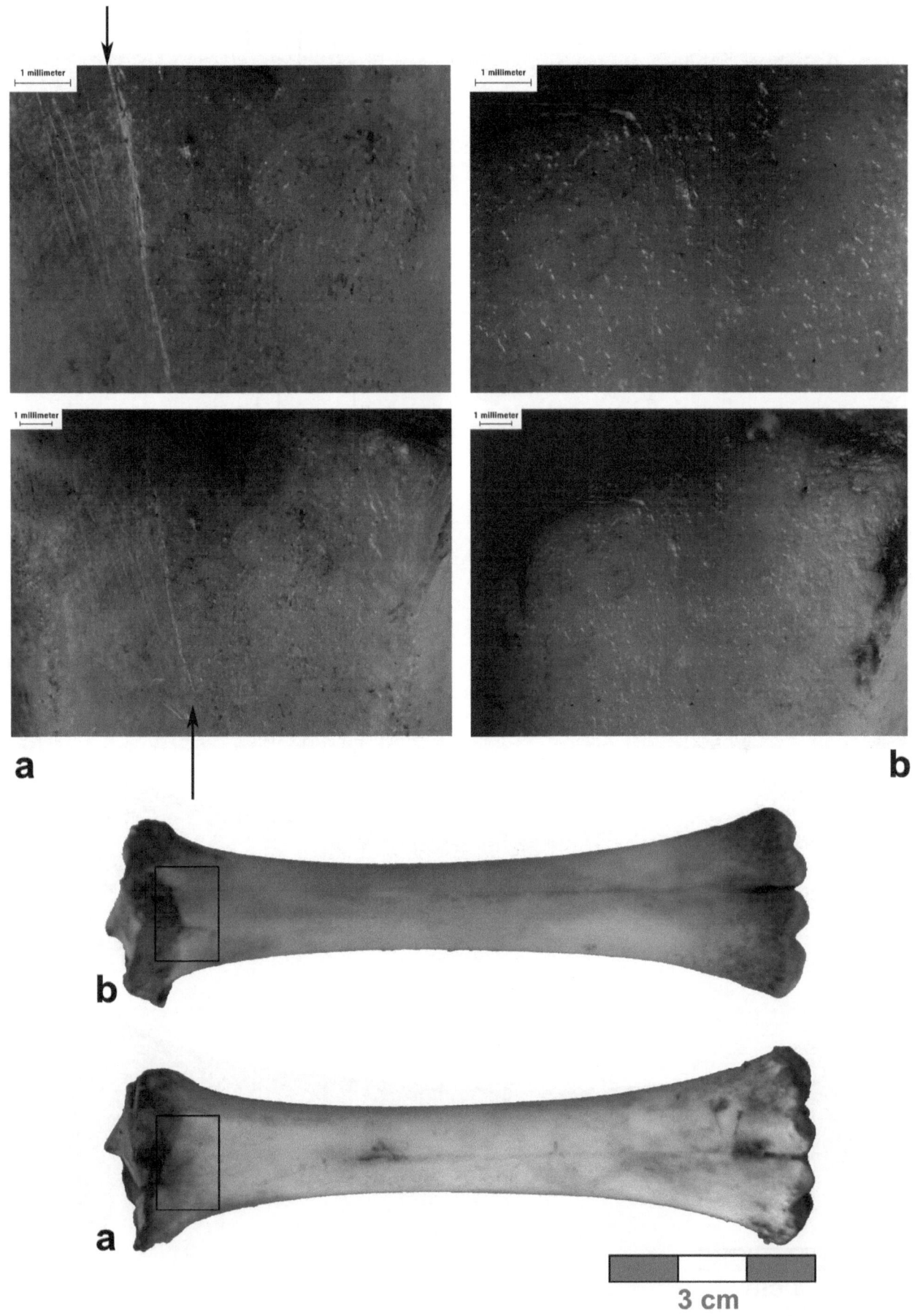

Fig. 6.46 Tumbling Experiment 17, sheep metatarsus. Bone-surface modification on the anterior face of the proximal metaphysis **a** before and **b** after tumbling. Removal of the cut mark, indicated by an *arrow*, was caused by tumbling. Both columns show an identical detail of the bone at different degrees of magnification

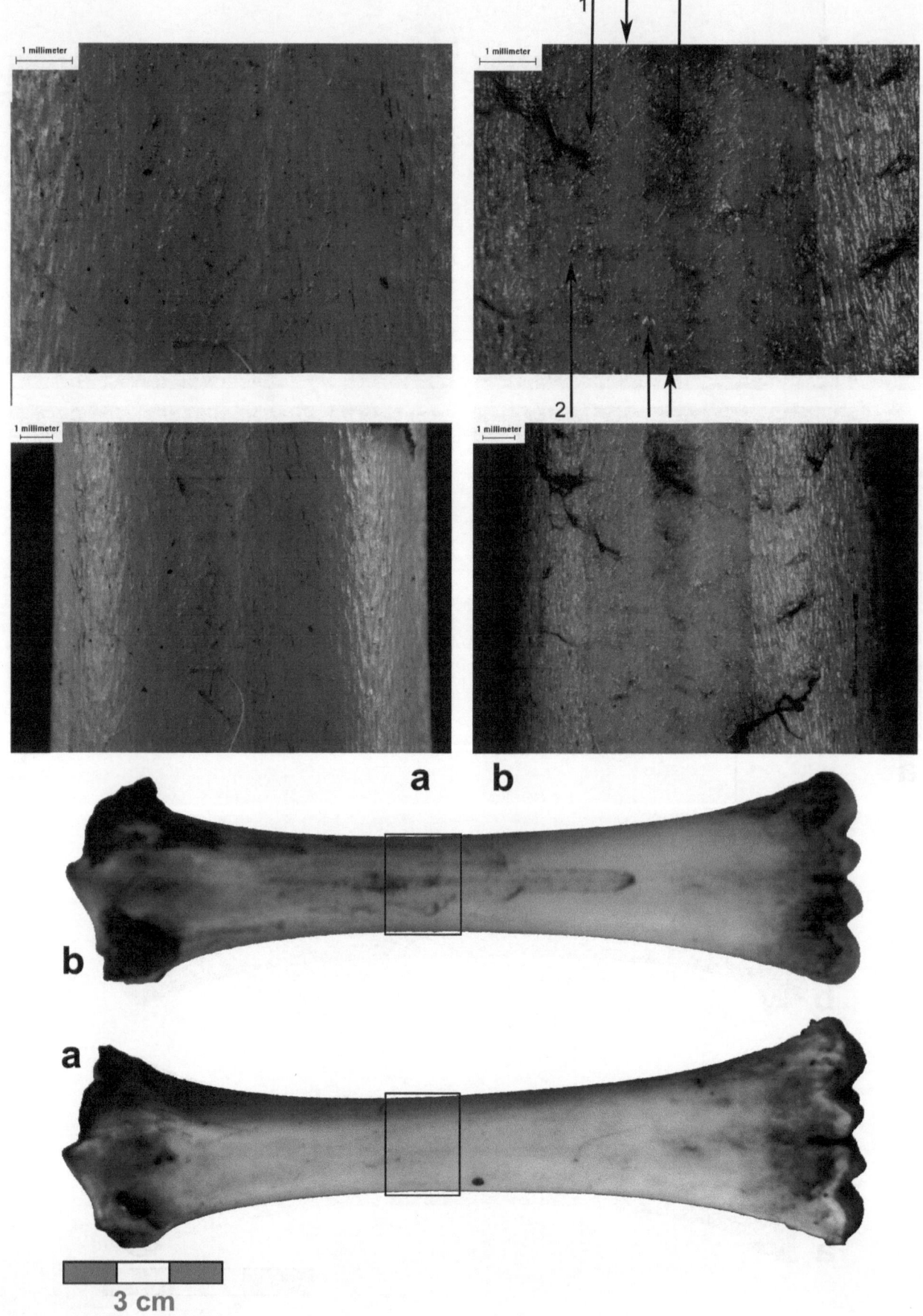

Fig. 6.47 Tumbling Experiment 17, sheep metatarsus. Bone-surface modification on the posterior face of the diaphysis **a** before and **b** after tumbling. Irregularly-shaped striations (1) and tiny round punctures (2) were created by tumbling. Both columns show an identical detail of the bone at different degrees of magnification

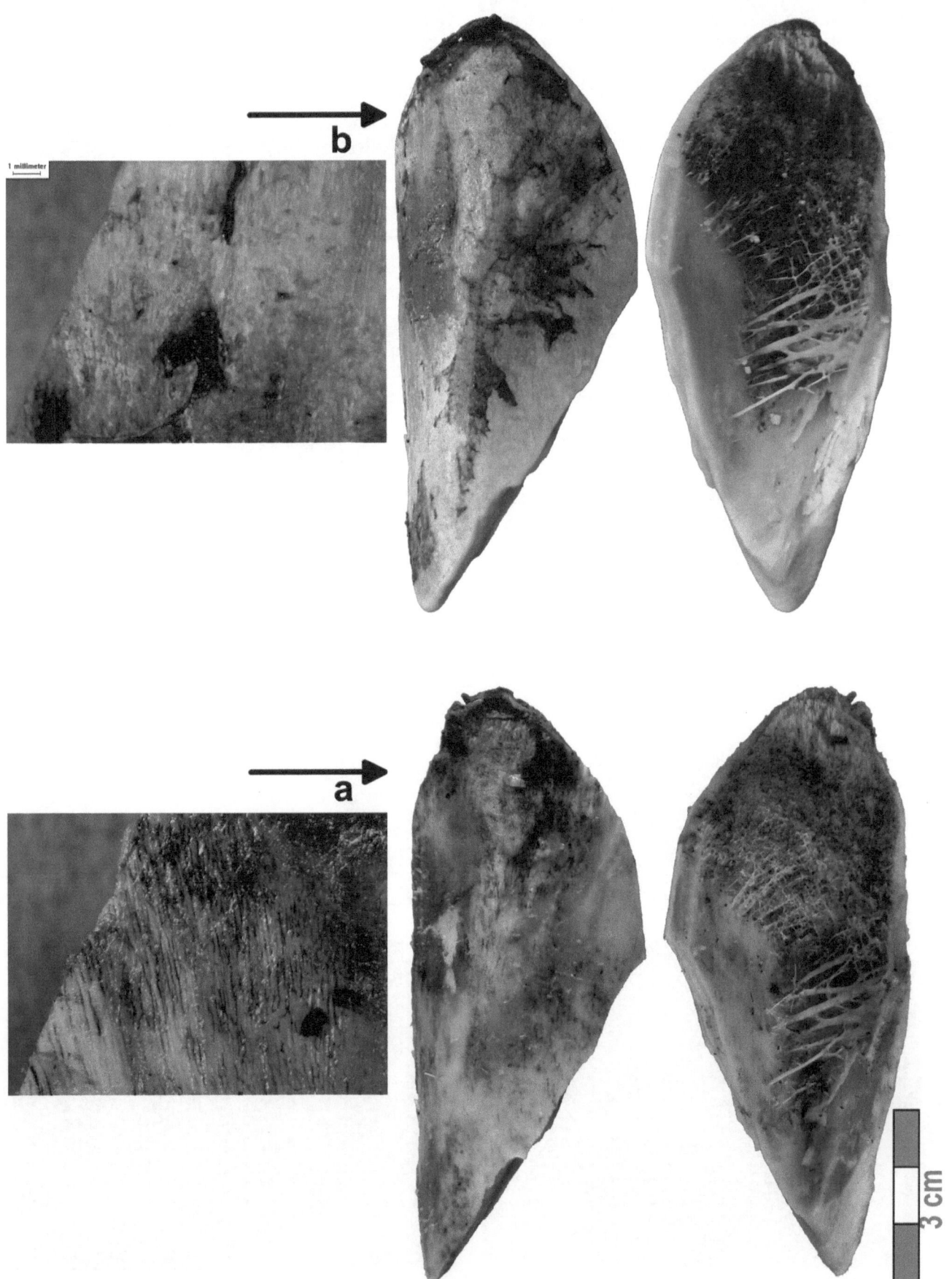

Fig. 6.48 Tumbling Experiment 18, domestic cow humerus fragment. Bone modifications **a** before and **b** after tumbling. An *arrow* indicates the area that was magnified. Removal of the periosteum and rounding of the fragment edges were caused by tumbling

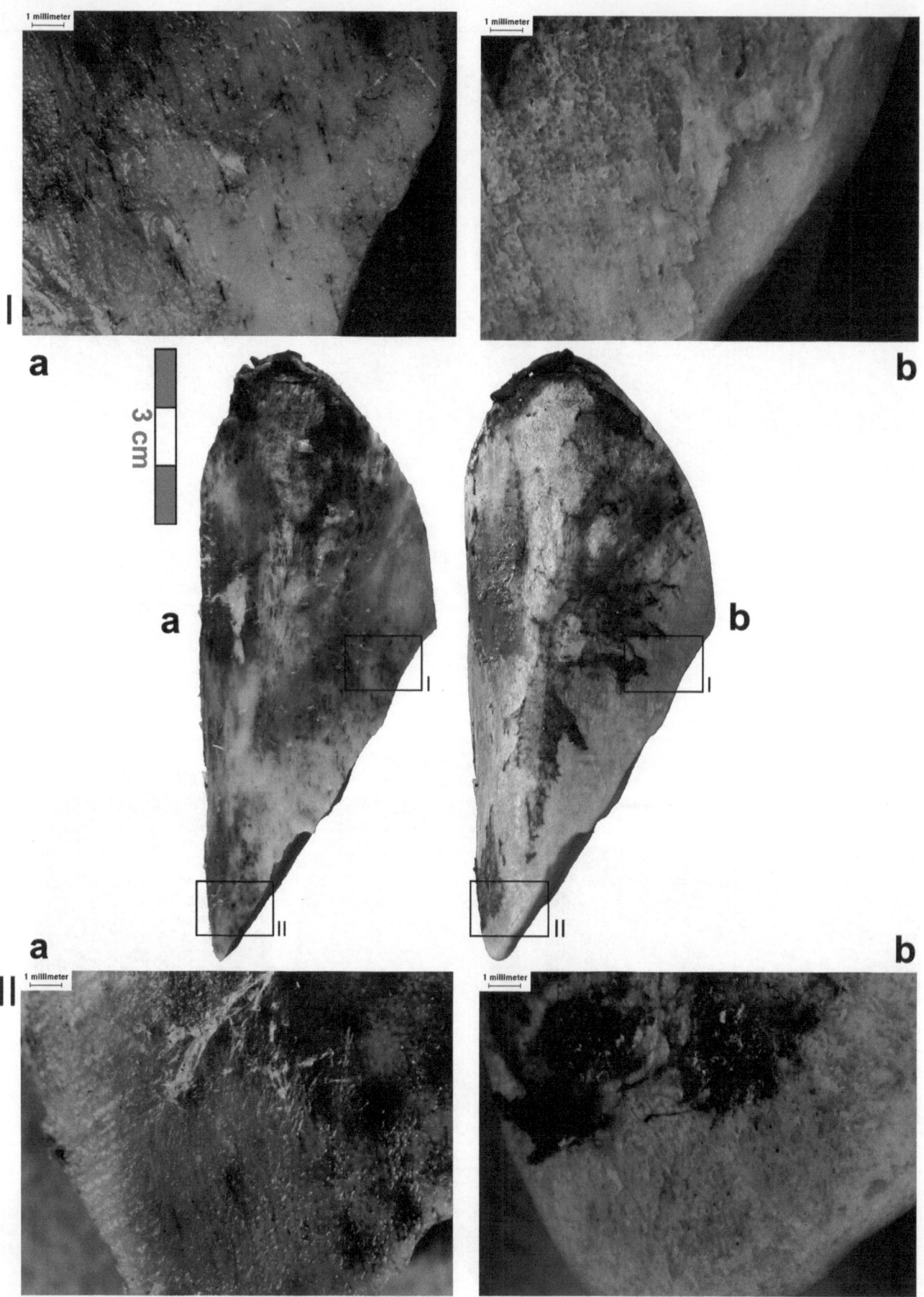

Fig. 6.49 Tumbling Experiment 18, domestic cow humerus fragment. Bone modifications **a** before and **b** after tumbling. (I) Exfoliation and (II) exposure of the porous structure of the bone surface were caused by tumbling

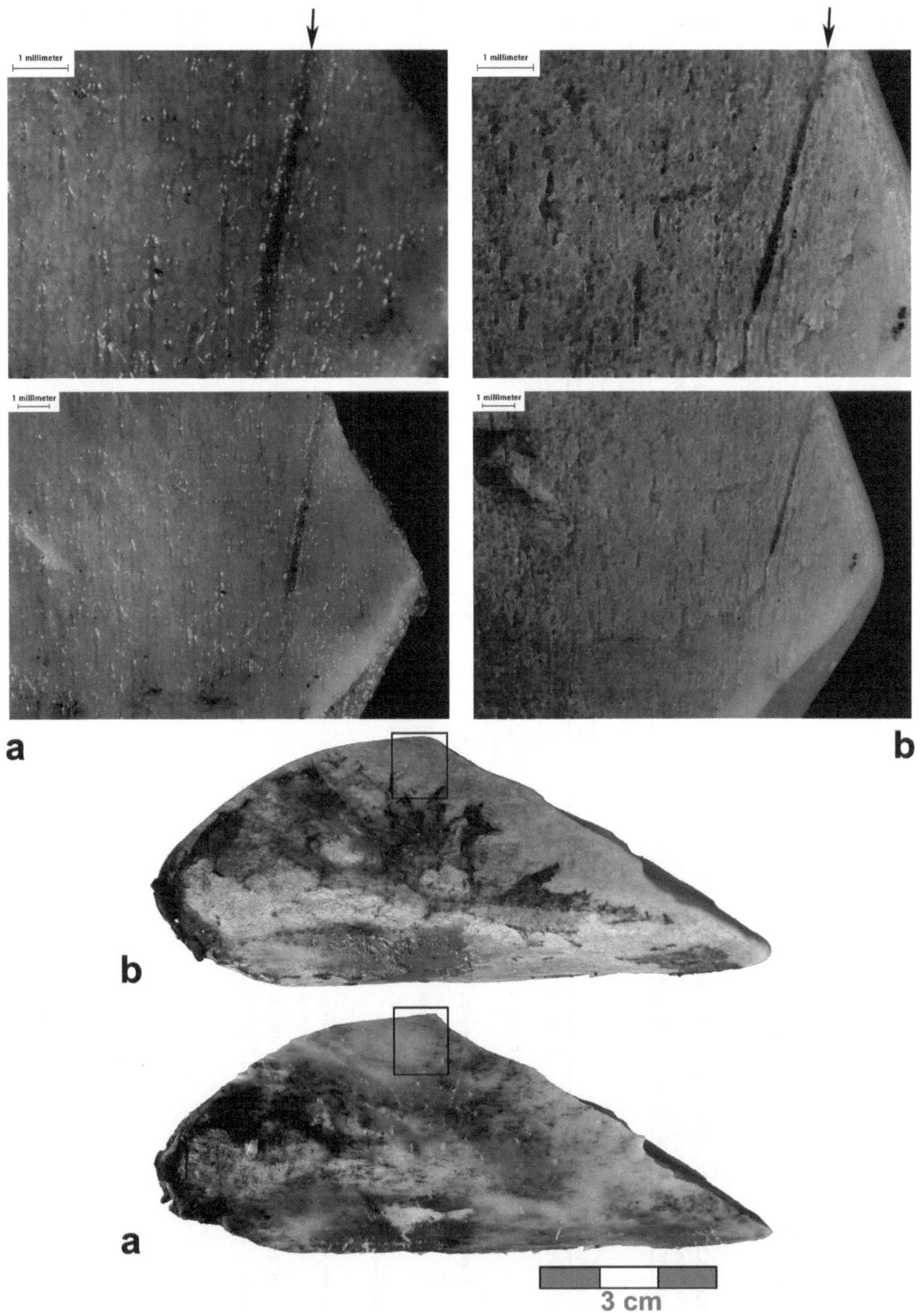

Fig. 6.50 Tumbling Experiment 18, domestic cow humerus fragment. Bone modification **a** before and **b** after tumbling. Modifications in the morphology of the cut mark, indicated by an *arrow*, were created by tumbling. Both columns show an identical detail of the bone at different degrees of magnification

Table 6.2 Sediment conditions–trampling experiment: (a) number (n) and volume (ml) of complete mollusks and of mollusk fragments before and after trampling; (b) ratio of complete mollusks (n) to mollusk fragments (n) before and after trampling; (c) ratio of complete mollusks (n) before and after trampling; (d) ratio of shell volume (ml) before and after trampling. The 500-ml sediment sample consisted of a 2:1 ratio of mollusks to clay. The mollusks were extracted by a 5-mm sieve

		Trampling experiments						
		2	2.1	3	4	5	6	7
Sediment conditions		Wet	Muddy	Dry	Muddy	Wet	Muddy	Muddy
(a) Before trampling	Complete	70 (n)	84 (n)	170 (n)	170 (n)	177 (n)	120 (n)	178 (n)
		45 ml	50 ml	135 ml	135 ml	180 ml	75 ml	190 ml
	Fragment	61 (n)	107 (n)	170 (n)	170 (n)	218 (n)	121 (n)	201 (n)
		30 ml	35 ml	70 ml	70 ml	100 ml	50 ml	100 ml
After trampling	Complete	62 (n)	62 (n)	215 (n)	80 (n)	163 (n)	59 (n)	132 (n)
		35 ml	30 ml	135 ml	40 ml	80 ml	25 ml	75 ml
	Fragment	101 (n)	109 (n)	184 (n)	145 (n)	403 (n)	141 (n)	353 (n)
		25 ml	40 ml	60 ml	60 ml	150 ml	60 ml	110 ml
(b) Complete/fragment (n)	Before	1.13	0.8	1	1.2	0.8	1	0.9
	After	0.6	0.57	1.2	0.6	0.4	0.4	0.4
(c) Before/after (n)	Complete	1.14	1.3	0.8	2.7	1.1	2	1.3
	Fragment	0.6	1	0.9	1.3	0.5	0.8	0.6
(d) Before/after (ml)	Complete	1.3	1.6	1	3.3	2.2	3	3
	Fragment	1.2	0.9	1	1	0.7	0.8	0.8

Periosteum remained attached to some parts of the proximal and distal metaphyses.

Although the cow bones were cleaned before fracturing, periosteum remained attached to their surfaces, especially the tibia. Prior to trampling, the remaining periosteum was removed with a surgical scalpel, during which the femur was seriously damaged. Horizontal, vertical, and diagonal cut marks were produced on the femur fragment, but no marks were produced on the tibia. Periosteum remained attached to parts of both long bone fragments.

This experiment utilized a mixture of 5.5 kg of mollusks (3,325 kg GBY 12/09/97; Layer V-6; 182/130b, 60.36/60.29–60.03/60.00; 2,250 kg GBY 15/09/97; Layer V-6; 181/130c, 60.00/59.88–59.84) and 2 kg of clay. Bones and sediment were placed in a large plastic container, to which 5 liters of water were added. Over the course of 10 days, the material was trampled for 10 minutes each day by a person with bare feet weighing approximately 80 kg. Since water is the most crucial variable in a shoreline environment, different amounts of water were added to the container in order to mimic flooded, soaked (muddy), and dry sediment conditions. Additional water was added only when the sediment

would become too dry due to evaporation: 2.5 liters on Day 2, 5 liters on Day 4, and 8 liters on Day 8 (Fig. 6.51).

Following trampling, the metacarpus displayed signs of abrasion and rounding. Most of the periosteum had been removed along the anterior proximal metaphysis, and the outermost bone surface was exfoliated (Fig. 6.52). The sulcus of the posterior face was flattened and a deep horizontal to diagonal striation appeared along the lateral diaphysis (Fig. 6.53).

The distal metaphysis of the metatarsus exhibited a high degree of abrasion and rounding. The periosteum of the proximal and distal metaphyses disappeared. The foramen on the anterior diaphysis became deeper and narrower, while the vertical striations became more pronounced with well-defined edges. The diagonal mark along one of the striations had disappeared (Fig. 6.54). Initial stages of horizontal striation formation were observed on the lateral proximal metaphysis (Fig. 6.55).

Abrasion and exfoliation altered the cow femur (Fem-2-2) in such a way that the porous (vascular) structure of the bone had become visible in several localized areas (Fig. 6.56). The fragment edges were rounded. A shallow

Fig. 6.51 Trampling Experiment 1, sediment composition: **a** mixture of mollusks and clay without water; **b** first day: sediment composition following the experiment; **c** second day: sediment composition prior to the experiment; **d** fourth day: sediment composition prior to the experiment; **e** fourth day: sediment composition following the experiment; **f** eighth day: sediment composition prior to the experiment; **g** eighth day: sediment composition following the experiment

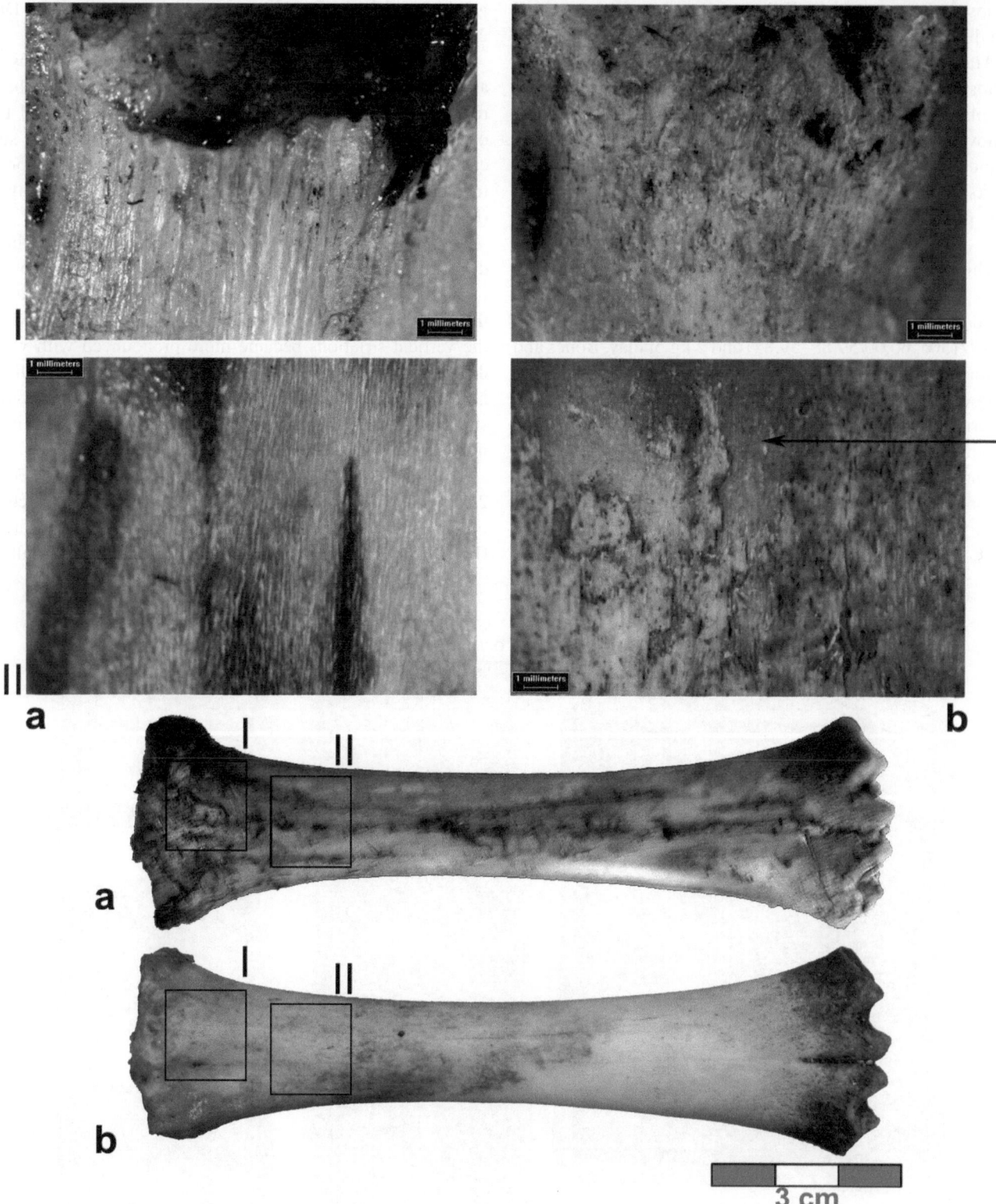

Fig. 6.52 Trampling Experiment 1, sheep metacarpus. The anterior face of the proximal metaphysis **a** after tumbling and before trampling and **b** after trampling. (I) Removal of the periosteum and (II) exfoliation of the bone surface, indicated by an *arrow*, were caused by trampling

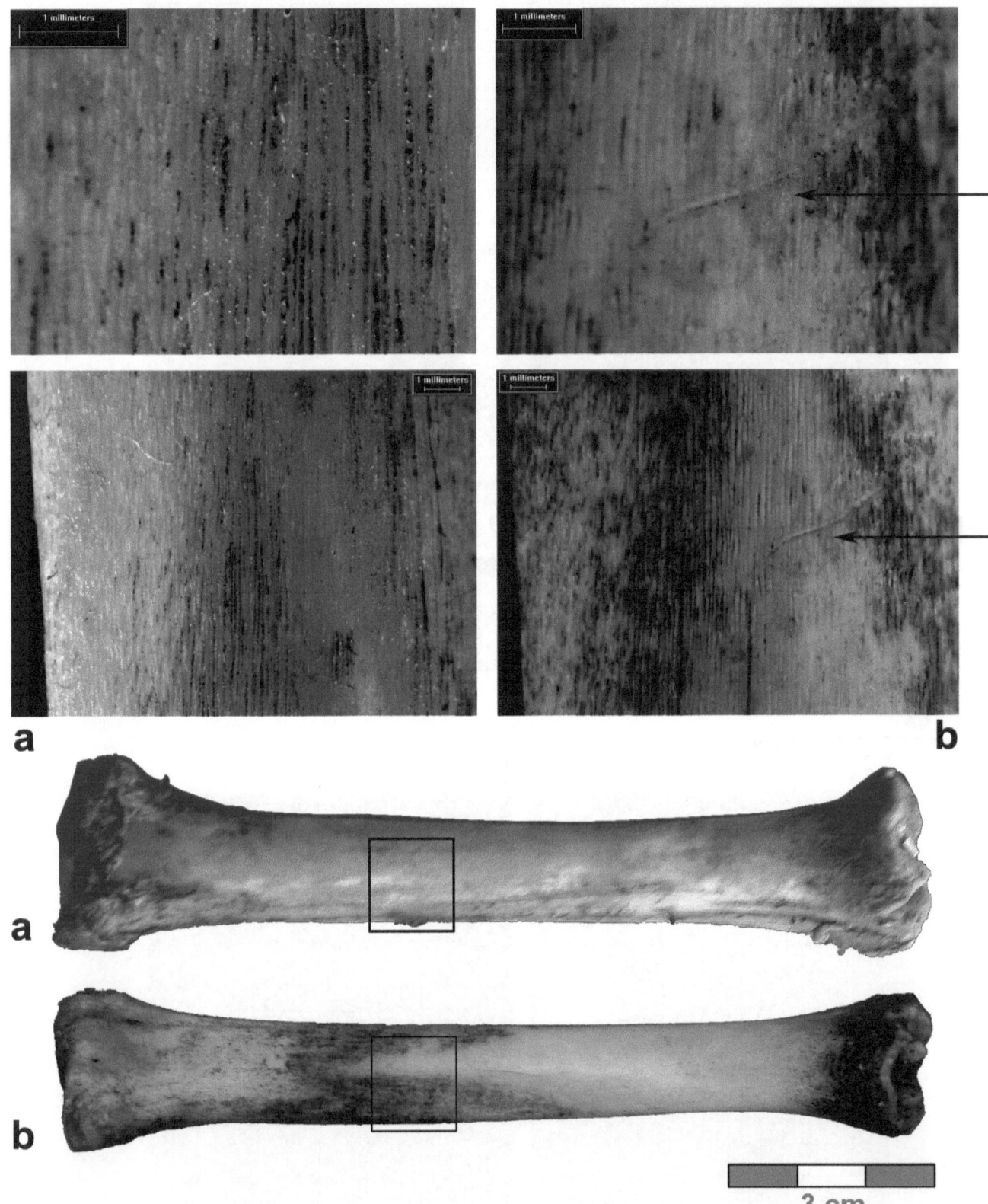

Fig. 6.53 Trampling Experiment 1, sheep metacarpus. The lateral diaphysis **a** after tumbling and before trampling and **b** after trampling. A striation, indicated by an *arrow*, was created by trampling. Both columns show an identical detail of the bone at different degrees of magnification

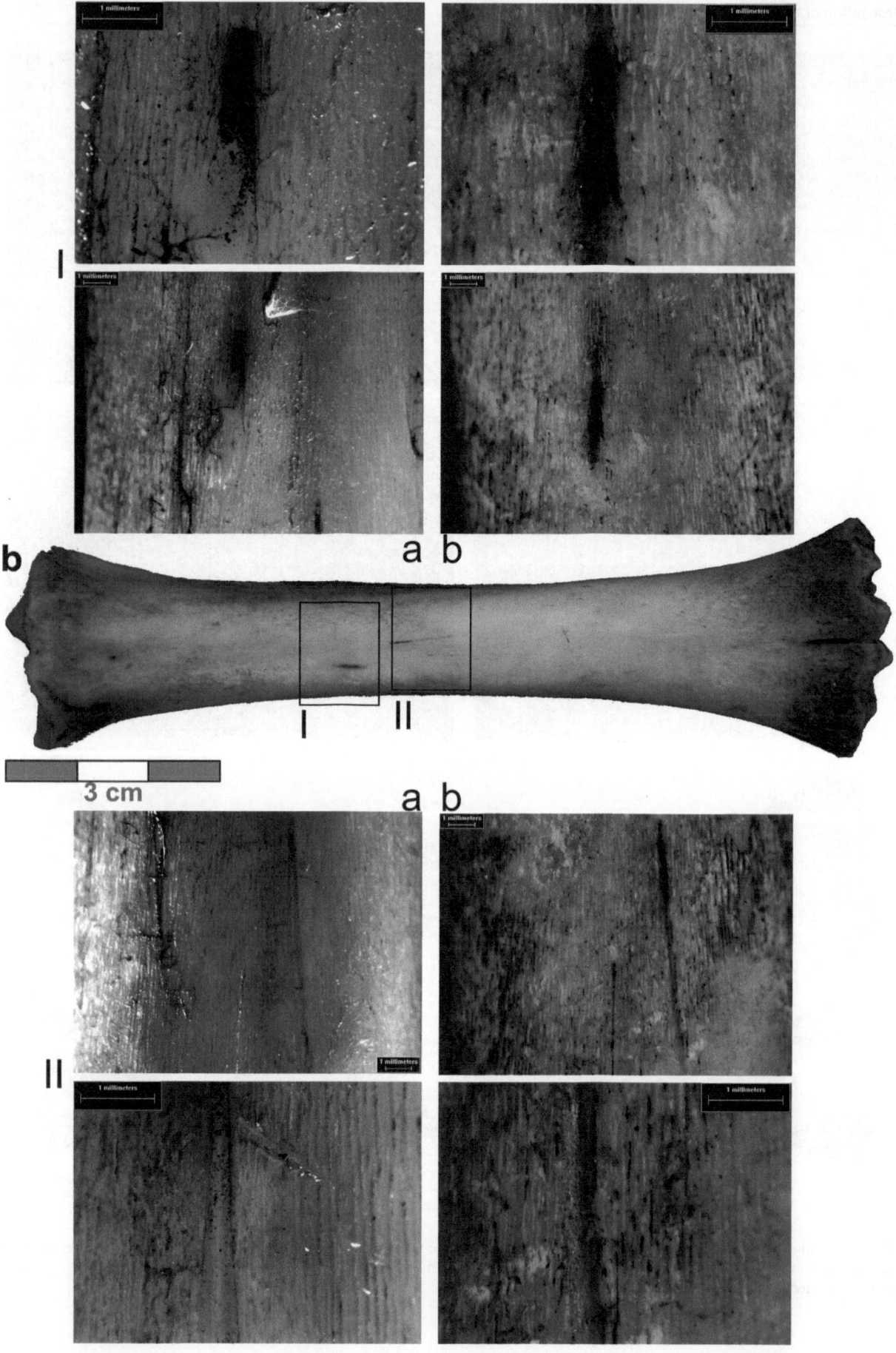

Fig. 6.54

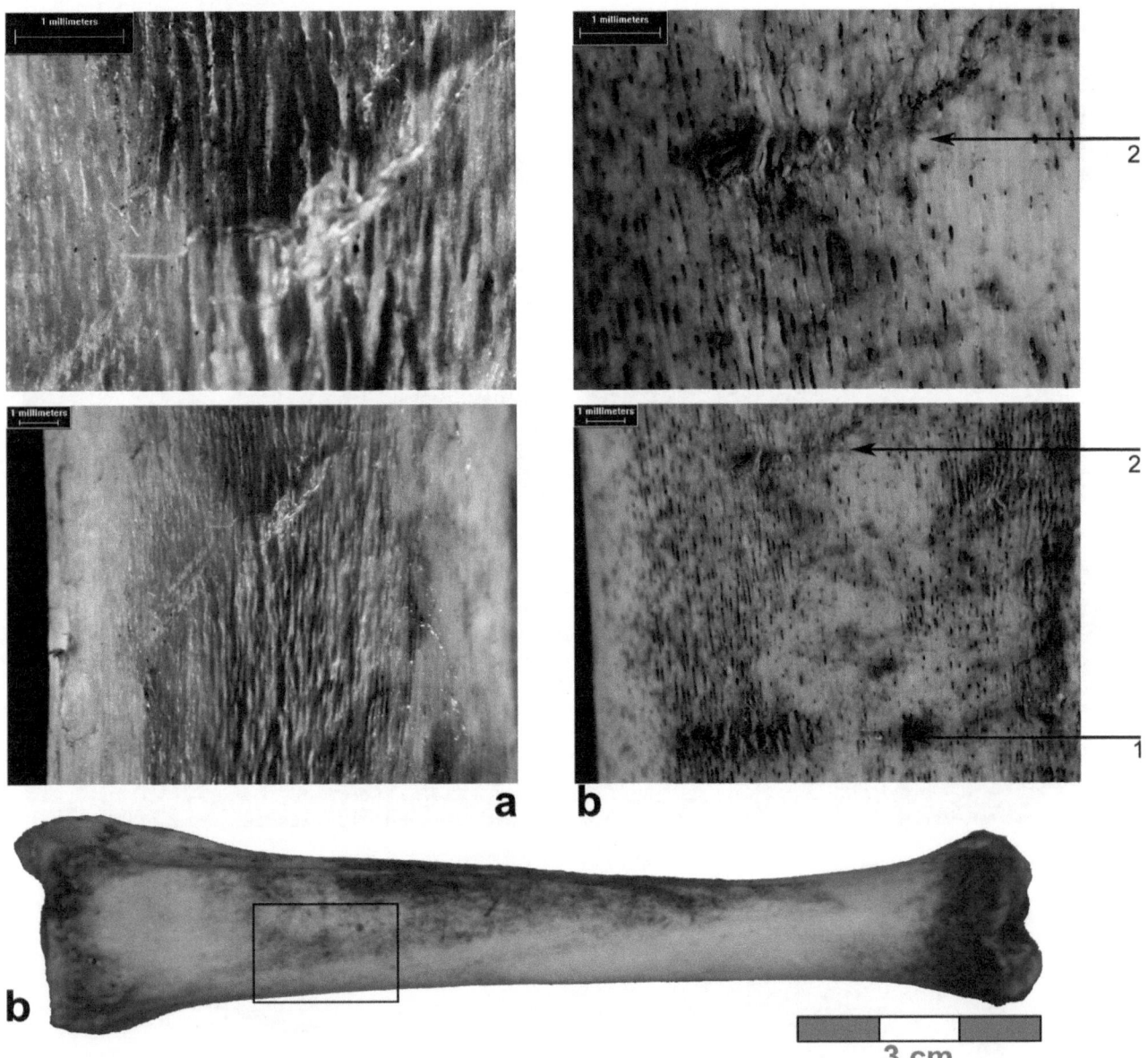

Fig. 6.55 Trampling Experiment 1, sheep metacarpus. The lateral proximal metaphysis **a** after tumbling and before trampling and **b** after trampling. Bone-surface modifications, indicated by *arrows* (1, 2), were created by trampling. Both columns show an identical detail of the bone at different degrees of magnification

Fig. 6.54 Trampling Experiment 1, sheep metacarpus. The anterior diaphysis **a** after tumbling and before trampling and **b** after trampling. (I) Morphological alteration of the foramen and (II) pronunciation of striations were caused by trampling. (I and II) Both columns show an identical detail of the bone at different degrees of magnification

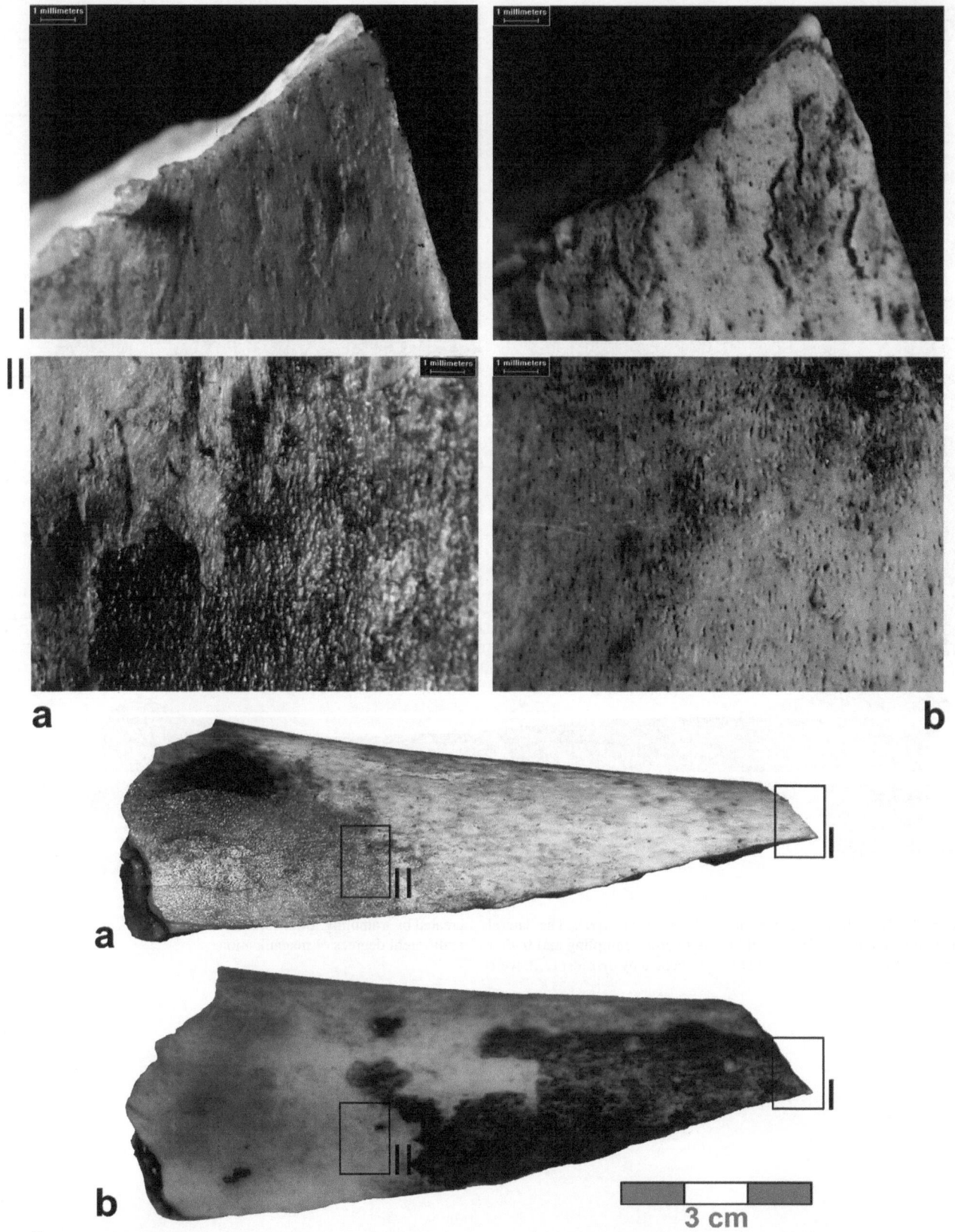

Fig. 6.56 Trampling Experiment 1, domestic cow femur fragment **a** before and **b** after trampling. Rounding of the edges of the bone fragment (I) and exfoliation (I and II) were caused by trampling

diagonal mark was visible on the exfoliated surface, near the broken proximal edge (Fig. 6.57). A diagonal cut mark in the center of the fragment, near the broken medial edge, was obliterated and replaced by diagonal and horizontal striations (Fig. 6.58). The distal part of the fragment displayed the same phenomenon whereby fine vertical striations present before trampling had disappeared and were replaced by diagonal striations (Fig. 6.59). Horizontal marks were produced along the broken lateral edge.

The edges of the cow tibia fragment were rounded. Abrasion and exfoliation had removed large parts of the bone surface so that the porous (vascular) structure of the bone had become visible (Fig. 6.60). Shreds of periosteum remained attached to the fragment's medial edge. A deep, flat-bottomed channel had appeared in this area (Fig. 6.61). No other striations or marks were produced in the course of trampling.

6.4.4.2 Trampling Experiment 2 and 2.1

A domestic cow radius and second rib from the National Natural History Collections, at the Hebrew University, comparative collection were used for the second trampling experiment. Isolated, fine diagonal striations of unknown origin were noted on the anterior face of the radius diaphysis. Slight abrasion and exfoliation had affected the entire bone surface. No preexisting marks were observed on the rib.

This experiment utilized a mixture of 1.6 kg of mollusks (GBY 14/09/97; Layer V-6; 183/130c, 60.11/59.86–60.07/59.80) and 0.4 kg of clay. Bone and sediment were placed in a plastic container, and were saturated with 6 liters of water. A person weighing approximating 50 kg and wearing rubber boots trampled the mixture for two hours.

Following trampling, the radius, especially the proximal epiphyses and the distal metaphysis, was heavily abraded. Protruding parts of the proximal epiphysis were considerably reduced. The bone surface of the distal metaphysis was heavily worn and the porous bone structure had become visible (Fig. 6.62). Abrasion of the diaphysis led to the comprehensive smoothing of the articular zone between the ulna and radius (Fig. 6.63). The anterior face of the radius exhibited several dry fractures, which appeared in an area that had already been exfoliated prior to trampling (Fig. 6.64). The fractures strongly resemble dry cracks caused by climatically induced weathering (according to Behrensmeyer 1978, Stage 1).

In addition to overall abrasion, striations occurred on the bone surface. Very shallow, irregular horizontal striations appeared on the anterior face of the distal metaphysis (Fig. 6.65). The diaphysis displayed striations mainly restricted to areas where the outermost bone surface was missing. It can be assumed that these striations are the result of bone-surface removal processes (Fig. 6.66). Numerous isolated areas of singular or parallel horizontal-to-diagonal striations were scattered across both faces of the diaphysis, accompanied by random traces of oval and round pit marks. Each striation emerges from a round pit and terminates in a striation with a v-shaped cross section. They strongly resemble carnivore teeth marks (Fig. 6.67).

Trampling seriously damaged the rib. The distal portion was completely destroyed, as was the tuberculum of the proximal portion. Dry longitudinal fractures were observed on the dorsal face (Fig. 6.68). The entire bone surface was marked by net-like sets of differently sized v-shaped and overlapping pit marks. The individual boundaries of the pit marks were not easily discernable (Fig. 6.69). When they appeared singularly, they shared the same characteristics as the v-shaped cross-sectioned striations seen on the radius.

Since the modifications visible on the radius and rib were produced in the same experiment, we conclude that their differing frequencies are not related to trampling duration. Form, shape and structure were obviously major factors in a bone's susceptibility to surface modification. The ventral face of the rib exhibited thin v-shaped cross-sectioned striations overlapped by large pit marks (Fig. 6.70). The fact that different generations of striations and pitting are observed on a single bone strengthens our hypothesis that the duration of trampling does not necessarily correlate with the intensity of bone-surface modification.

The domestic cow radius used in Trampling Experiment 2 was used again for a follow-up experiment, Trampling Experiment 2.1.

The sediment mixture comprised 4.5 kg of mollusks (GBY 15/09/97; Layer V-6; 183/129b, 60.02/59.96–59.80/59.75) and 2.35 kg of clay. Bone and the sediment were placed in a plastic container filled with two liters of water, which would mimic muddy conditions. A person weighing approximately 50 kg and wearing rubber boots trampled on the mixture for two hours.

Following trampling, the abrasion noted from Experiment 2 had increased considerably. Several dry fractures from the previous experiment were observed on the anterior face of the radius, an area of the bone that had already been exfoliated prior to the first trampling. These fractures were not affected by the additional trampling, but new, fine dry fractures were noted (Fig. 6.71).

The most significant effect of prolonged trampling was the increase in the amount of striations. The striations attributed to bone-surface removal in Experiment 2 had disappeared and were replaced by unidirectional striations. They are not as numerous as the "peeling striations," but altered the appearance of the bone surface in a much more serious way (Fig. 6.72).

The abrasion caused by Experiment 2.1 intensified the damage already caused to the bone surface, which had

Fig. 6.57 Trampling Experiment 1, domestic cow femur fragment **a** before and **b** after trampling. Rounding of the edges and a diagonal mark on the exfoliated bone surface, indicated by an *arrow*, were caused by trampling. Both columns show an identical detail of the bone at different degrees of magnification

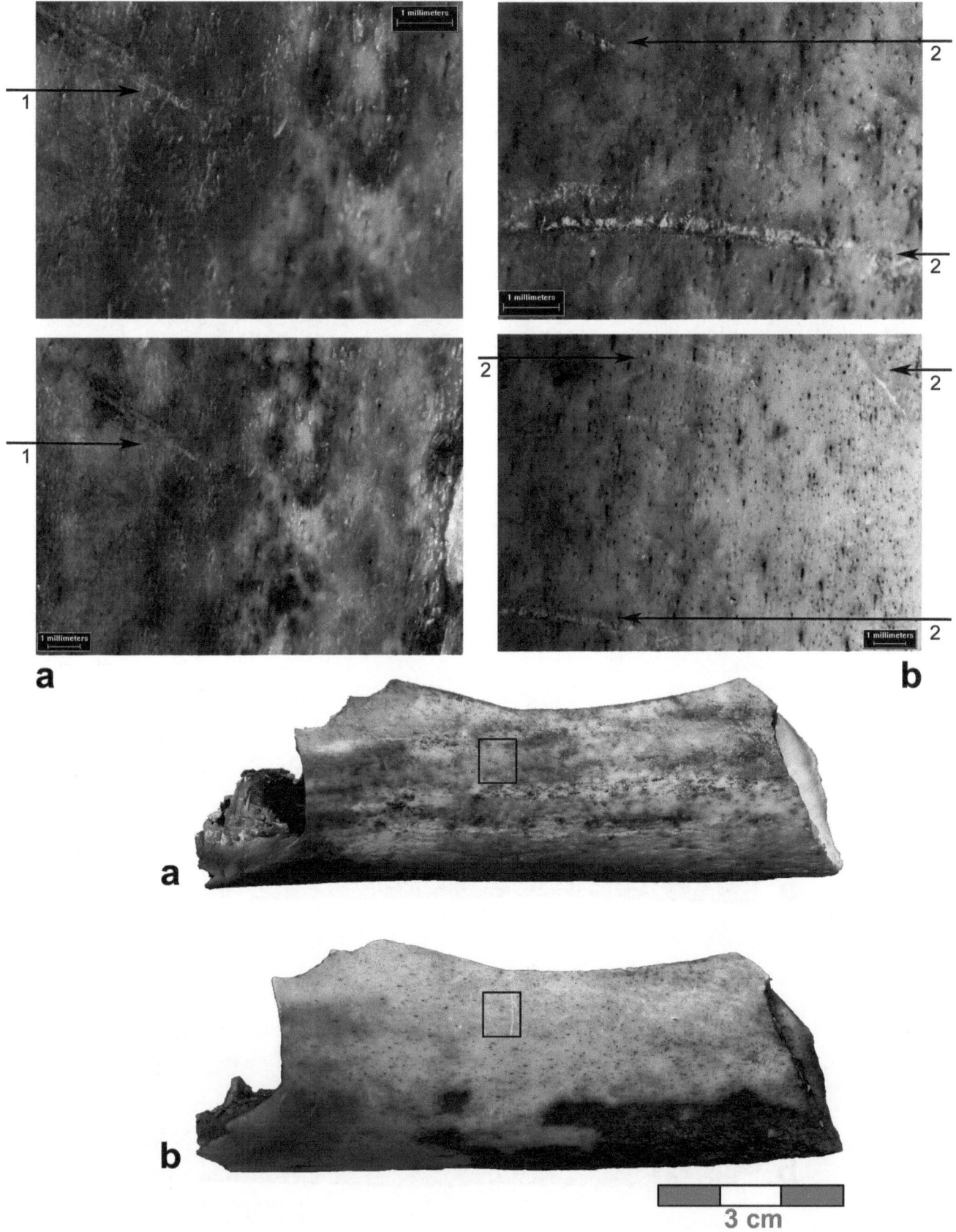

Fig. 6.58 Trampling Experiment 1, domestic cow femur fragment **a** before and **b** after trampling. After trampling, the cut mark (1) was obliterated by striations (2). Both columns show an identical detail of the bone at different degrees of magnification

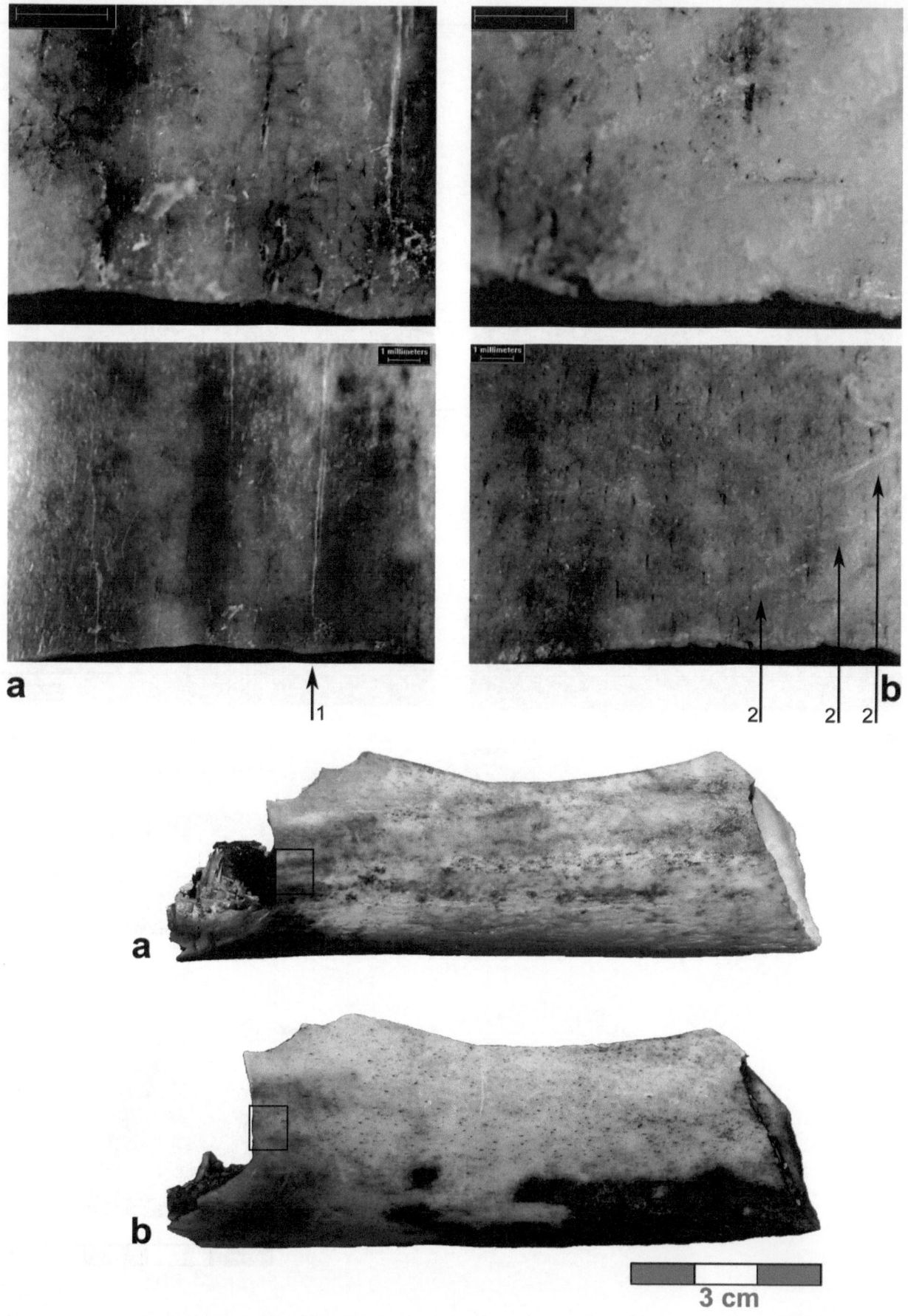

Fig. 6.59 Trampling Experiment 1, domestic cow femur fragment **a** before and **b** after trampling. After trampling, the cut mark (1) was obliterated by striations (2). Both columns show an identical detail of the bone at different degrees of magnification

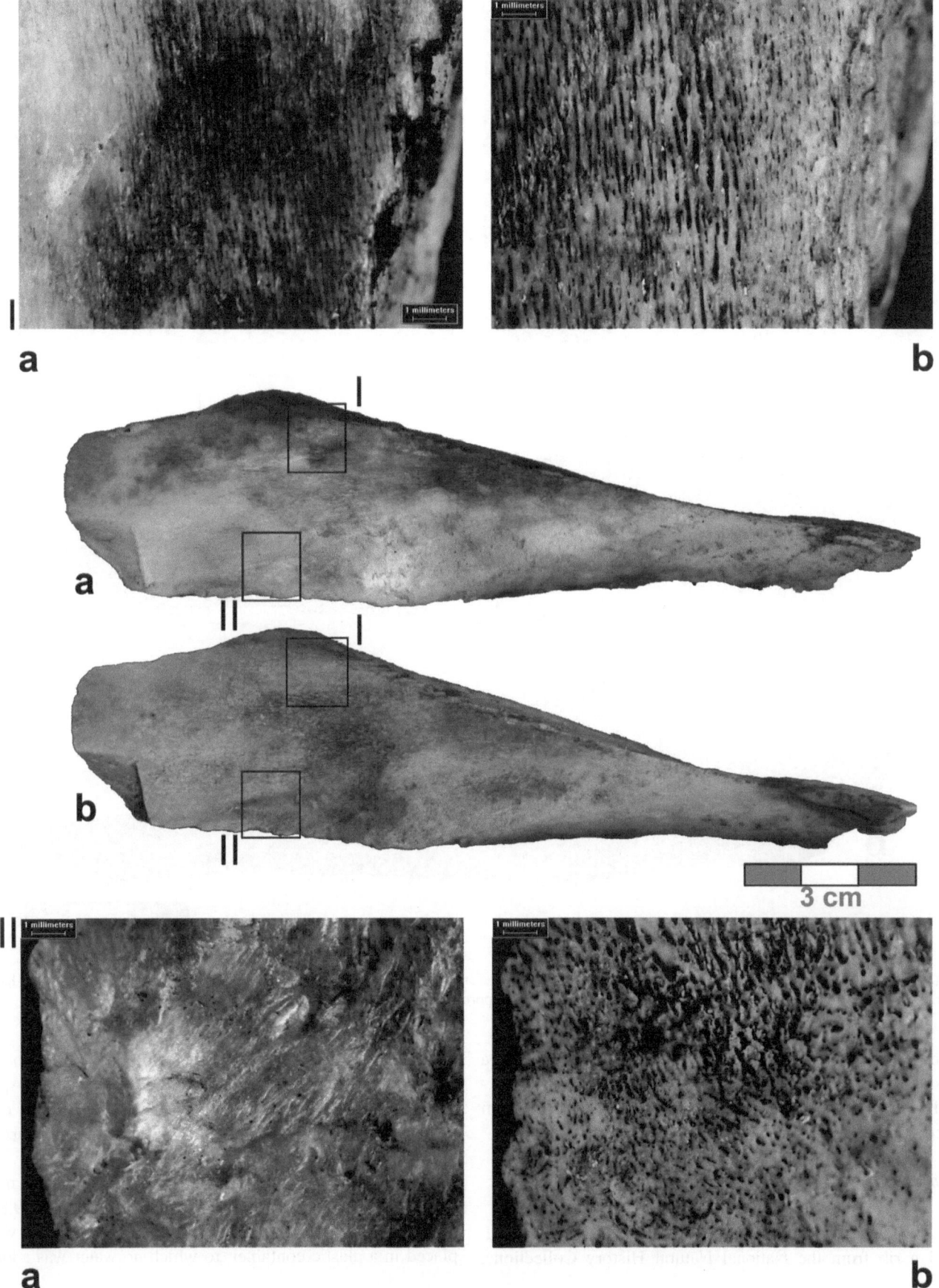

Fig. 6.60 Trampling Experiment 1, domestic cow tibia fragment **a** before and **b** after trampling. (I and II) Removal of the periosteum was caused by trampling. The porous structure of the bone became visible after trampling

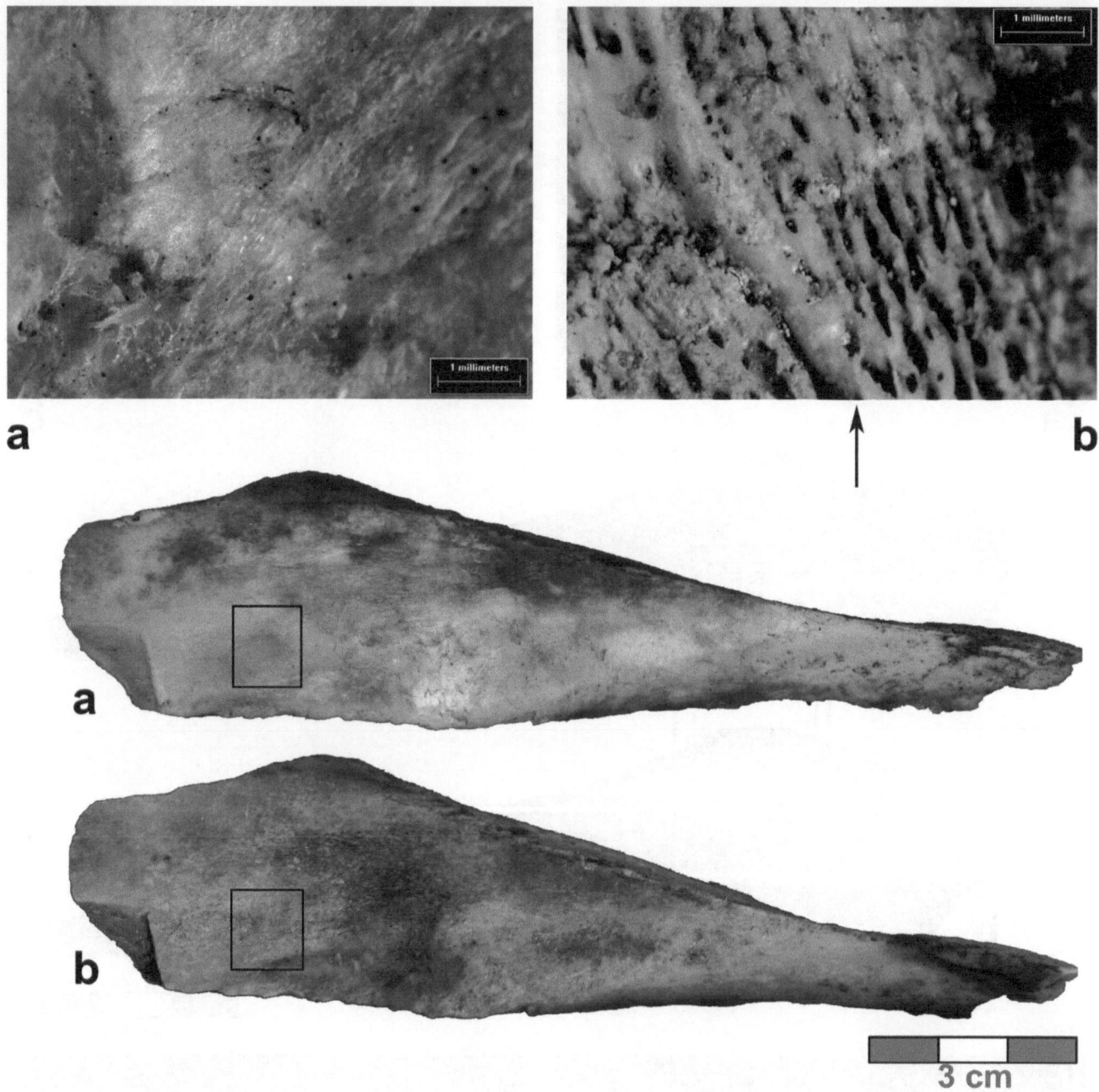

Fig. 6.61 Trampling Experiment 1, domestic cow tibia fragment **a** before and **b** after trampling. Removal of periosteum during trampling led to the formation of a flat-bottomed channel, indicated by an *arrow*. The porous bone structure became visible after trampling

become much more vulnerable to striation agents. As a result, portions of the bone surface exhibited a considerably higher number of striations than those resulting from Experiment 2.

6.4.4.3 Trampling Experiment 3

A *Hyaena hyaena* tibia, ulna, pelvis, metatarsus (MT) III, and a rib from the National Natural History Collections,

at the Hebrew University of Jerusalem comparative collection were used for the third trampling experiment. Apart from slight weathering traces and exfoliation, especially near the long bone epiphyses and the pelvis rims, no surface modifications were observed.

The experiment utilized a mixture of 3.65 kg of mollusks (GBY 14/09/97; Layer V-6; 183/130c, 60.11/59.86–60.07/59.80) and 2.15 kg of clay. Bones and sediment were placed in a plastic container, to which no water was added.

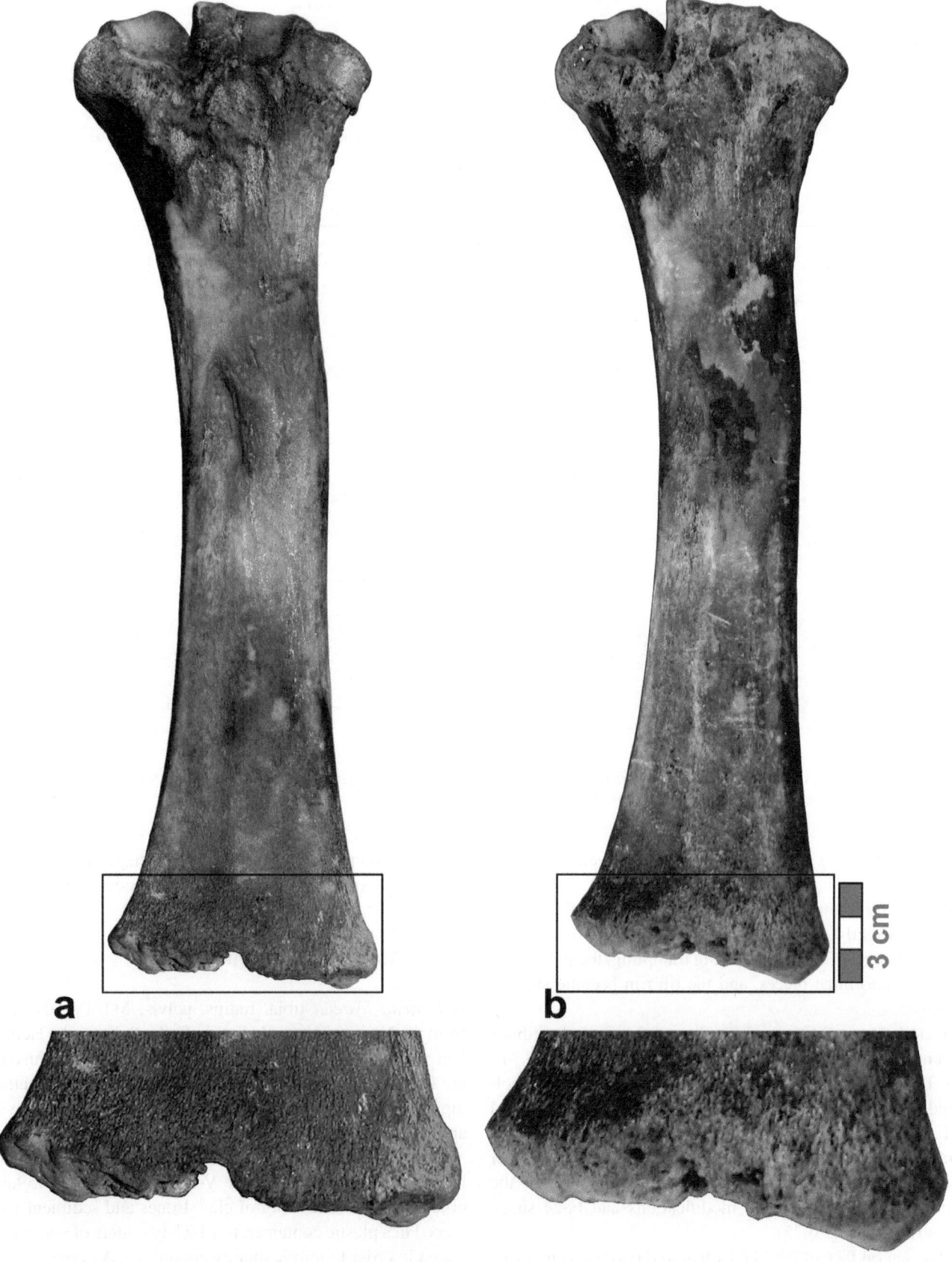

Fig. 6.62 Trampling Experiment 2, domestic cow radius. The posterior face of the distal metaphysis **a** before and **b** after trampling. Morphological changes of the bone due to abrasion were created by trampling

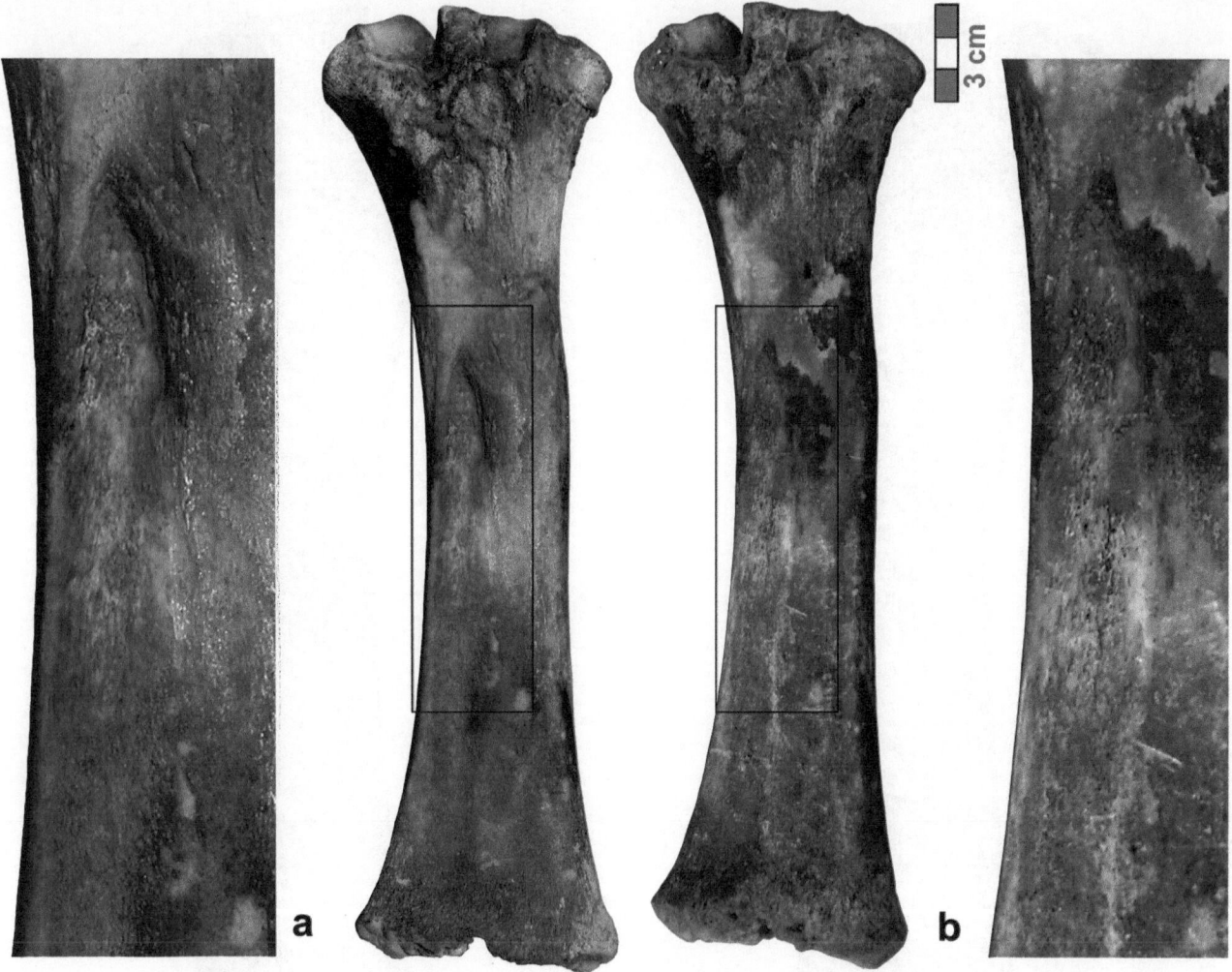

Fig. 6.63 Trampling Experiment 2, domestic cow radius. The posterior face of the bone **a** before and **b** after trampling. Smoothing of the articular zone between the ulna and the radius was caused by trampling

A person weighing approximately 50 kg and wearing rubber boots trampled the mixture for an hour.

Following trampling, the tibia, ulna, pelvis, MT III, and rib were slightly abraded, resulting in only minor morphological changes (Fig. 6.73). During trampling, the pelvis was fractured into three pieces, and the rib rim became slightly damaged.

Detailed examination of the morphology of the tibia's foramen nutricium revealed only minor abrasion (Fig. 6.74). All bone surfaces appeared to have been slightly polished. Exfoliation was observed only slightly. The tibia and MT III display random and localized areas with shallow round pit marks of insignificant depth, which reinforces our above-mentioned hypothesis of a correlation between the appearance of bone-surface modifications and bone shape and structure (Fig. 6.75).

The lateral face of the MT III featured two horizontal striations on the diaphysis. They were accompanied by random oval and round pit marks (Fig. 6.76). Each striation emerges from a round pit and terminates in a v-shaped cross-section. Again, they strongly resemble carnivore tooth marks.

6.4.4.4 Trampling Experiment 4

A *Hyaena hyaena* tibia, radius, pelvis, MT IV, and a rib from the National Natural History Collections, at the Hebrew University of Jerusalem comparative collection were used in the fourth trampling experiment. Apart from slight weathering and exfoliation, especially near the long bone epiphyses and the pelvis rims, no surface modifications were observed.

This experiment utilized a mixture of 3.65 kg of mollusks (GBY 14/09/97; Layer V-6; 183/130c, 60.11/59.86–60.07/59.80) and 2.15 kg of clay. Bones and sediment were placed in a plastic container, to which two liters of water were added in order to mimic muddy conditions. A person weighing approximately 50 kg and wearing rubber boots trampled on the mixture for two hours.

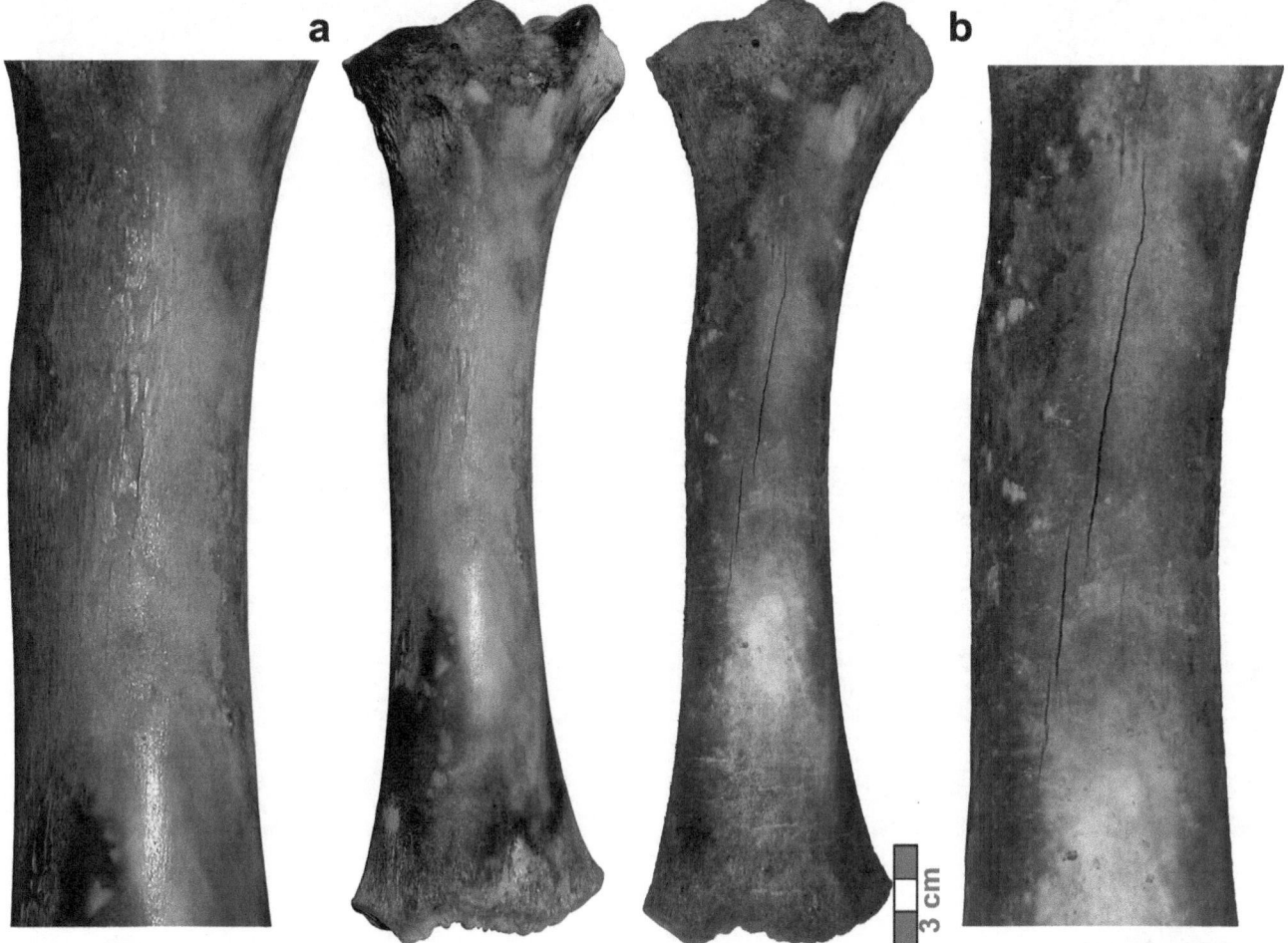

Fig. 6.64 Trampling Experiment 2, domestic cow radius. The anterior face of the bone **a** before and **b** after trampling. Dry fractures were created by trampling

Following trampling, heavy abrasion had reduced the tibia and radius to rod-like cylinders. The proximal and distal metaphyses and the crista of the tibia diaphysis were heavily abraded, yet the overall bone surfaces were only slightly abraded and not polished. In addition, several dry fracture lines appeared on the tibia and radius diaphyses. In contrast to the tibia and radius, the MT IV remained unaffected by abrasion (Fig. 6.77).

During trampling, the pelvis was fractured into three pieces. Heavy abrasion was observed, especially along the pelvis rims and the acetabulum, resulting in significant morphological changes. The bone surface was also heavily exfoliated. Polish was not observed (Fig. 6.78).

The rib underwent the most extreme morphological changes. Its proximal and distal ends were completely destroyed. The bone was compressed, straightened, and dissolved into a fibrous structure, as in the case of damage resulting from severe climatically induced weathering.

The anterior and posterior faces of the tibia and radius, all faces of the metatarsus, and the rare intact sections of the rib exhibited scattered horizontal striations. They were accompanied by random traces of oval and round pit marks (Fig. 6.79). These striations emerge from a round pit mark and terminate in a v-shaped cross-section. As mentioned above, they strongly resembled carnivore tooth marks. Very few such traces were noted on the pelvis.

6.4.4.5 Trampling Experiment 5

A *Hyaena hyaena* proximal ulna, MT II, and two ribs from the National Natural History Collections, at the Hebrew University of Jerusalem comparative collection were used for the fifth trampling experiment. Slight weathering and exfoliation, especially near the long bone epiphyses, as well as tiny shallow scratches were observed prior to the experiment.

This experiment utilized a mixture of 2.5 kg of mollusks (GBY 14/09/97; Layer V-6; 184/130, 60.39/60.10–60.19/60.01; 182/129d, 59.91/59.79–59.85/59.72; 182/130c, 60.12/59.85–60.02/59.86) and 2 kg of clay. Bones and

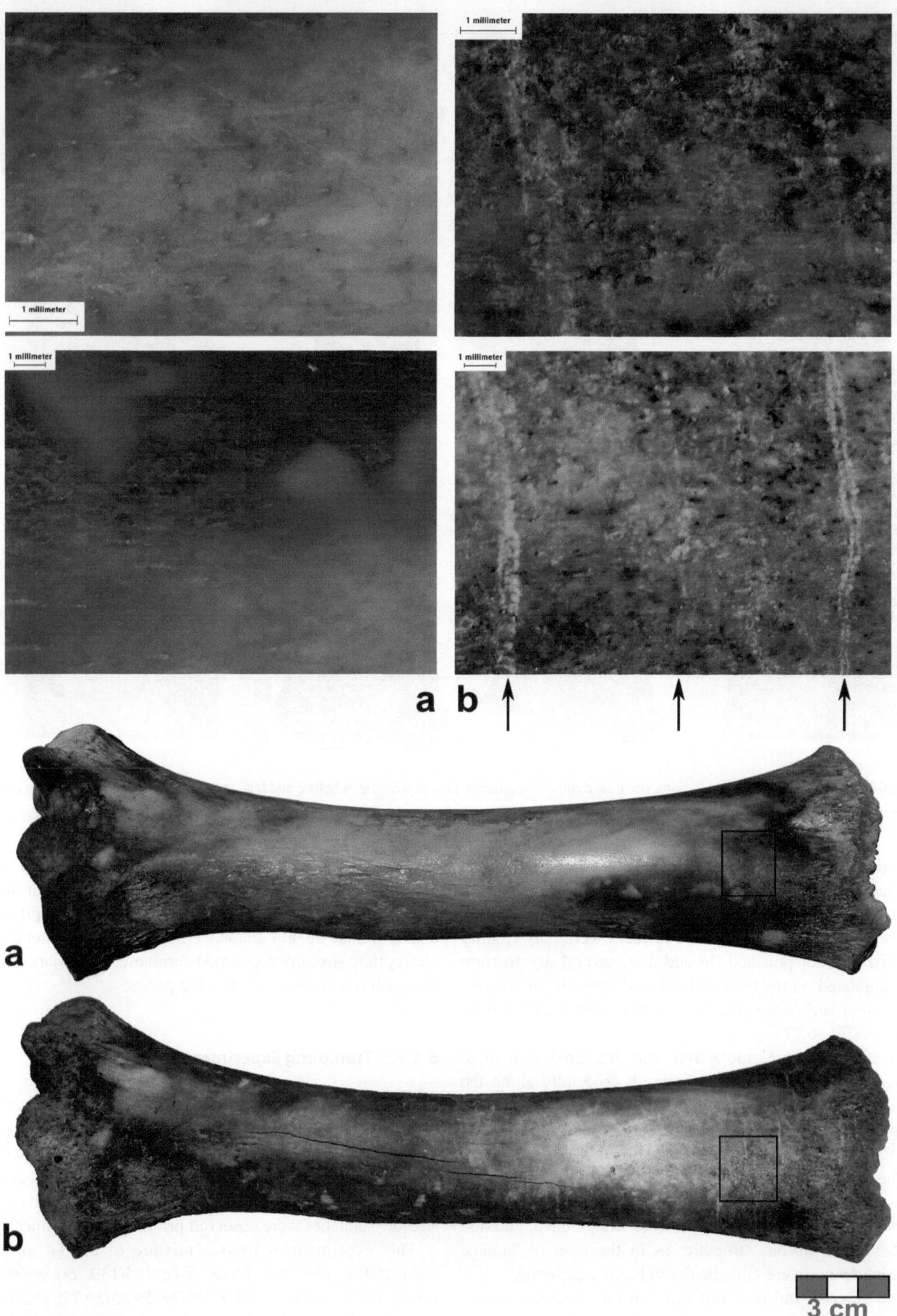

Fig. 6.65 Trampling Experiment 2, domestic cow radius. The anterior face of the distal metaphysis **a** before and **b** after trampling. Striations, indicated by *arrows*, were created by trampling. Both columns show an identical detail of the bone at different degrees of magnification

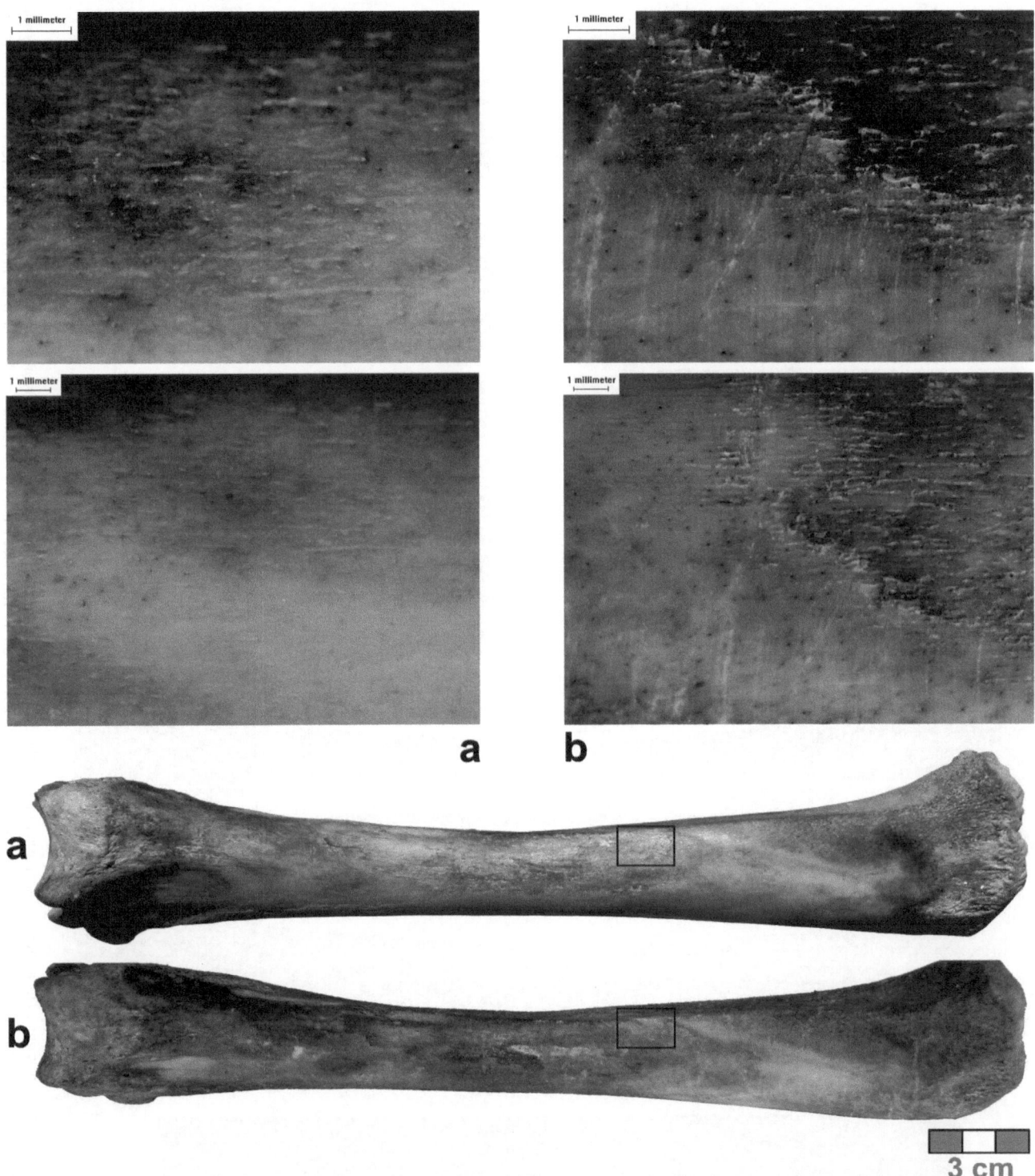

Fig. 6.66 Trampling Experiment 2, domestic cow radius. The lateral diaphysis of the bone **a** before and **b** after trampling. Exfoliation of the bone surface was caused by trampling. Both columns show an identical detail of the bone at different degrees of magnification

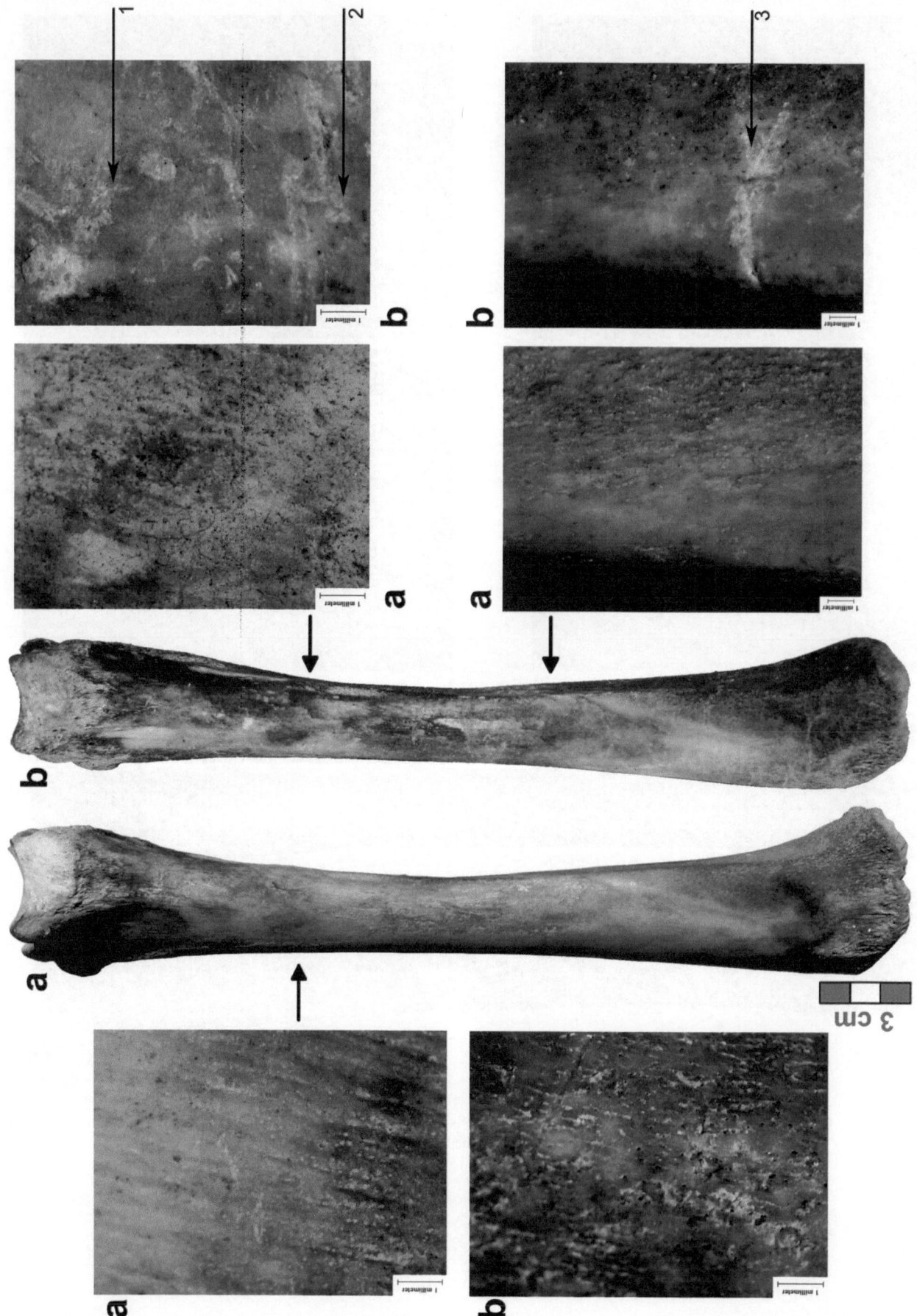

Fig. 6.67 Trampling Experiment 2, domestic cow radius. The posterior and anterior faces of the diaphysis **a** before and **b** after trampling. The *arrows* indicate the areas of the bone that were magnified. After trampling, striations (1–3) were seen to emerge from a round pit with a v-shaped cross-section. Rows on the right side show an identical detail of the bone at different degrees of magnification

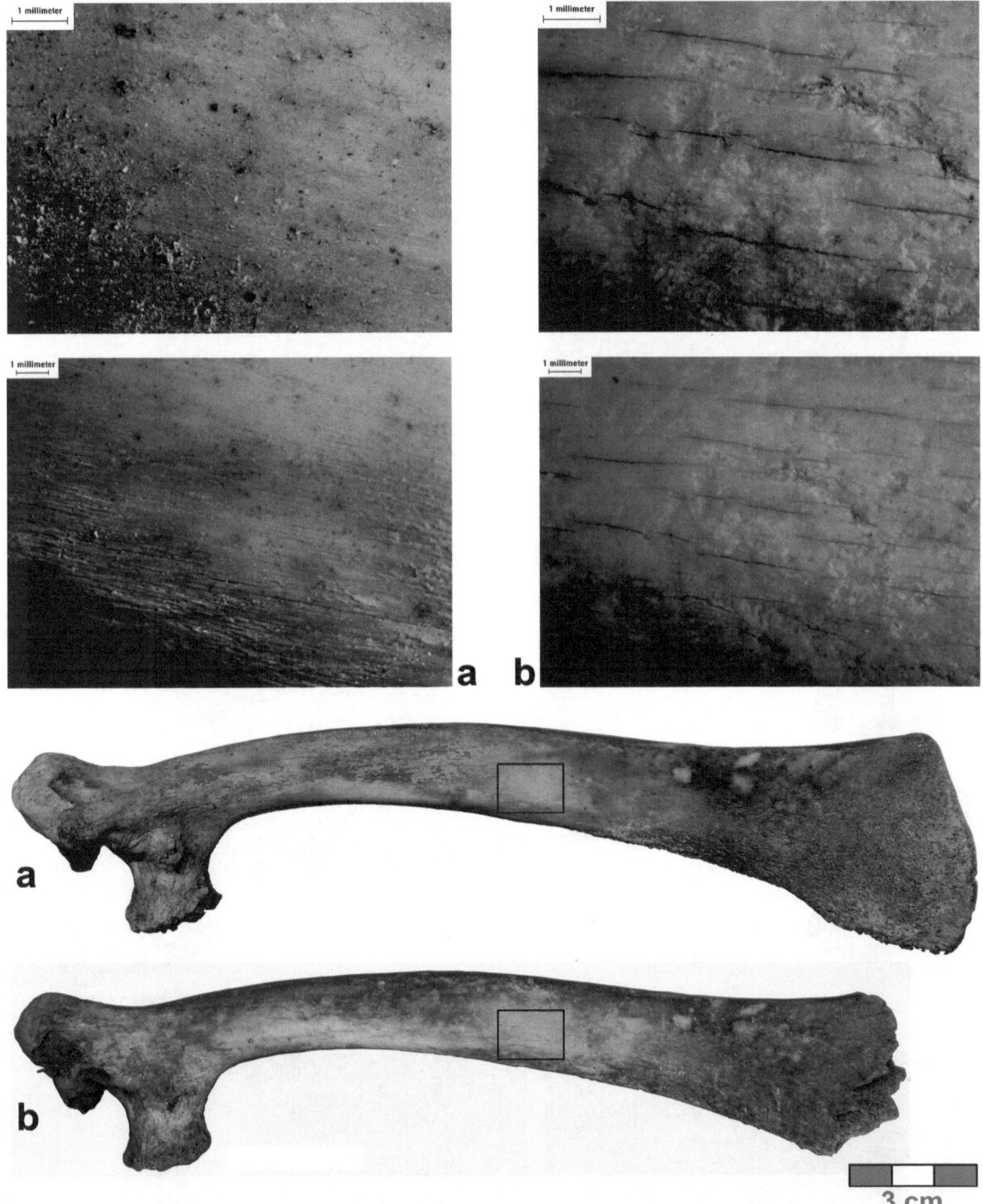

Fig. 6.68 Trampling Experiment 2, domestic cow rib. The dorsal face **a** before and **b** after trampling. Dry fractures were created by trampling. Both columns show an identical detail of the bone at different degrees of magnification

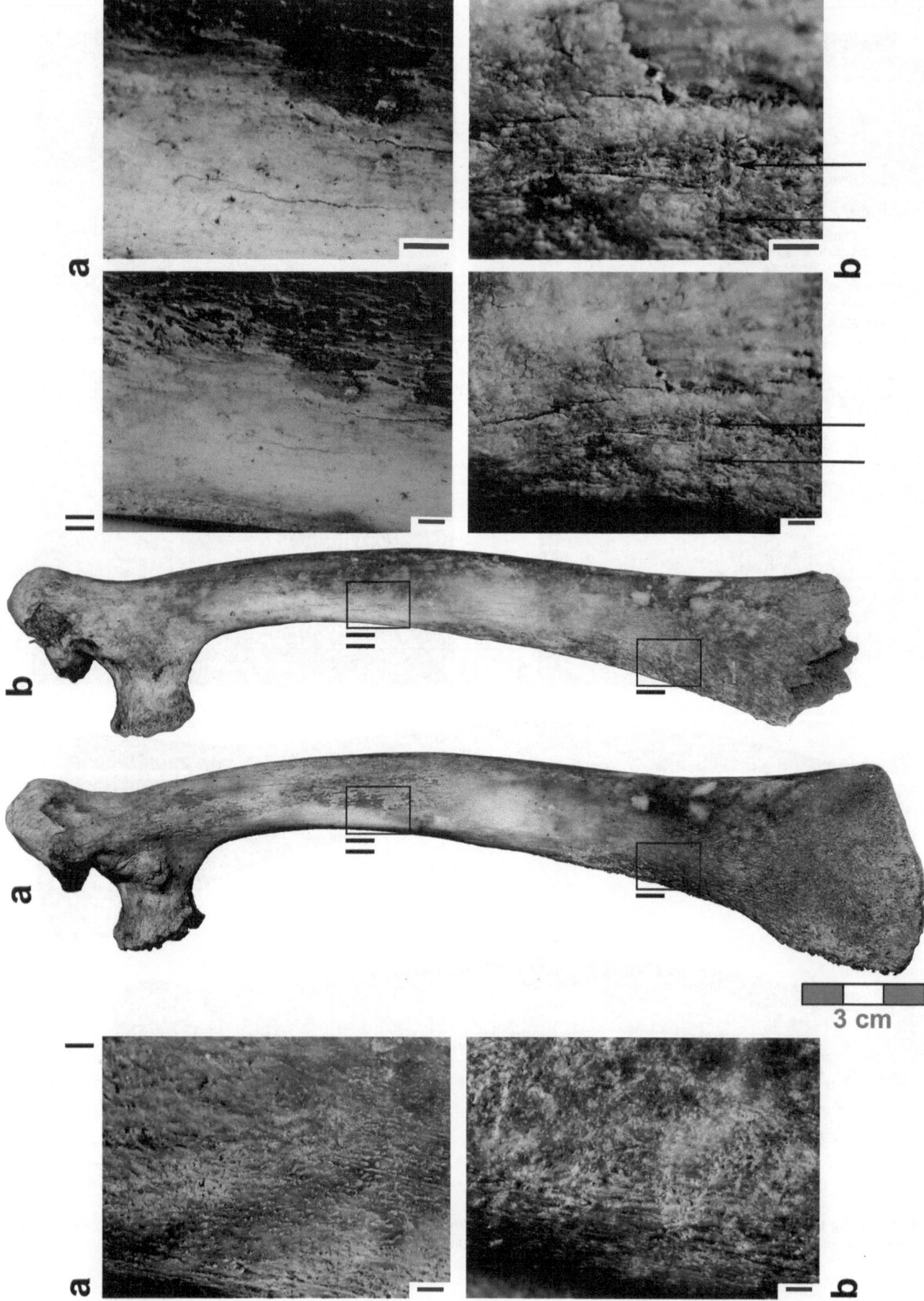

Fig. 6.69 Trampling Experiment 2, domestic cow rib. The dorsal face **a** before and **b** after trampling. V-shaped pit marks, indicated by *arrows*, were created by trampling. (II) Both columns show an identical detail of the bone at different degrees of magnification. The scale bar of the microscope images is 1 mm

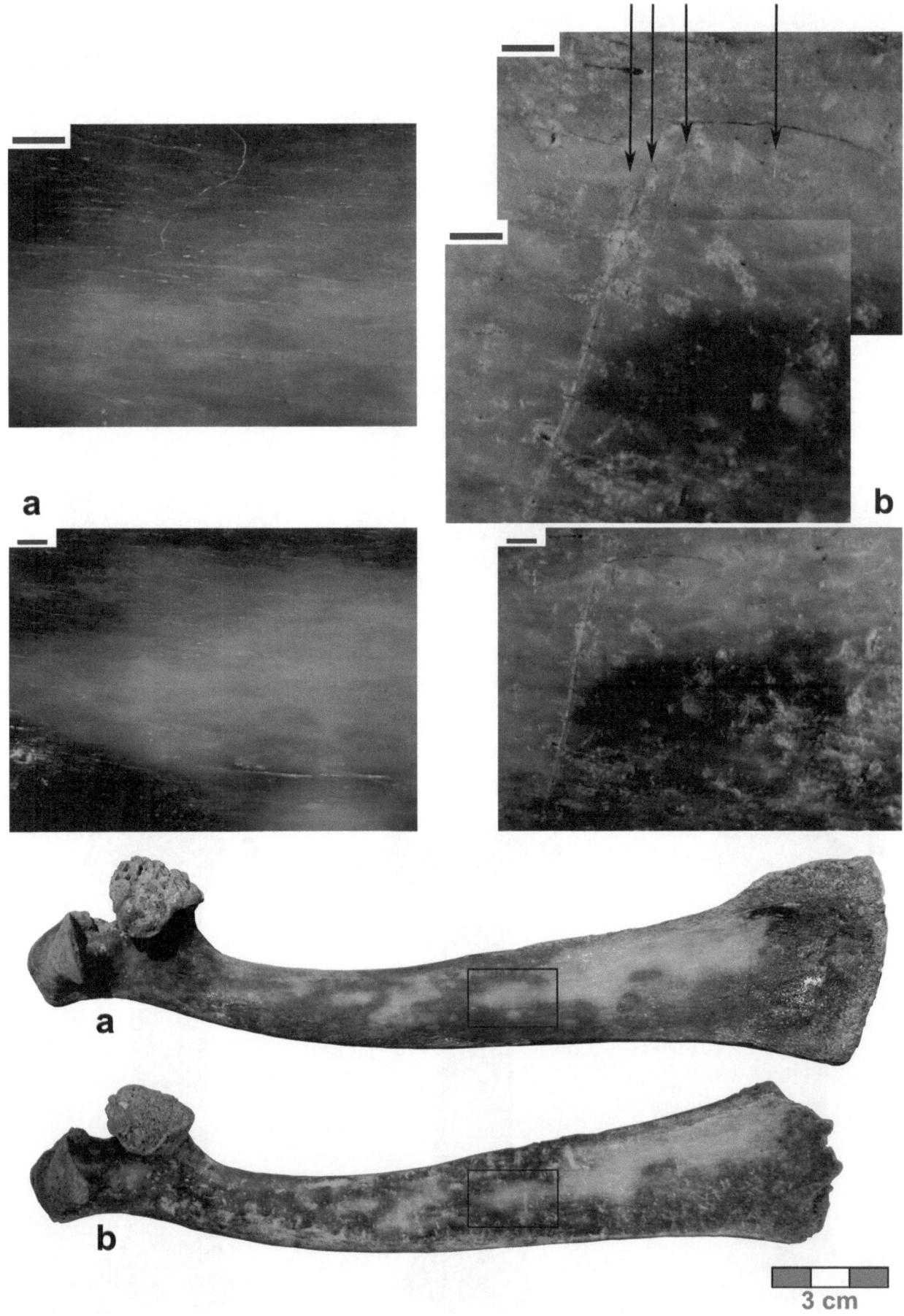

Fig. 6.70 Trampling Experiment 2, domestic cow rib. The ventral face **a** before and **b** after trampling. V-shaped, cross-sectioned striations, indicated by *arrows*, superimposed by large pit marks, were created by trampling. Both columns show an identical detail of the bone at different degrees of magnification. The scale bar of the microscope images is 1 mm

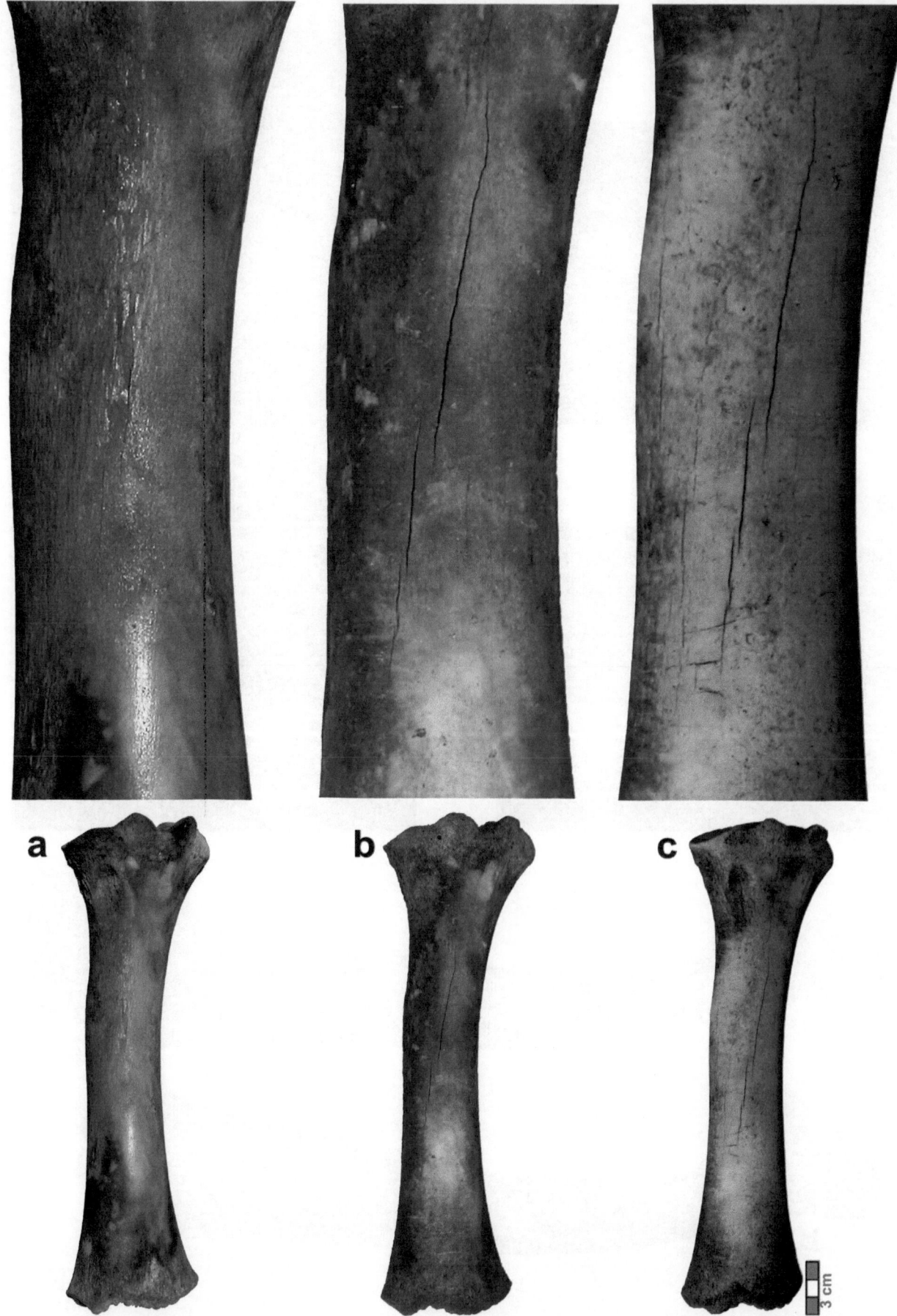

Fig. 6.71 Trampling Experiment 2.1, domestic cow radius. The anterior face **a** before trampling, **b** after two hours of trampling, and **c** after four hours of trampling. Dry fractures were created by trampling

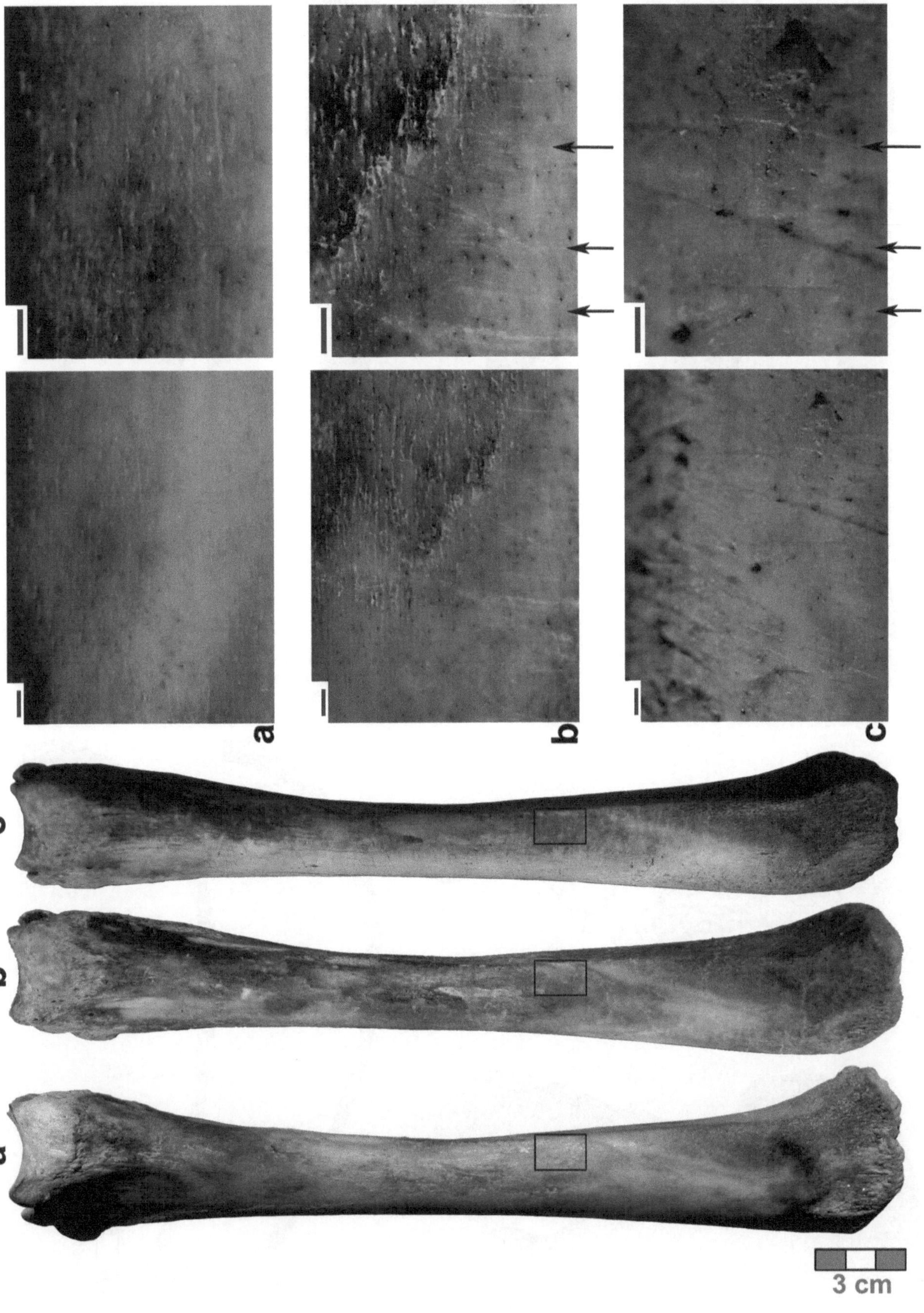

Fig. 6.72 Trampling Experiment 2.1, domestic cow radius. The lateral face of the bone **a** before trampling, **b** after two hours of trampling, and **c** after four hours of trampling. Superimposed striations, indicated by *arrows*, were created by prolonged trampling. The scale bar of the microscope images is 1 mm

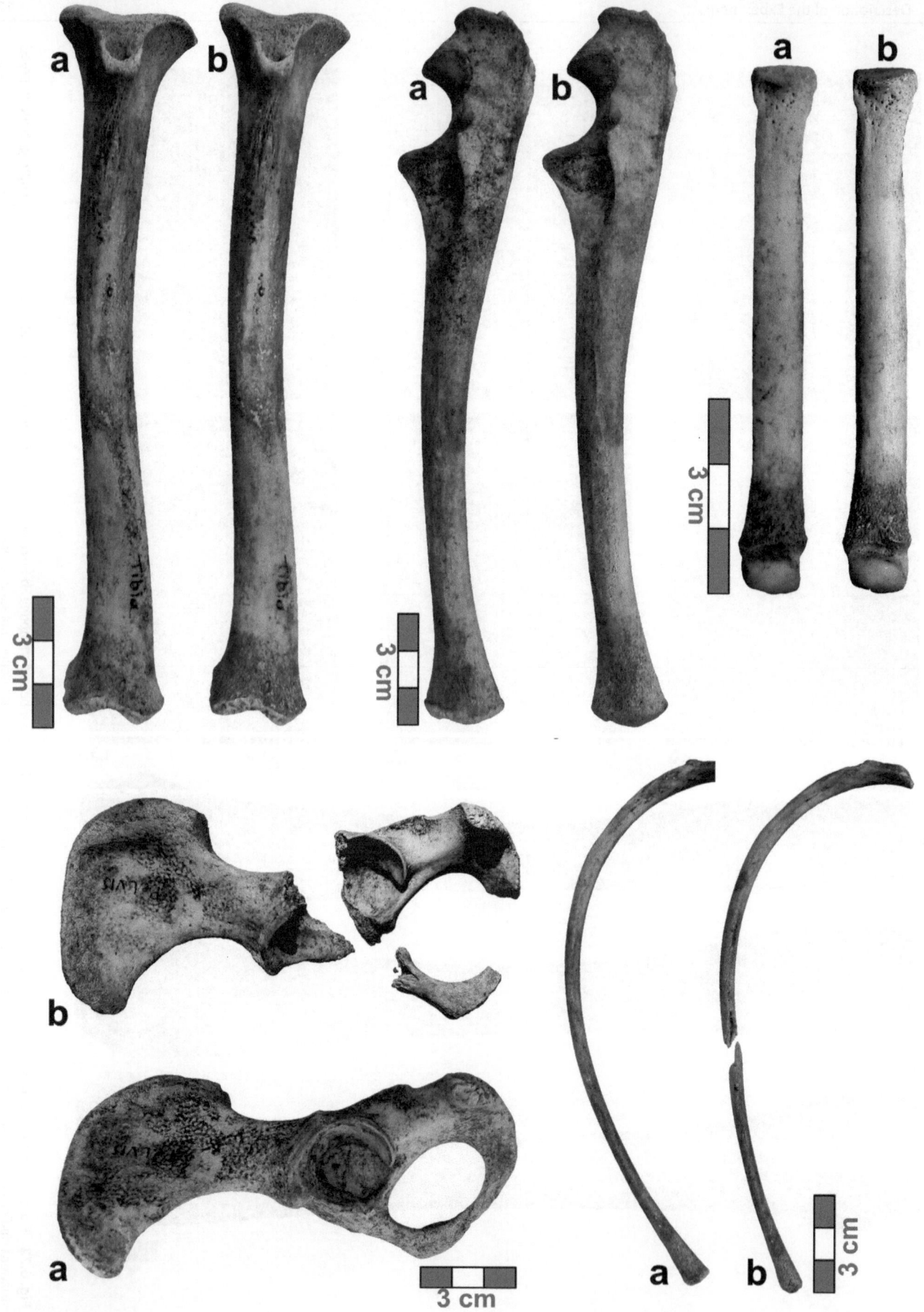

Fig. 6.73 Trampling Experiment 3. hyena tibia, ulna, pelvis, MT III, and rib **a** before and **b** after trampling. Only slight morphological changes were created by trampling

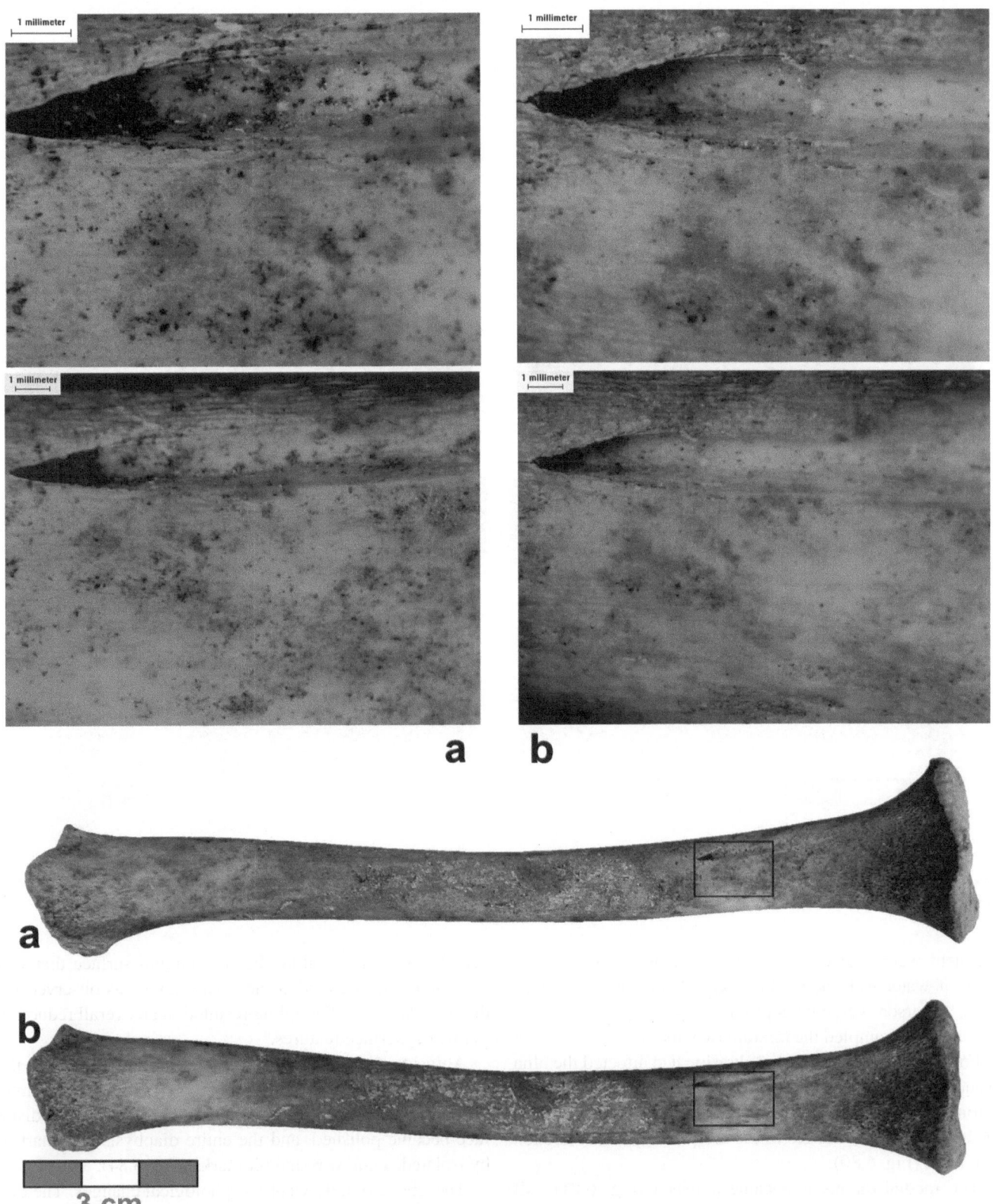

Fig. 6.74 Trampling Experiment 3, hyena tibia. The foramen nutricium **a** before and **b** after trampling. Only minor abrasions were observed after trampling. Both columns show an identical detail of the bone at different degrees of magnification

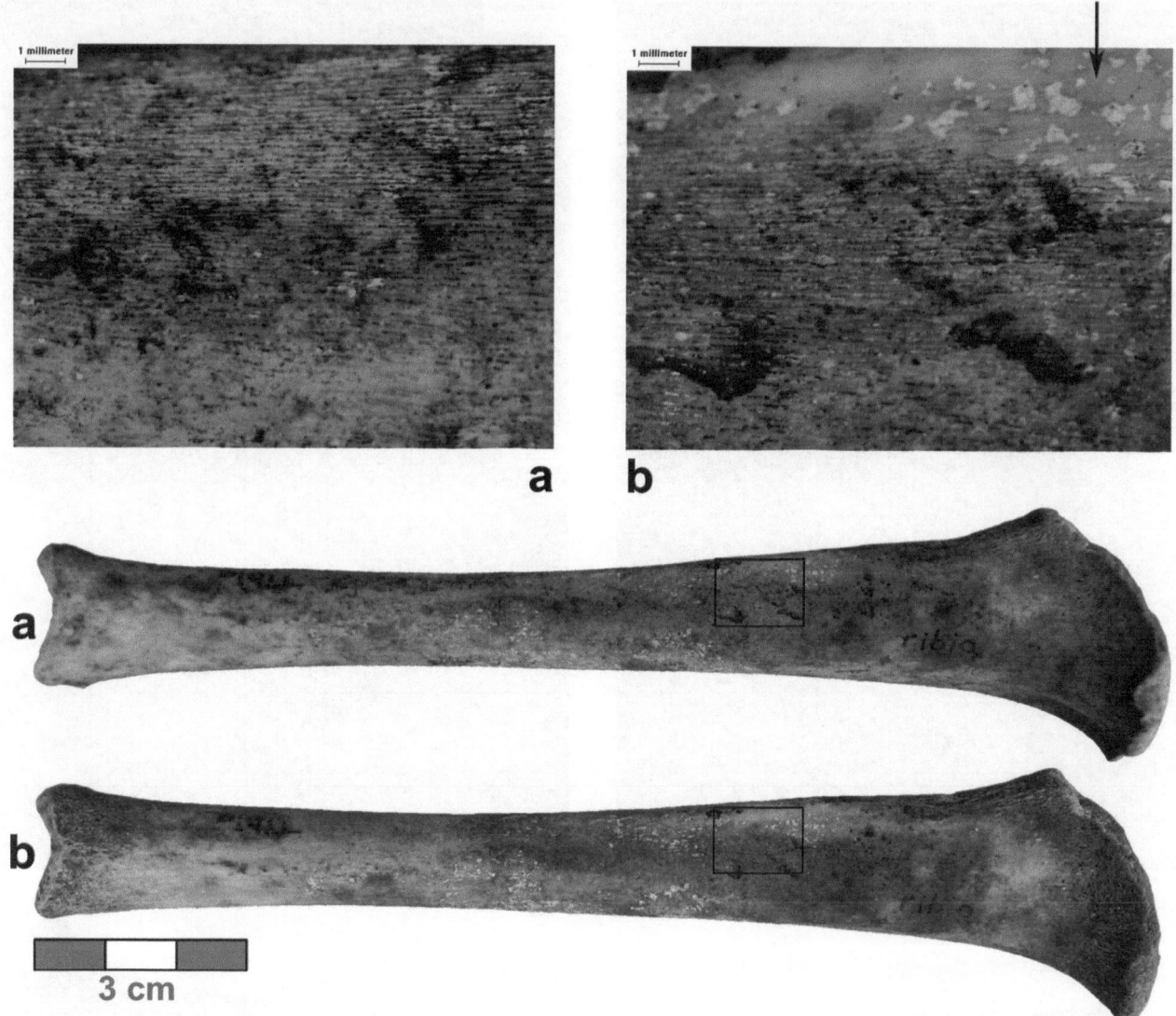

Fig. 6.75 Trampling Experiment 3, hyena tibia. The lateral face of the diaphysis **a** before and **b** after trampling. Shallow pit marks, indicated by an *arrow*, were created by trampling

sediment were placed in a plastic container, to which five liters of water were added in order to simulate wet conditions. A person weighing approximately 50 kg and wearing rubber boots trampled the mixture two hours.

Following trampling, heavy abrasion had affected the ulna in such a way that protruding sections of the proximal articulation almost vanished from the lateral and medial bone face. We also observed changes in the morphology of the broken distal end (Fig. 6.80).

The medial diaphysis became polished (Fig. 6.81). All traces present before trampling, such as slight exfoliation and scratches, were obliterated. The bone surface structure of the cranio/lateral face had not been homogeneously preserved, especially when comparing the area of articulation with the radius with the rest of the bone (Fig. 6.82). This area was

heavily exfoliated, unlike the surrounding surface that was comparable to the medial face. The same was observed for the posterior face. Trampling resulted in an overall reduction in distinct surface features.

Abrasion did not significantly affect the MT II, but the bone's prominent features became more pronounced, as may be seen on the lateral face (Fig. 6.83). The anterior distal face became polished, and the entire diaphysis was marked by isolated, shallow round pit marks (Fig. 6.84).

The ribs also underwent morphological changes. The distal ends were destroyed, and the bones were compressed and partially dissolved into a fibrous consistency, as in the case of bone damage resulting from severe climatically induced weathering (Fig. 6.85). The lateral and medial faces also displayed differences in bone-surface preservation. The lateral

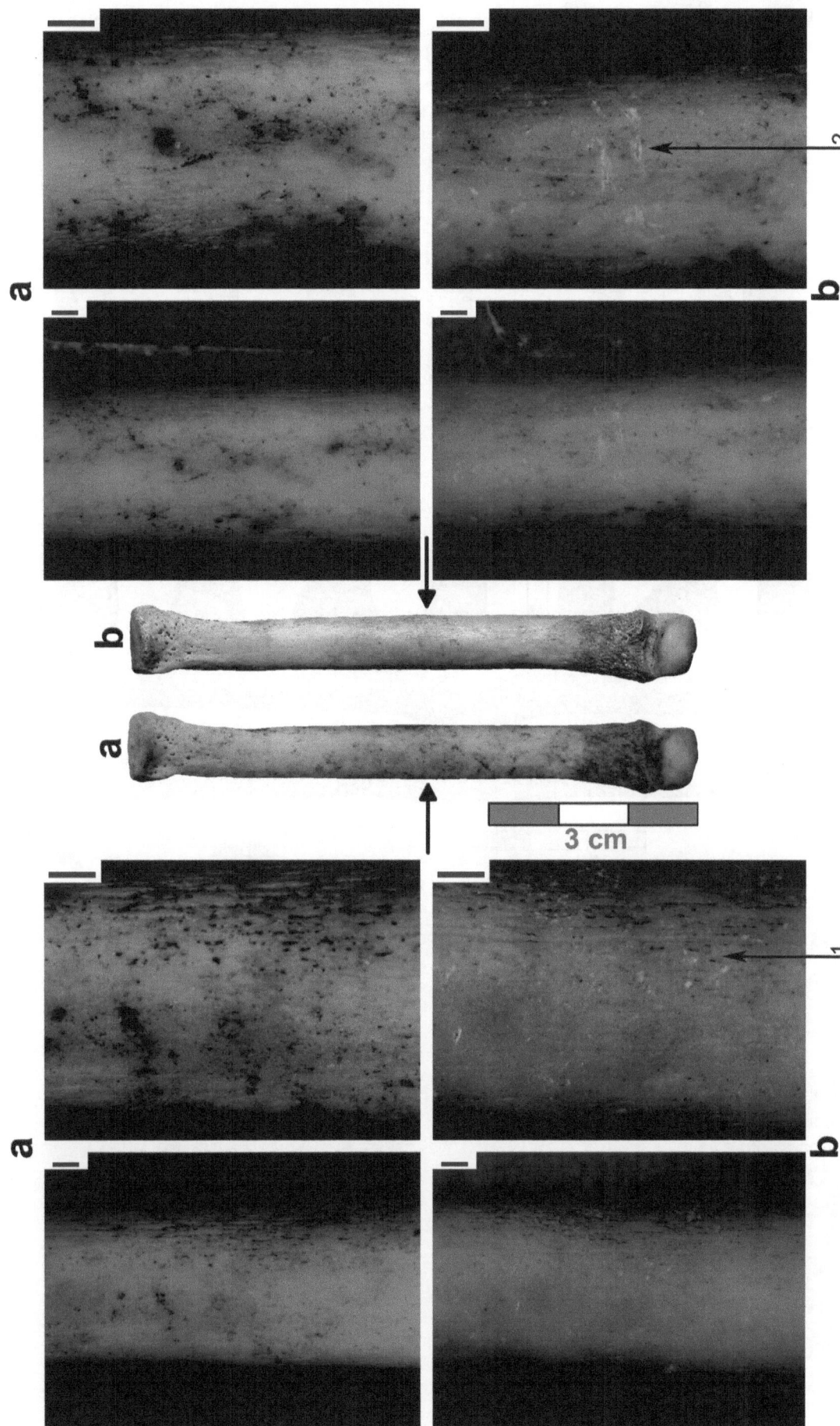

Fig. 6.76 Trampling Experiment 3, hyena MT III. The lateral and medial face of the diaphysis **a** before and **b** after trampling. Striations and pit marks (1, 2) were created by trampling. *Arrows* indicate the area of the bone that was magnified. Rows show identical details of the bone at different degrees of magnification. The scale bar of the microscope images is 1 mm

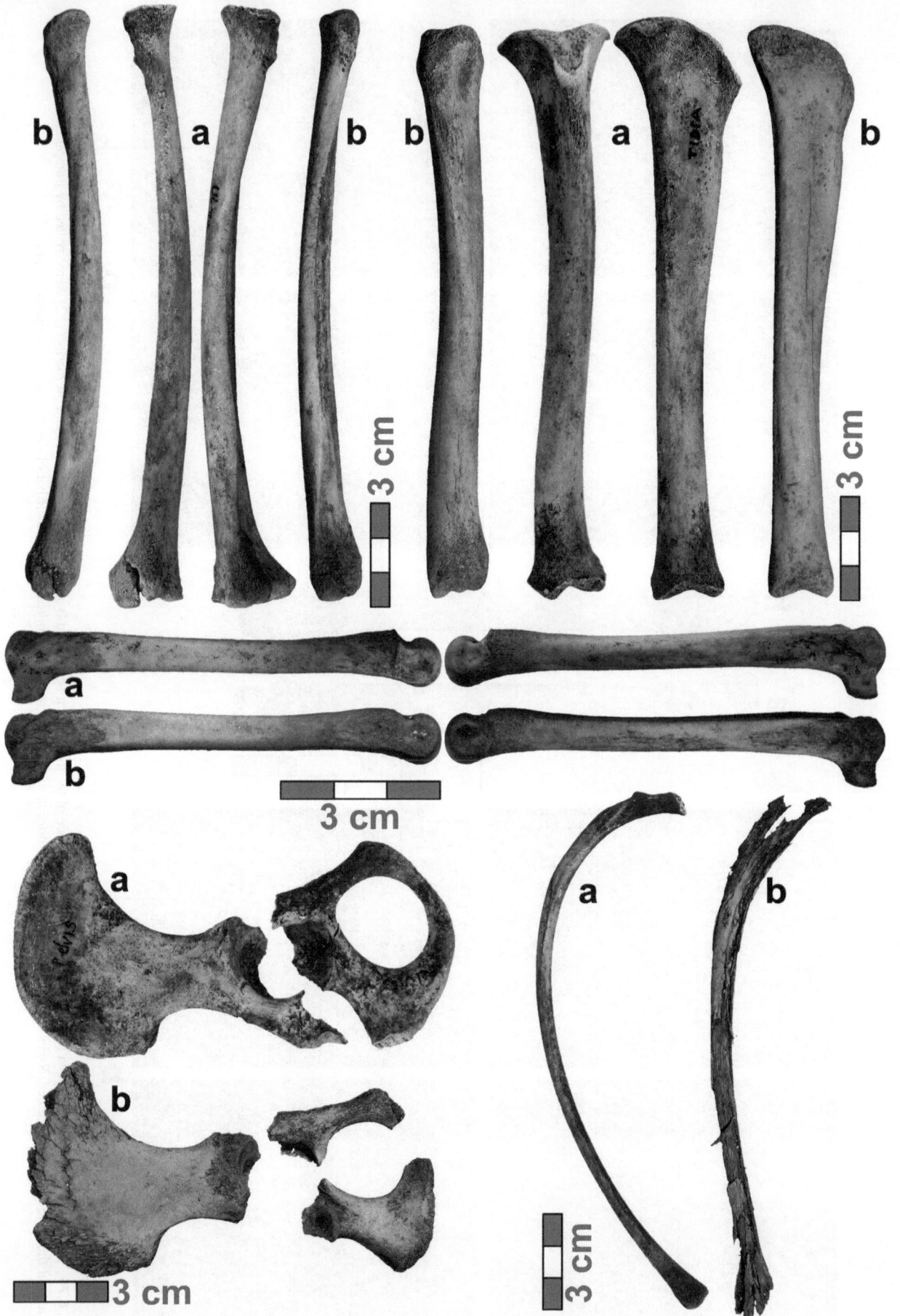

Fig. 6.77 Trampling Experiment 4: hyena tibia, radius, pelvis, MT IV, and rib **a** before and **b** after trampling. Morphological reduction of the tibia and radius was caused by trampling. The morphology of MT IV remained almost unchanged. Heavy abrasion affected the pelvis, resulting in exfoliation. The rib was straightened, compressed, and dissolved into a fibrous structure

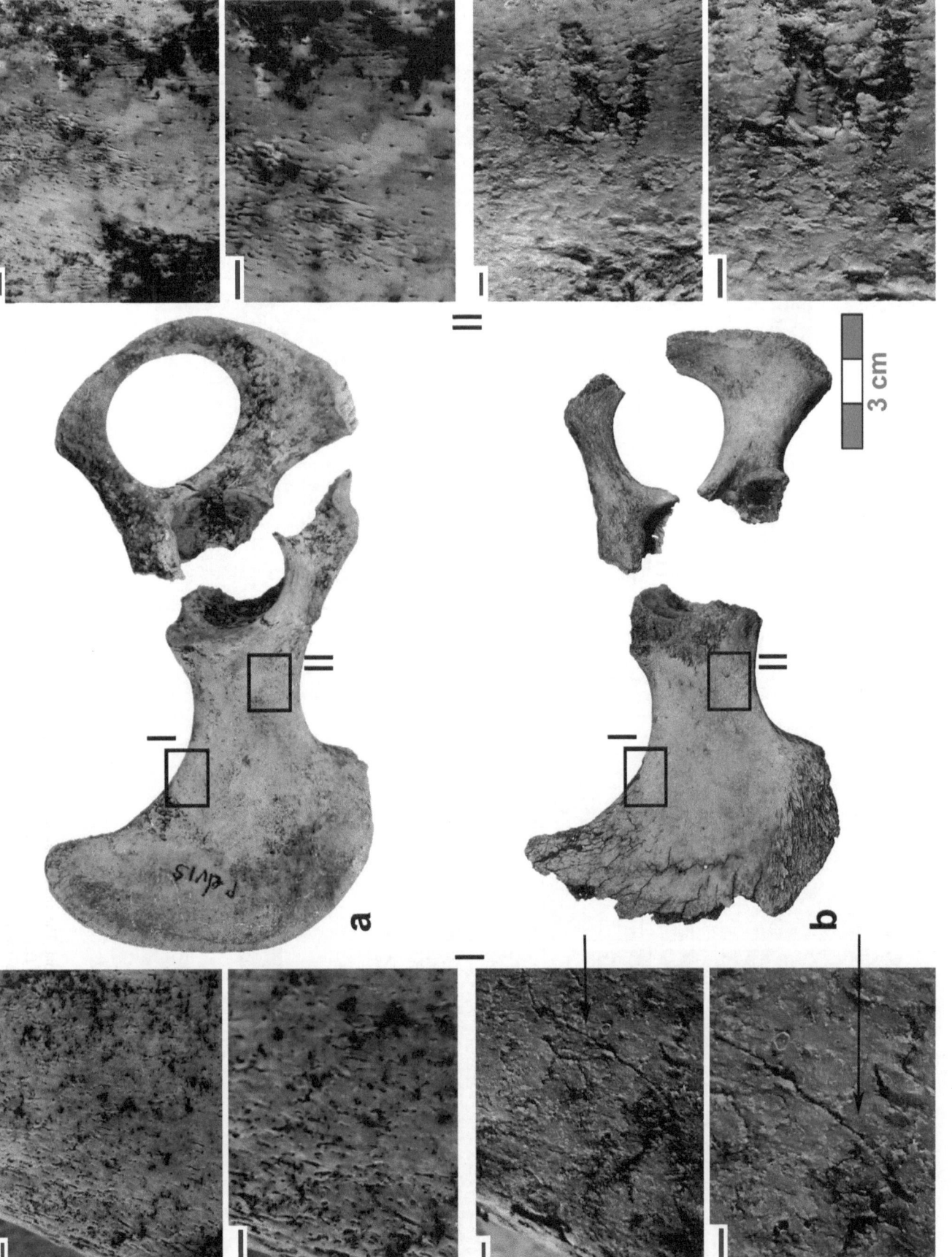

Fig. 6.78 Trampling Experiment 4, hyena pelvis **a** before and **b** after trampling. Morphological changes and exfoliation and dry fractioning of the bone surface, indicated by an *arrow*, were caused by trampling. (I) On the *left* is an enlarged view of the ilium; (II) on the *right* is an enlarged view of the pelvis near the rim of the acetabulum. Both columns show an identical detail of the bone at different degrees of magnification. The scale bar of the microscope images is 1 mm

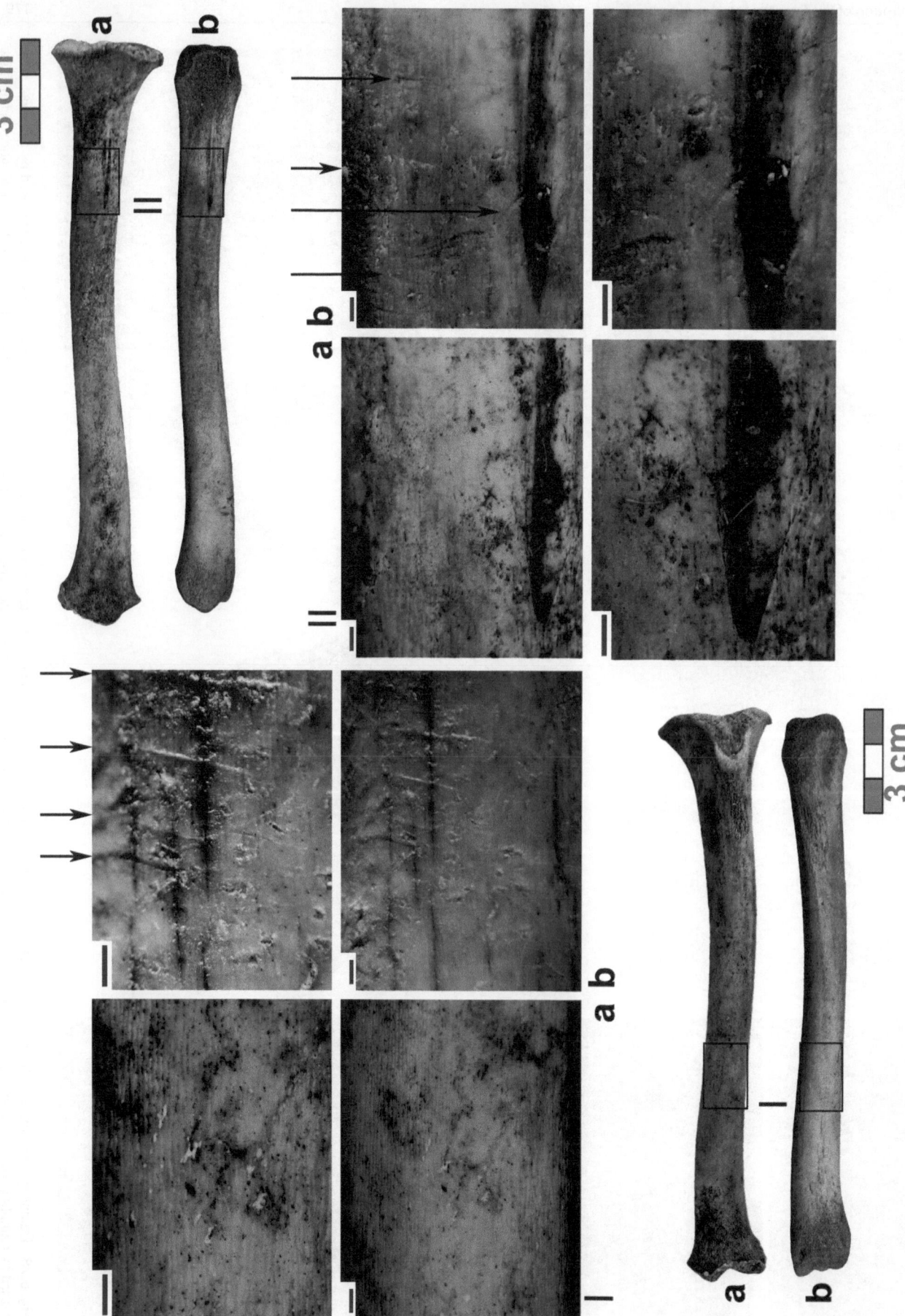

Fig. 6.79 Trampling Experiment 4, hyena tibia. The anterior and posterior face of the diaphysis **a** before and **b** after trampling. Scattered horizontal striations, indicated by *arrows*, seen to emerge from pit marks, were created by trampling. (I) Detail of the anterior face of the tibia; (II) detail of the posterior face of the tibia. All columns show an identical detail of the bone at different degrees of magnification. The scale bar of the microscope images is 1 mm

Fig. 6.80 Trampling Experiment
5, hyena ulna **a** before and **b** after
trampling. Heavy abrasion,
resulting in the reduction of
protruding sections of the
proximal articulation and the
broken distal end, was caused by
trampling

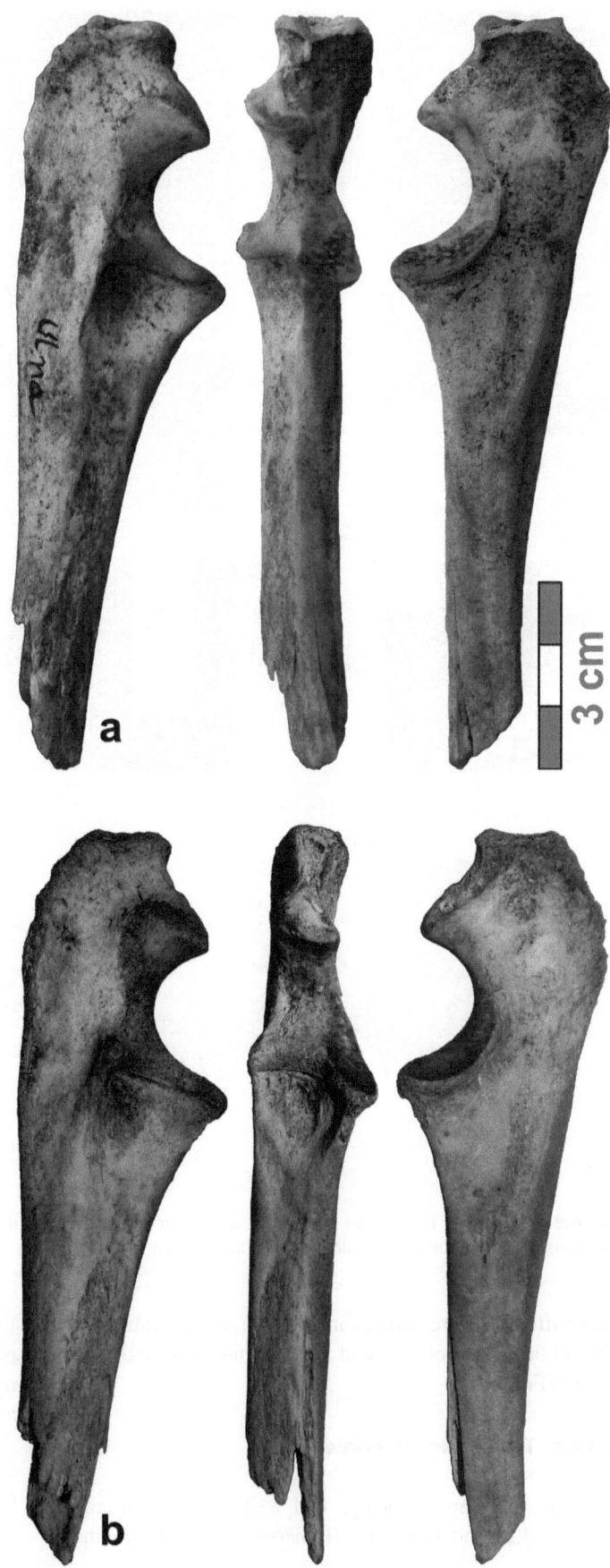

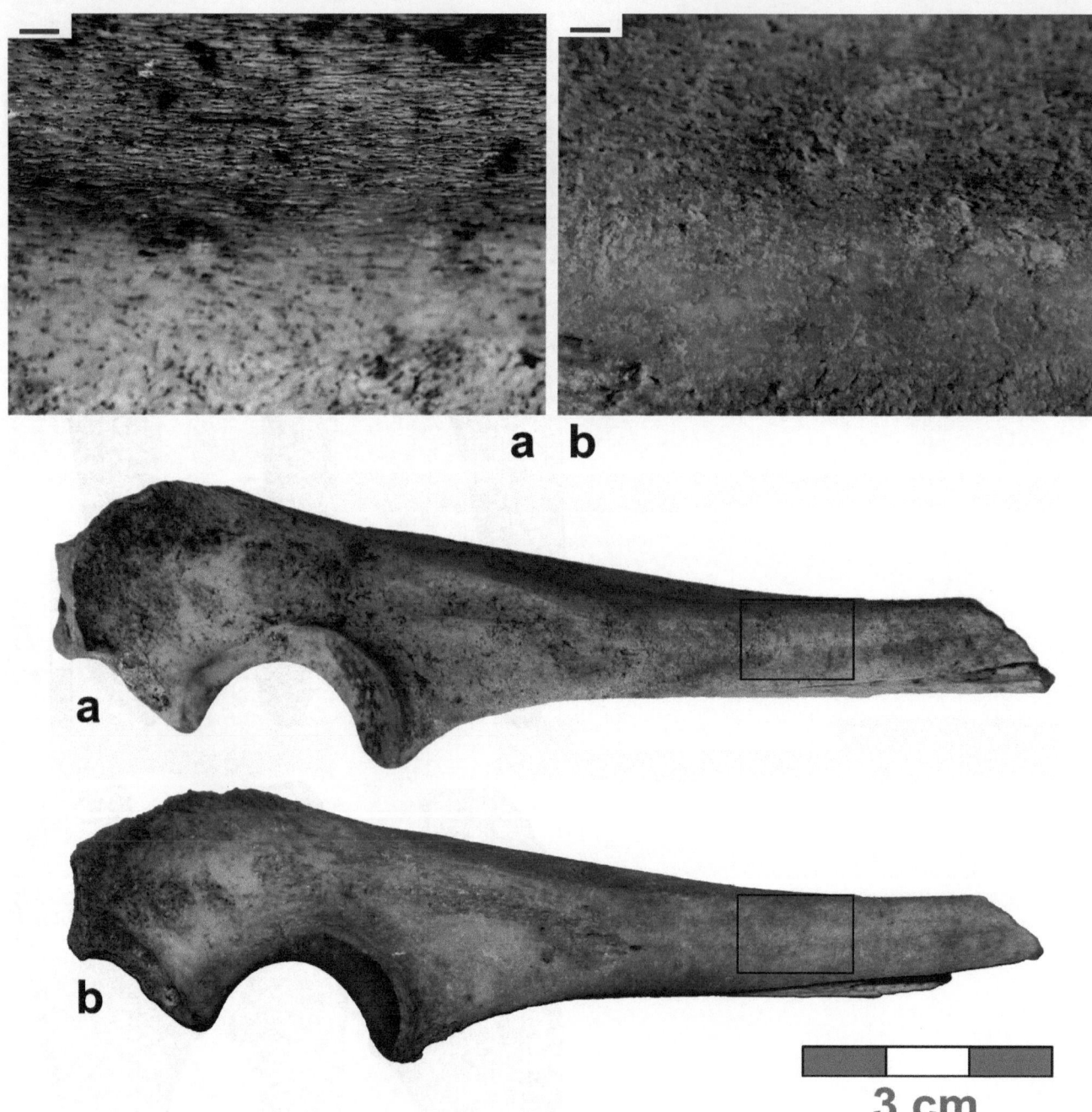

Fig. 6.81 Trampling Experiment 5, hyena ulna. The medial face of the diaphysis **a** before and **b** after trampling. Polishing of the bone surface was caused by trampling. The scale bar of the microscope images is 1 mm

face suffered severe damage in the form of perforations. The ventral face was polished and featured numerous isolated pit marks (Fig. 6.86).

6.4.4.6 Trampling Experiment 6

Two fresh sheep metacarpi and a domestic cow radius (Radius 3.2) and humerus (Humerus 1.1) shaft fragments were used for the sixth trampling experiment. One metacarpus displayed fine vertical striations along the anterior face resulting from periosteum removal. Periosteum remained attached to all bones.

This experiment utilized a mixture of 3.75 kg of mollusks (GBY 16/1997; Layer V-6; 184/130a, 60.35/60.19–60.10/60.01) and 1.75 kg of clay. Bones and sediment were placed in a plastic container, to which two liters of water

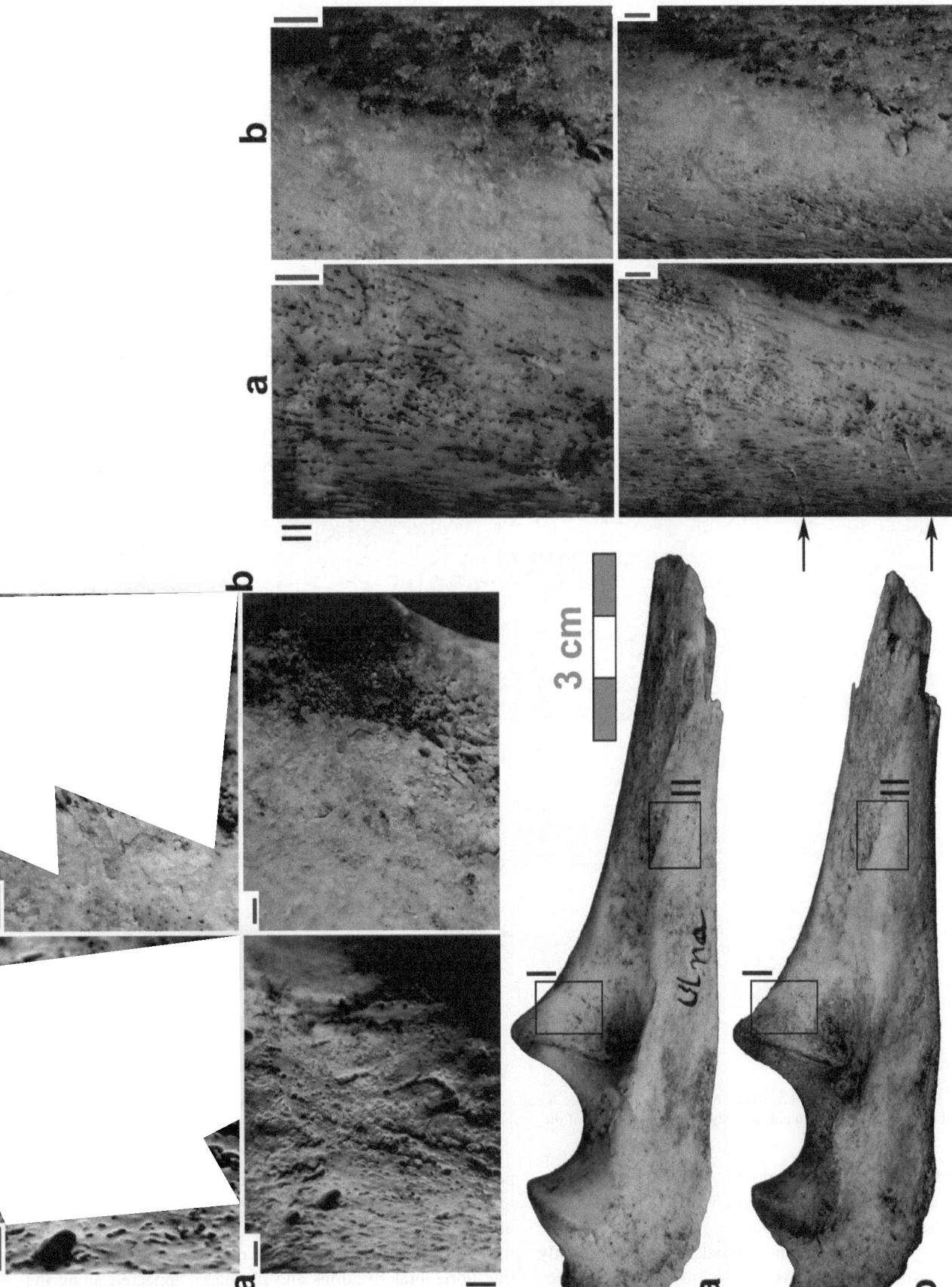

Fig. 6.82 Trampling Experiment 5, hyena ulna. The anterior/lateral face **a** before and **b** after trampling. Non-homogeneous bone-surface preservation resulted from trampling. Polishing of the rough parts of the bone and the removal of striations, indicated by an *arrow*, were also caused by trampling. (I) Detail of the proximal articulation. Both columns show an identical detail of the bone at different degrees of magnification. (II) Detail of the anterior/lateral face. Both columns show an identical detail of the anterior/lateral face. The scale bar of the microscope images is 1 mm

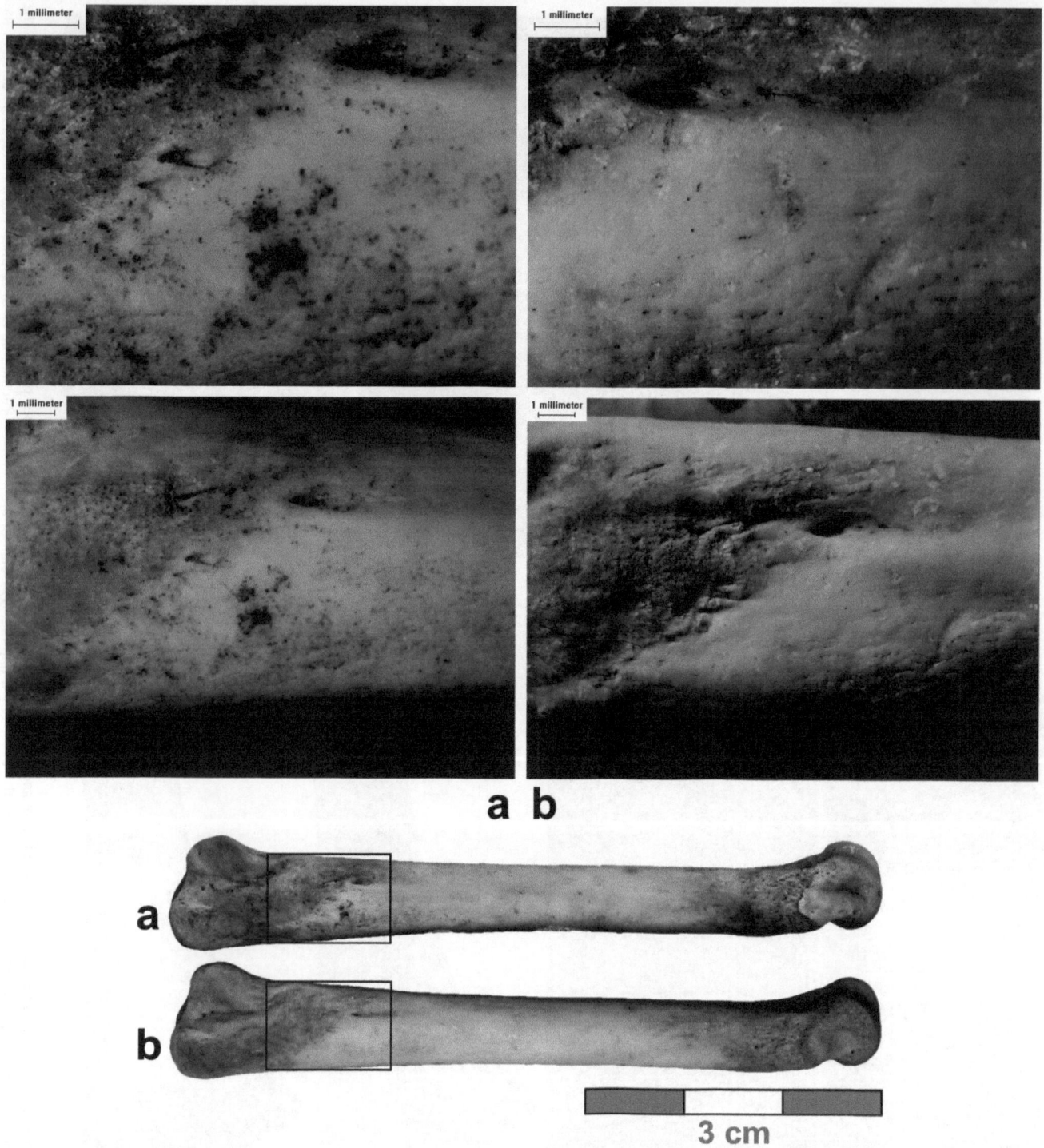

Fig. 6.83 Trampling Experiment 5, hyena MT II. The lateral face **a** before and **b** after trampling. The bone's prominent features became more pronounced after trampling. Both columns show the same detail of the bone at different degrees of magnification

was added in order to mimic muddy conditions. A person weighing approximately 50 kg and wearing rubber boots trampled the material for two hours.

Following trampling, the metapodials were only slightly abraded; however, changes were noted in the bone-surface morphology due to the removal of the remaining scraps of periosteum, which, in turn, led to higher bone susceptibility to abrasion. For example, the foramina became more pronounced (Fig. 6.87). On the other hand, abrasion obliterated the striations on the anterior side of the metapodials. Instead, fine horizontal striations and pit marks appeared along the anterior diaphysis rims (Fig. 6.88).

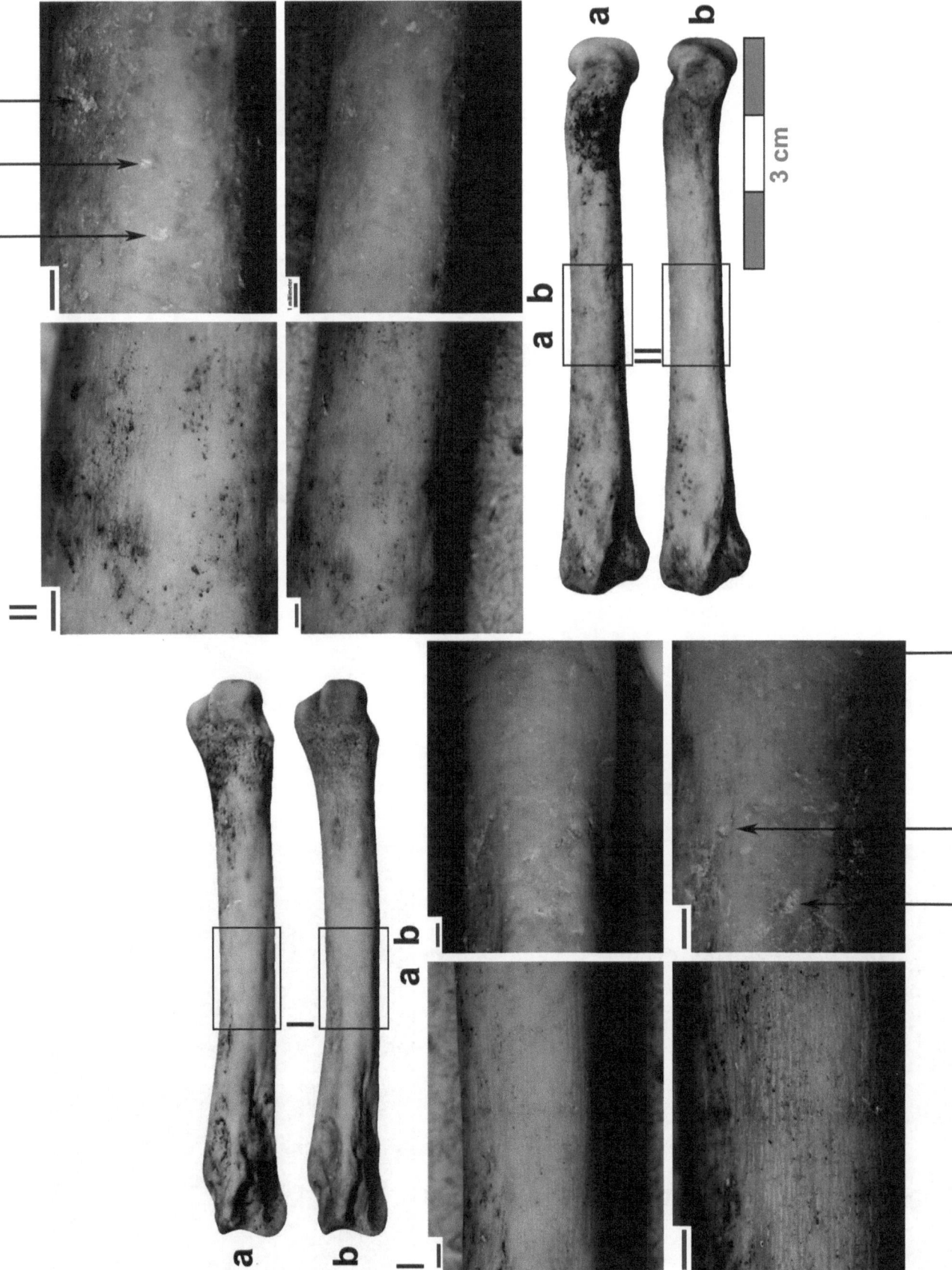

Fig. 6.84 Trampling Experiment 5, hyena MT II. The posterior and lateral face of the diaphysis **a** before and **b** after trampling. Isolated, shallow round pit marks, indicated by *arrows*, were created by trampling. (I) Detail of the posterior face. Both columns show an identical detail of the bone at different degrees of magnification. (II) Detail of the lateral face. Both columns show an identical detail of the bone at different degrees of magnification. The scale bar of the microscope images is 1 mm

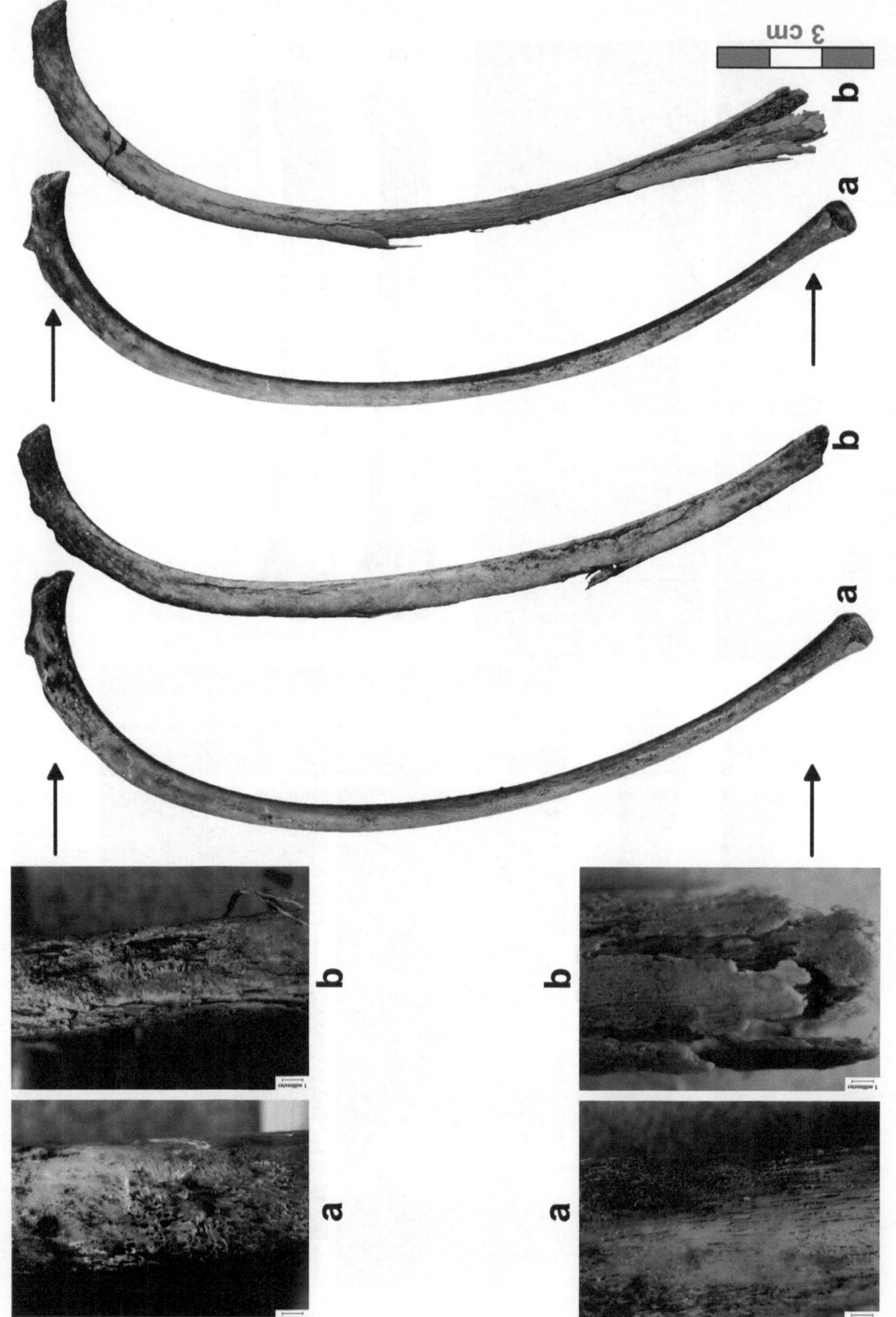

Fig. 6.85 Trampling Experiment 5, hyena ribs **a** before and **b** after trampling. Both the reduction of the bones into a fibrous consistency and bone disintegration were caused by trampling. *Arrows* indicate the areas of the bone that were magnified

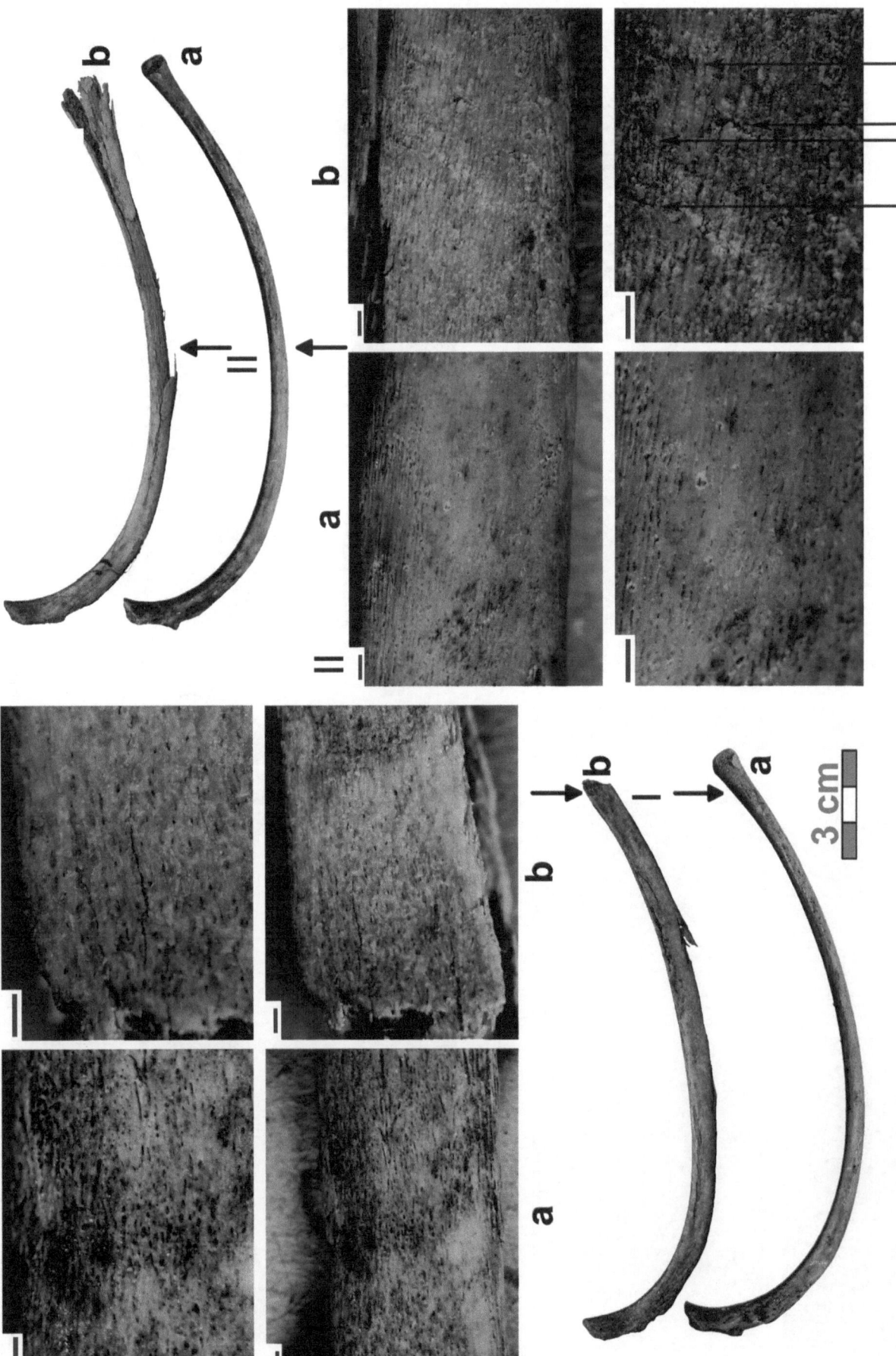

Fig. 6.86 Trampling Experiment 5, hyena rib. Bone-surface preservation **a** before and **b** after trampling. (I) Detail of the polished ventral face. Both columns show an identical detail of the bone at different degrees of magnification. (II) Detail of the lateral face showing isolated pit marks indicated by *arrows*. Both columns show an identical detail of the bone at different degrees of magnification. *Arrows* indicate the areas of the bone that were magnified. The scale bar of the microscope images is 1 mm

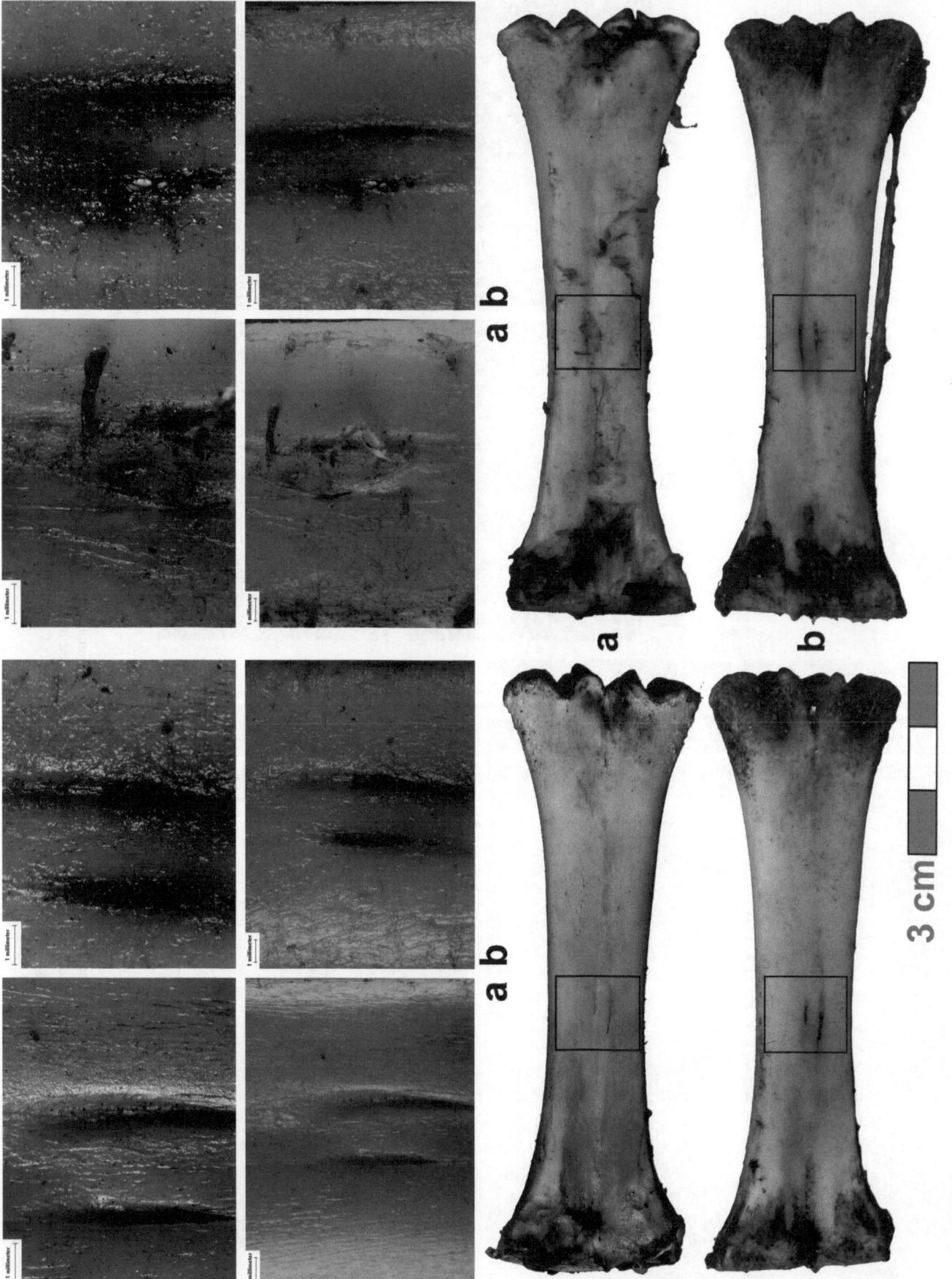

Fig. 6.87 Trampling Experiment 6, sheep metacarpi. The foramina on the posterior face of the diaphysis **a** before and **b** after trampling. Pronunciation of the foramina was caused by trampling. All columns show an identical detail of the bone at different degrees of magnification

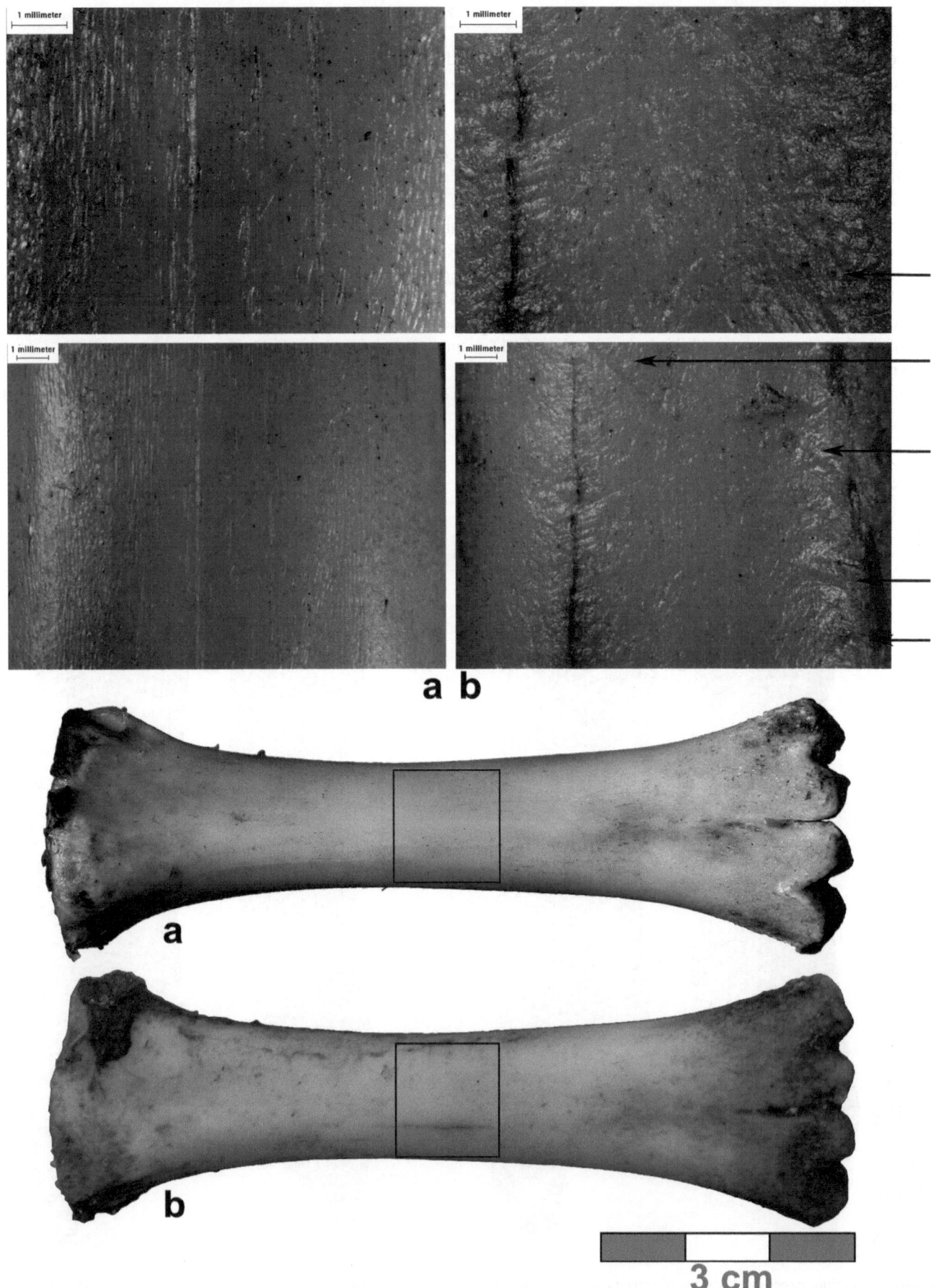

Fig. 6.88 Trampling Experiment 6, sheep metacarpus. The anterior face of the diaphysis **a** before and **b** after trampling. Fine striations and pit marks, indicated by *arrows*, were created by trampling. Both columns show an identical detail of the bone at different degrees of magnification

In contrast to the metapodials, the entire radius shaft fragment was rounded and smoothened (Fig. 6.89). The bone flakes still adhering to the bone were dislodged and destroyed, leaving conical impact marks on the radius fragment. The bone surface was heavily polished, as illustrated by the changed morphology of the distal edge that, prior to trampling, was characterized by irregular fracture planes along distinct crests. During trampling, these crests were erased, leaving a smooth round surface. Striations were not observed (Fig. 6.90).

The humerus underwent the same changes as the radius. Removal of the cancellous bone along the proximal and distal edges of the fragment produced striking changes (Fig. 6.91). The conical fracture plane changed in shape. Fine, shallow horizontal parallel striations were observed near the proximal

edge of the fragment, which had also been smoothened and polished (Fig. 6.92).

6.4.4.7 Trampling Experiment 7

A fresh sheep tibia, two sheep metacarpi with attached distal epiphyses and a proximal fragment of a bovid radius (Radius 4.1) were used for the seventh trampling experiment. Periosteum remained attached to the epiphyses and diaphyses of all bones, but the posterior diaphyses of the metacarpi exhibited two vertical striations caused by periosteum removal.

This experiment utilized a mixture of 3.75 kg of mollusks (GBY 16/1997; Layer V-6; 184/130a, 60.35/60.19–60.10/

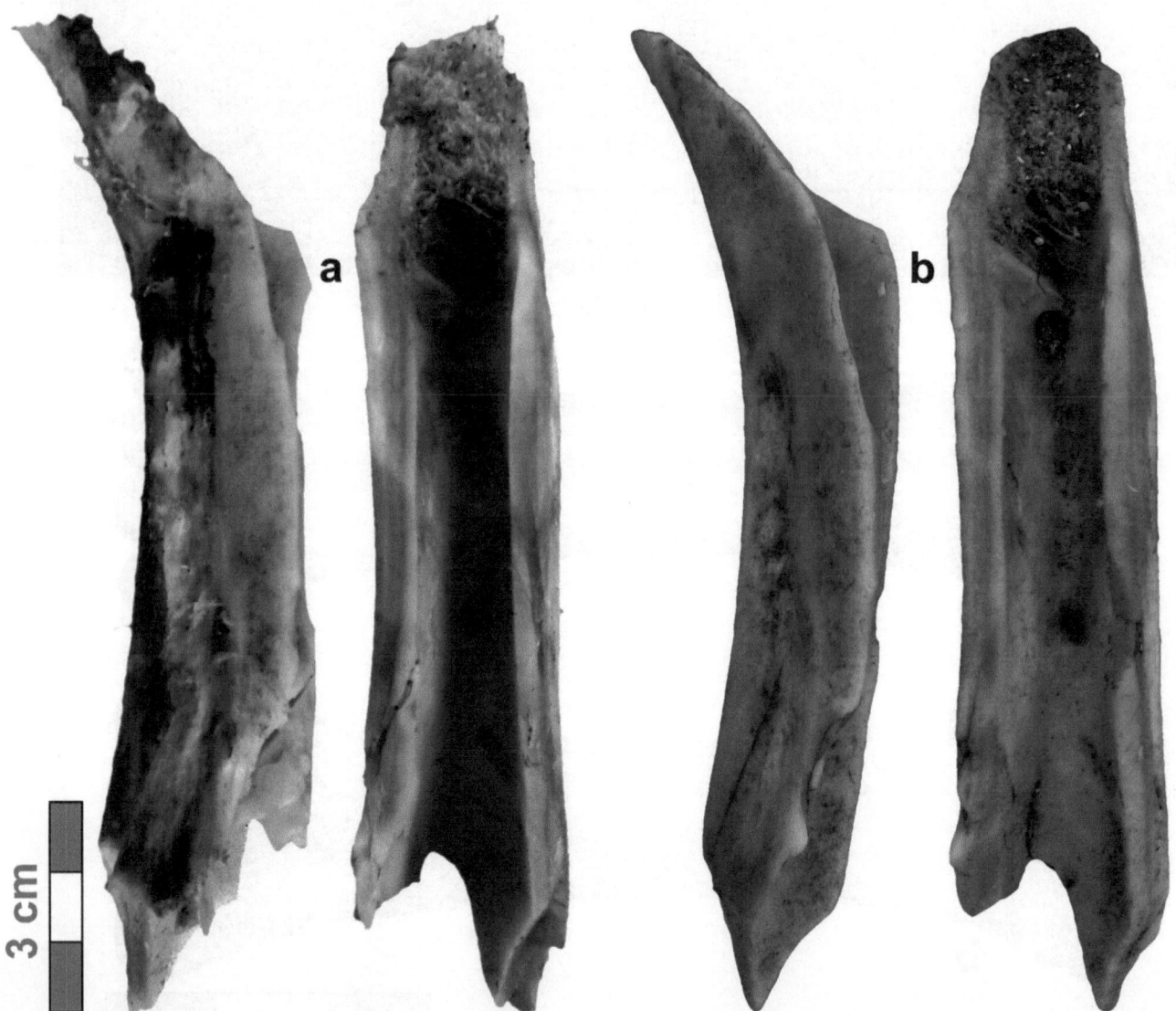

Fig. 6.89 Trampling Experiment 6, domestic cow radius fragment **a** before and **b** after trampling. Rounding and smoothing of the shaft fragment were caused by trampling

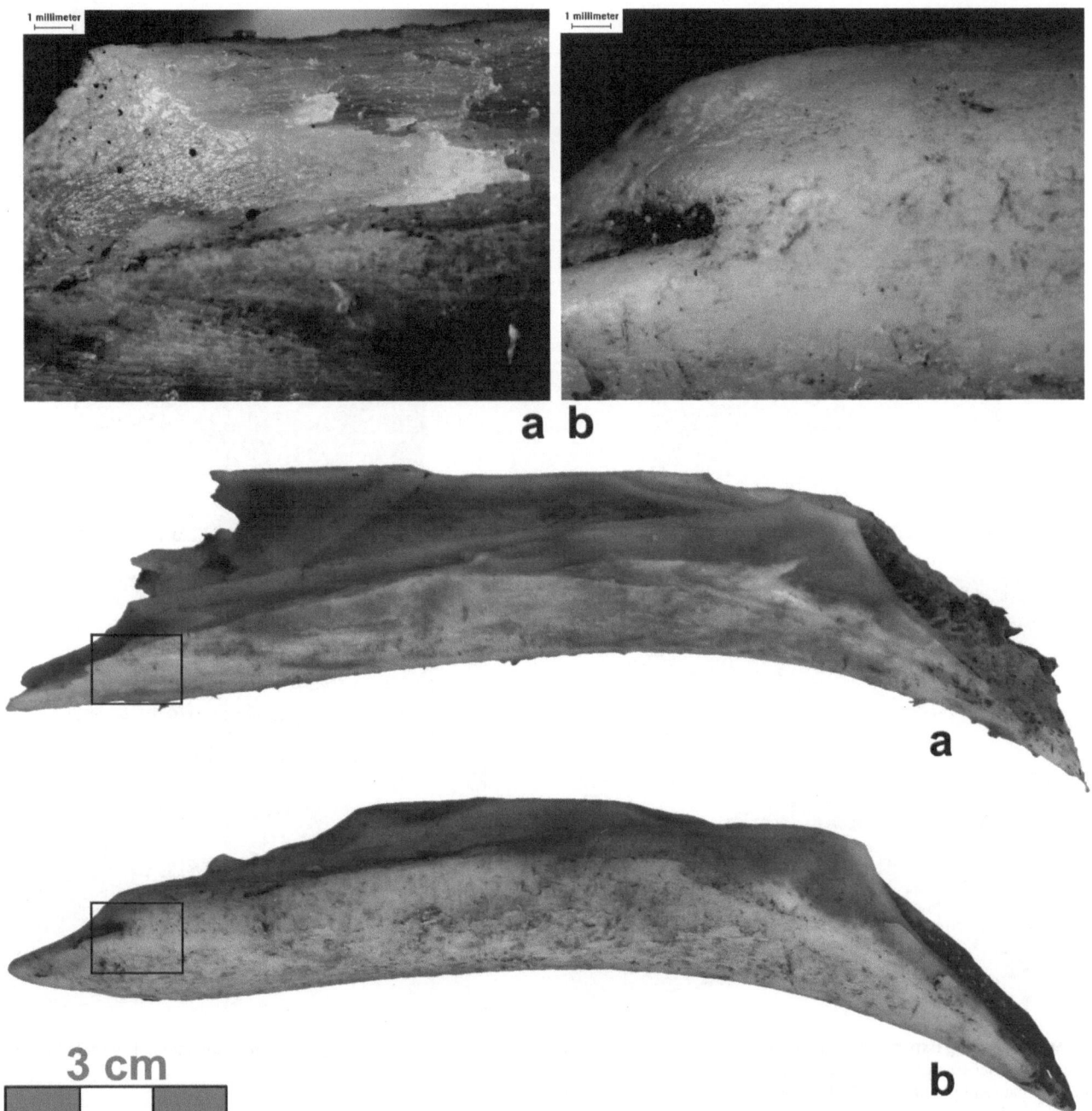

Fig. 6.90 Trampling Experiment 6, domestic cow radius fragment **a** before and **b** after trampling. Polishing and rounding of the shaft fragment were caused by trampling

60.01) and 1.75 kg of clay. Bones and sediment were placed in a plastic container, to which two liters of water were added in order to mimic muddy conditions. A person weighing approximately 50 kg and wearing rubber boots trampled the mixture for two hours.

Following trampling, we noted only a few morphological changes that had resulted from abrasion, but most of the periosteum had been removed. The anterior and posterior

faces of the tibia diaphysis displayed slight abrasion in those areas where periosteum was still attached prior to trampling. Pit marks and fine horizontal striations appeared only on the lateral and cranio/lateral diaphysis (Fig. 6.93).

Periosteum remained attached to parts of the epiphyses and to the lateral and medial face of one of the metacarpi diaphysis. Partial loss of the epiphysis was noted where periosteum was missing. Those parts of the posterior diaphysis

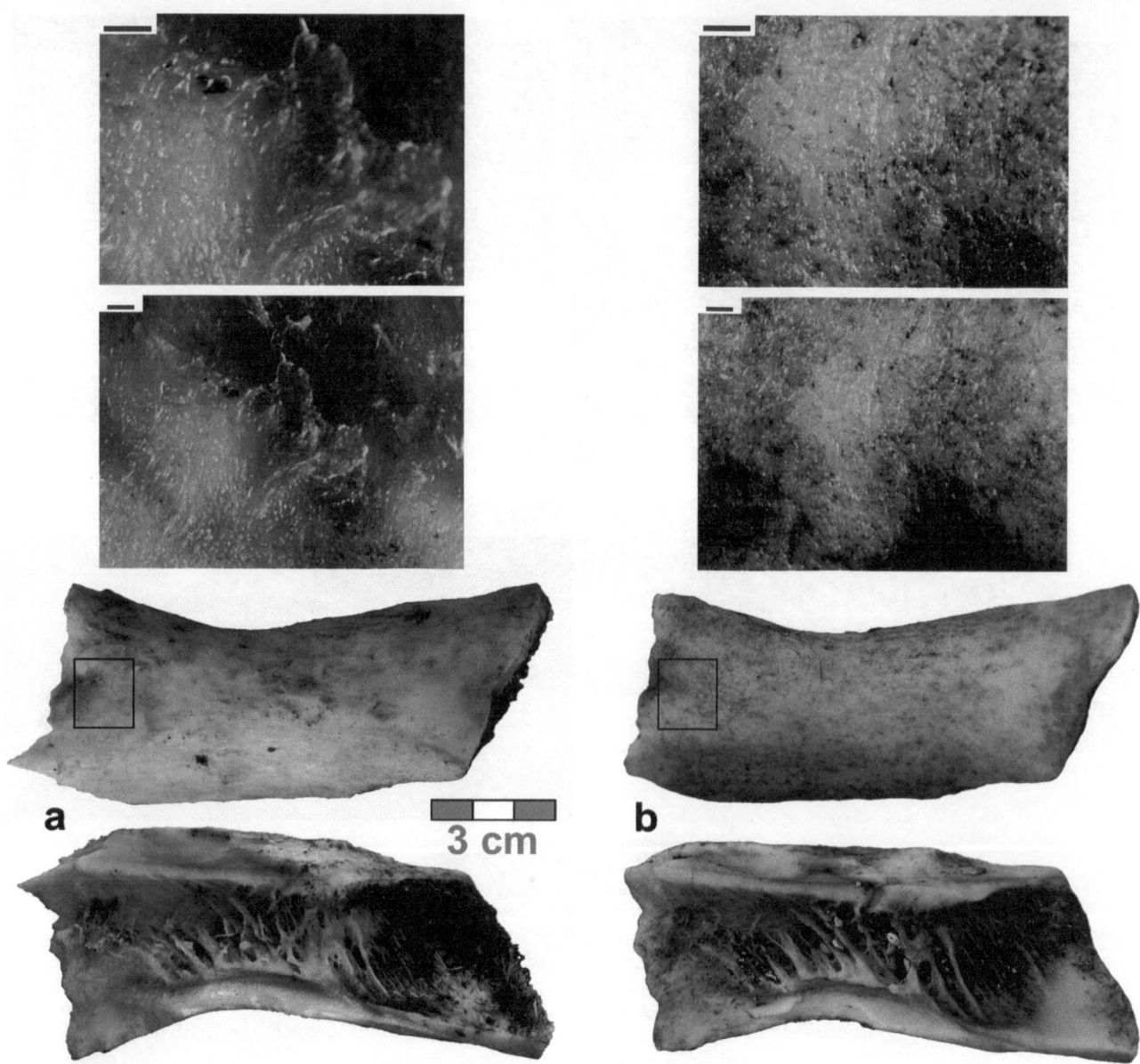

Fig. 6.91 Trampling Experiment 6, domestic cow humerus fragment **a** before and **b** after trampling. Polishing and rounding of the shaft fragment were caused by trampling. Both columns show an identical detail of the bone at different degrees of magnification. The scale bar of the microscope images is 1 mm

missing periosteum were marked by oval and round pits (Fig. 6.94). In addition, the posterior sulcus was more pronounced and the vertical striations produced during initial periosteum removal were smoothened.

A thick layer of periosteum still covered the other metacarpus, especially on its proximal and distal epiphyses. Surface modifications occurred only in areas where periosteum had been completely removed. The posterior face displayed traces of pit marks (Fig. 6.95). Some of

the marks created during periosteum removal became more pronounced, while others disappeared.

No periosteum remained on the medial distal face of the radius; instead, polish, tiny pit marks, and fine oval and round puncture traces were noted (Fig. 6.96). The posterior face of the diaphysis was abraded and displayed a variety of striations: fine horizontal striations; fine, overlapping vertical striations originating from puncture marks; and a deep, broad tapering striation (Fig. 6.97).

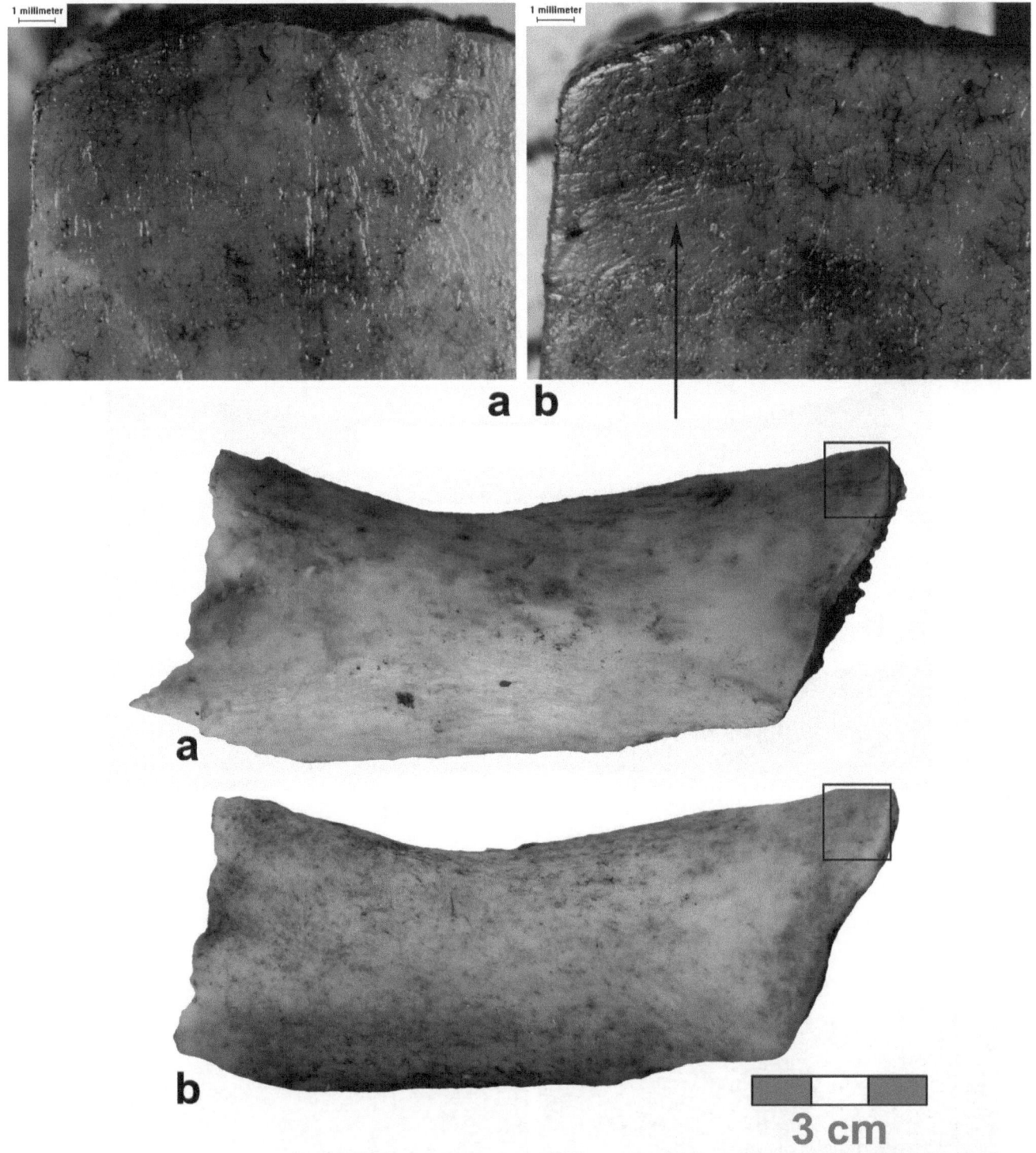

Fig. 6.92 Trampling Experiment 6, domestic cow humerus fragment. The proximal edge of the fragment **a** before and **b** after trampling. Fine parallel striations, indicated by an *arrow*, as well as the polishing and smoothing of the bone surface, were caused by trampling

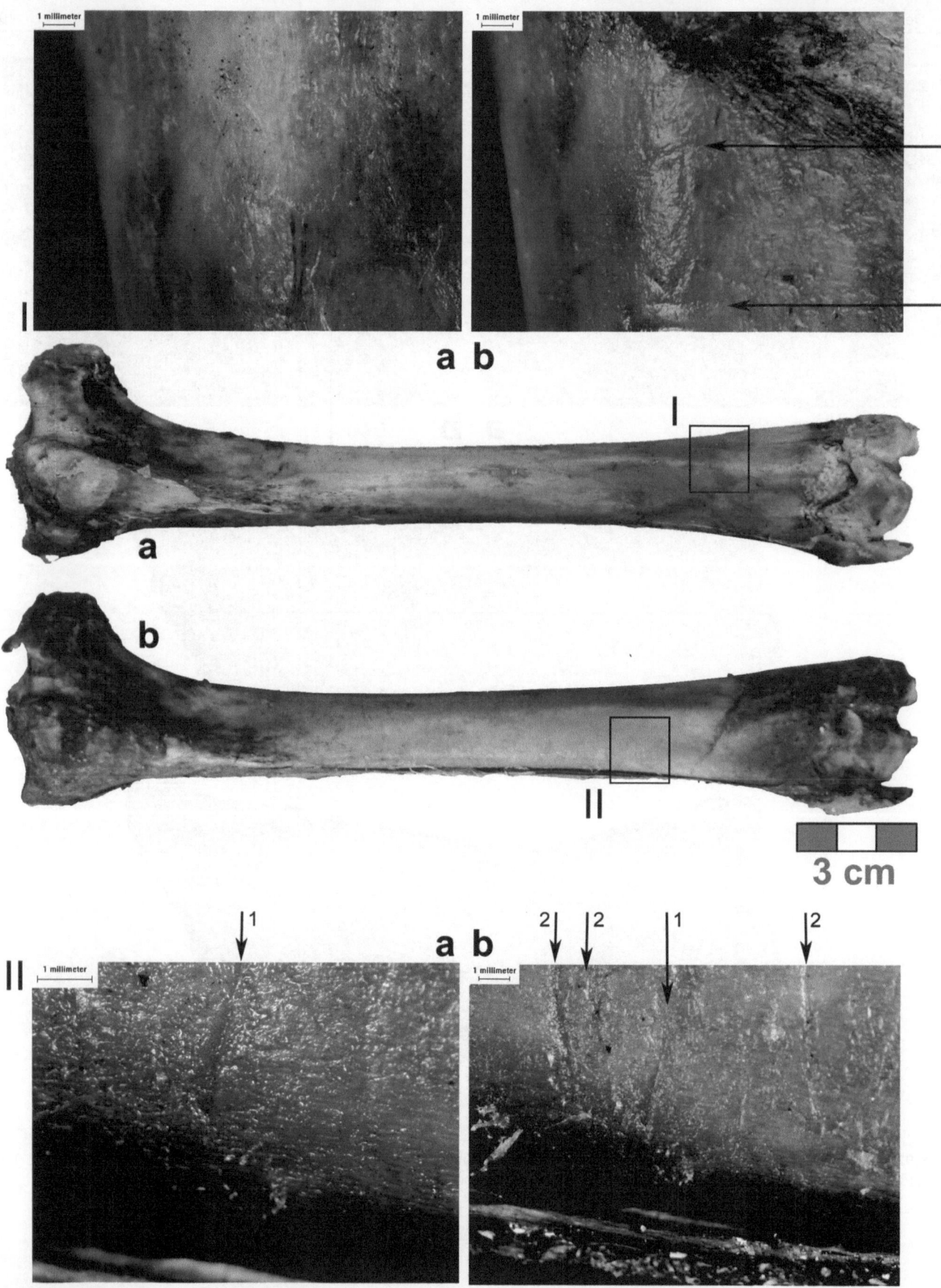

Fig. 6.93 Trampling Experiment 7, sheep tibia. The lateral and anterior diaphysis **a** before and **b** after trampling. (I) Horizontal striations on the anterior face were created by trampling. (II) Smoothing of the striations produced prior to trampling (1), and the formation of new pit marks and striations (2) after trampling

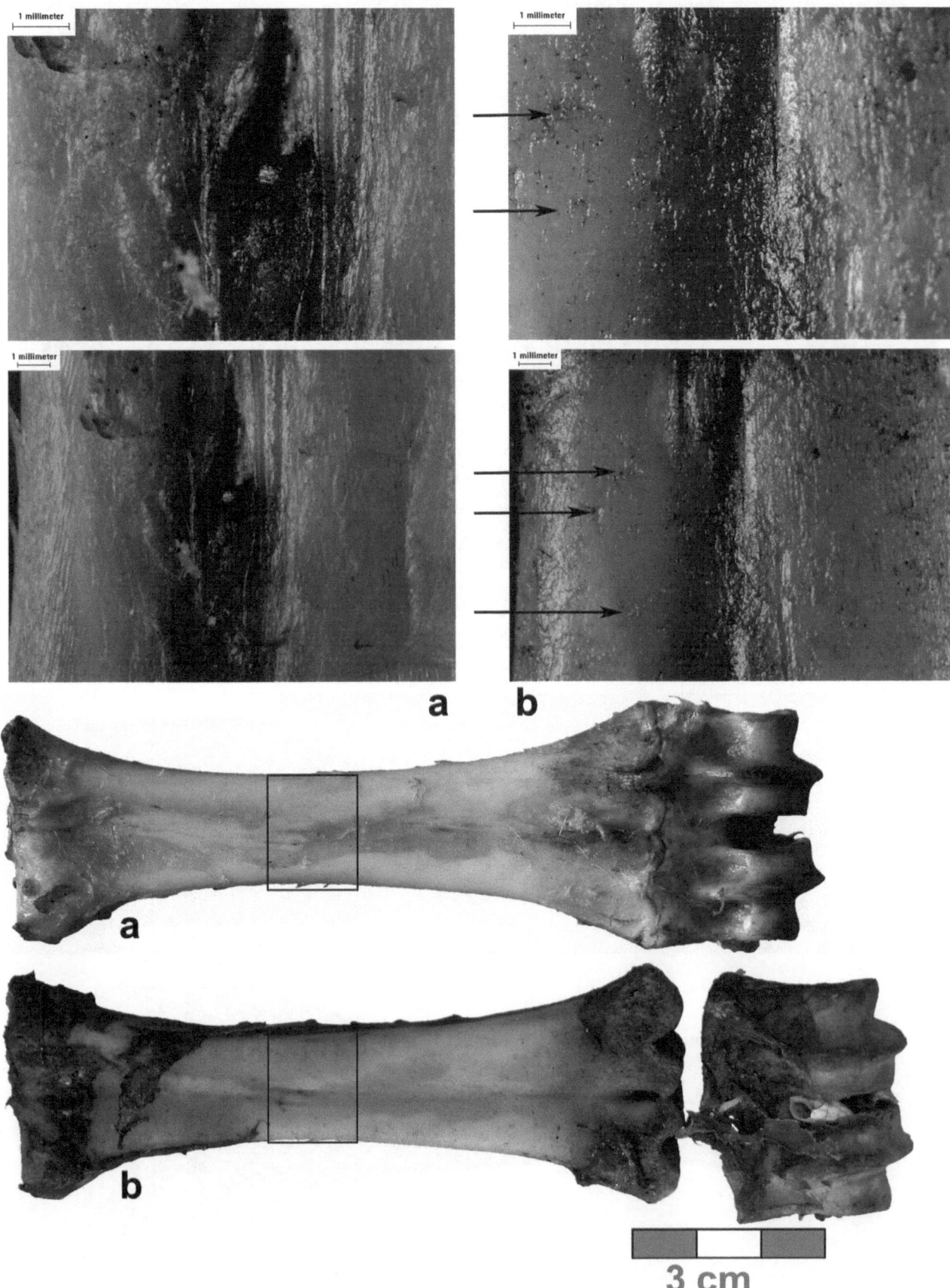

Fig. 6.94 Trampling Experiment 7, sheep metacarpus. The posterior diaphysis **a** before and **b** after trampling. Oval- and round-shaped pits, indicated by *arrows*, were created by trampling. Both columns show an identical detail of the bone at different degrees of magnification

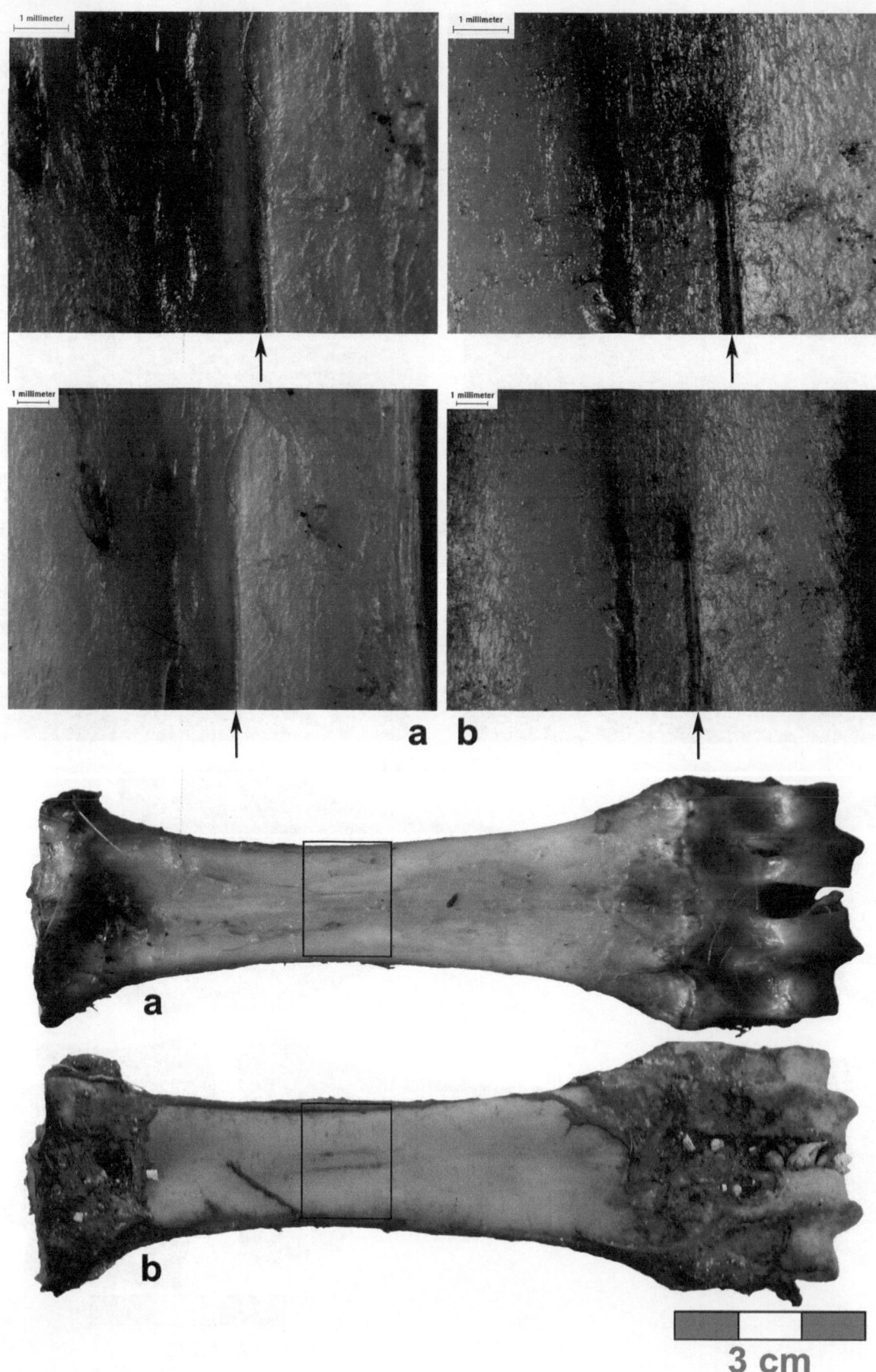

Fig. 6.95 Trampling Experiment 7, sheep metacarpus. The posterior diaphysis **a** before and **b** after trampling. Pronunciation of the posterior sulcus and the smoothing of the striation, indicated by an *arrow*, were caused by trampling. Both columns show an identical detail of the bone at different degrees of magnification

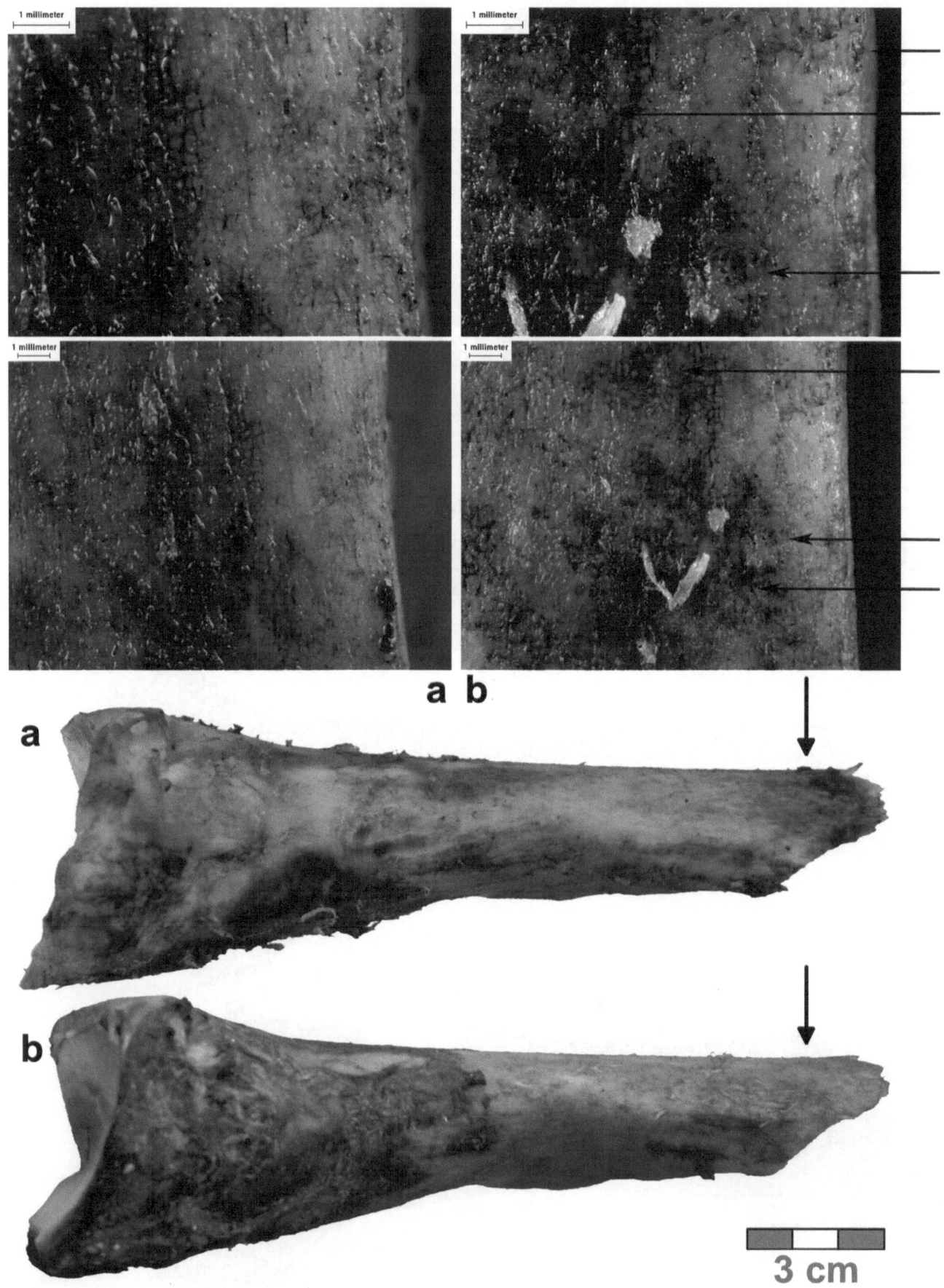

Fig. 6.96 Trampling Experiment 7, domestic cow radius fragment. The medial/distal diaphysis **a** before and **b** after trampling. Polishing and pit marks, indicated by *arrows*, were caused by trampling. Both columns show an identical detail of the bone at different degrees of magnification. *Arrows* indicate the area of the bone that was magnified

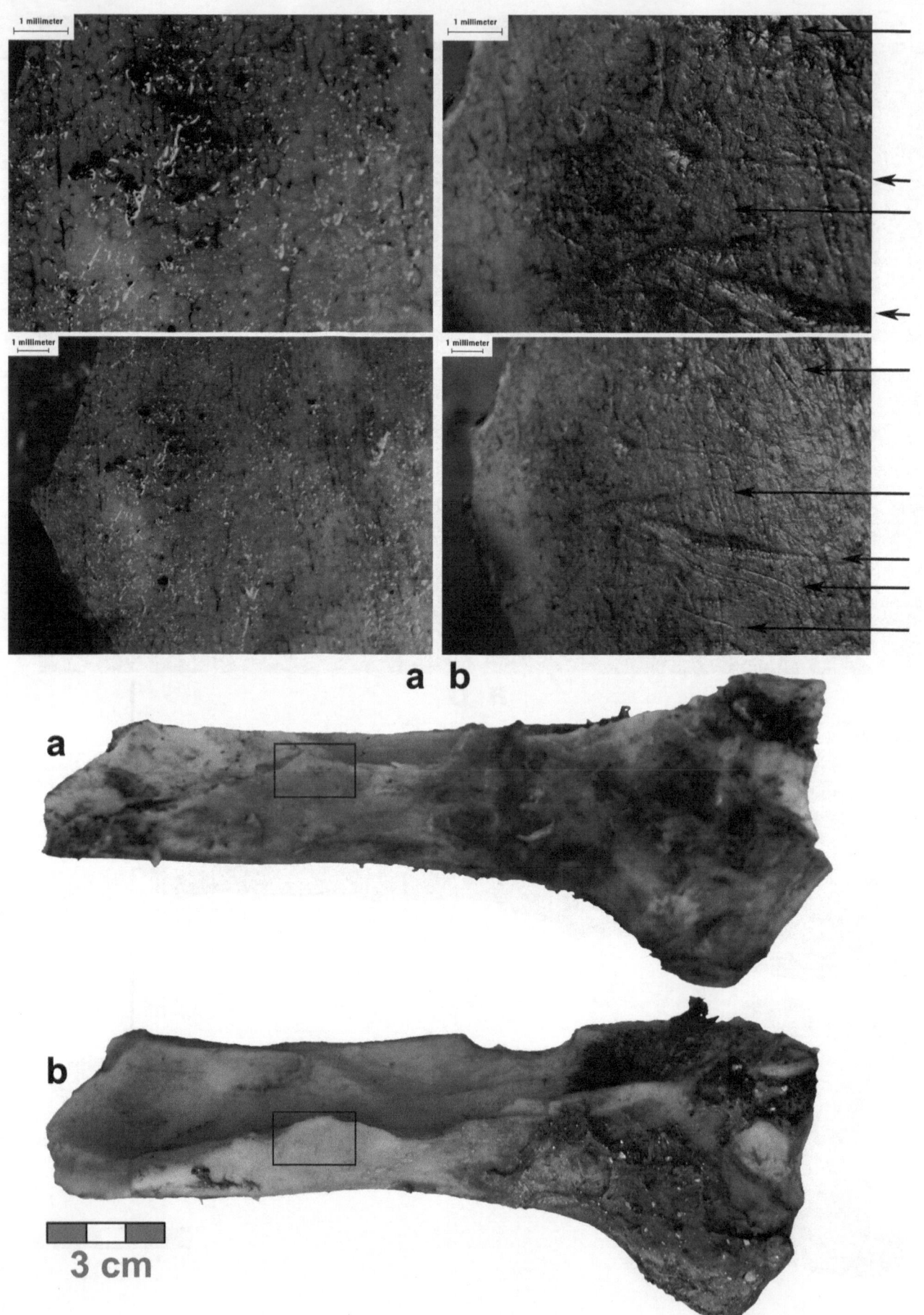

Fig. 6.97 Trampling Experiment 7, domestic cow radius fragment. The posterior face of the diaphysis **a** before and **b** after trampling. Abrasion and various striations, indicated by *arrows*, were caused by trampling. Both columns show an identical detail of the bone at different degrees of magnification

6.5 Summary and Results of the Experiments

In the following the results of the experiments undertaken will be summarized.

The scratching experiments illustrated that considerable force is needed to produce scratches and striations on bone surfaces using the pointed helix of mollusks and fragmented mollusks.

The burial experiment illustrated that striations on bone surfaces at GBY are the result of specific environmental or sedimentological conditions at the site. In order to substantiate this hypothesis detailed trampling and tumbling experiments were undertaken to mimic the conditions at GBY. Detailed experimental protocol is described in Section 6.4.

6.5.1 Results of the Tumbling Experiments

The tumbling experiments were designed to mimic unidirectional (Tumbling Experiments 1–8) and multidirectional (Tumbling Experiments 9–18) water movements (see also Gaudzinski-Windheuser et al. 2010). A summary of the tumbling experiments is provided in Table 6.3. Our experiments have demonstrated that the dominant taphonomic effect caused by both uni- and multidirectional water movements is abrasion. As suggested, the results of Tumbling Experiments 17 and 18 indicate that fragmented and complete mollusks were the agents mainly responsible for the surface alterations of the bones.

It should be noted that uni- and multidirectional settings each produce distinctly different abrasion characteristics on bone surfaces. Unidirectional water movement led to abrasion, which is best documented and assessed at the bones' fractured edges and on distal metaphyses. Three main factors appeared to control abrasion, of which the first is the amount of water involved. This became apparent when we compared the results of dry tumbling (Tumbling Experiment 4) with those of tumbling with 70 liters of water (Tumbling Experiments 5–8) and with 100 liters of water (Tumbling Experiments 1–3).

The second variable is bone morphology; i.e., species, skeletal element, and anatomical landmark. This is reflected in the tumbling of sheep metacarpi and metatarsi (Tumbling Experiments 2–3), as compared to the tumbling of cow long bone fragments (Tumbling Experiments 5–6).

The third, and most critical, variable is the state of periosteum preservation, since it determines the degree of abrasion on the bone surface underneath (see the results of Tumbling Experiment 8). It is important to note that the effects of time (i.e., duration of tumbling) did not operate in a simple linear manner. No differences in bone-surface modifications were observed when the duration of tumbling was increased (e.g., compare Tumbling Experiments 7 and 8).

Even though the matrix used for the experiments usually contained numerous sharp-edged shell fragments, not a single significant striation resulted from unidirectional water movement. However, the abrasive effects had a very distinct effect on the morphology of surface modifications that were present prior to tumbling.

Longitudinal striations resulting from periosteum removal and scraping marks on sheep bones consisting of horizontal striations overlain by fine vertical striations became more pronounced following tumbling. The effects of tumbling on anatomical landmarks, such as foramina and sulci, were minimal for time intervals up to six hours (compare Tumbling Experiment 2, 5, and 8). Significant changes in the character of bone-surface damage were observed only after six hours of unidirectional tumbling. These changes revealed that the destruction and formation of preexisting modifications accelerated exponentially over time.

Abrasion affected preexisting surface modifications differently on cow bones. Unidirectional tumbling led to the disappearance of longitudinal cut marks, whereas horizontal cut marks remained present, albeit somewhat altered (compare Tumbling Experiments 5 and 6). In addition, we noted that the same effects that caused the disappearance of longitudinal cut marks also led to the complete alteration of other cut- and scraping marks (compare Tumbling Experiments 5 and 6). Comparison of the effects of unidirectional tumbling on sheep and cow bones has shown that the specific bone structure was a crucial variable in the survival and morphology of the surface modifications.

As mentioned above, multidirectional tumbling also led to abrasion. As in the case of unidirectional tumbling, both the state of periosteum preservation, which determines the degree of abrasion, and the specific bone structure comprise the major contributing variables. In contrast to unidirectional movement, the amount of water involved in multidirection movement does not play a significant role; rather, it is the presence or absence of water that influences abrasion. However, the most crucial difference between the two tumbling types is that the results of multidirectional water movement depended on the duration of tumbling. This became obvious when comparing the results of Tumbling Experiments 9 and 11, in which a higher degree of abrasion was found to result from the increased duration of tumbling. In short, compared to unidirectional water movement, multidirectional water movement produced a larger variety of bone-surface modifications. Multidirectional water movement also accelerated the formation of bone-surface modifications.

Our experiments have demonstrated that striation-formation processes and the individual contributing variables involved can be studied in detail. Against this background,

Table 6.3 Results of the tumbling experiments. Observed modifications after tumbling are summarized as modification of outer bone tissue, bone morphology, and bone surface. (For identification of the bone specimens involved in the experiments, see Table 6.1)

Tumbling experiment number	Bone	Duration (hours)	Treatment and observed modifications before tumbling	Observed modifications after tumbling		
				Alterations of outer bone tissues	Alterations of bone morphology	Bone-surface modifications
1	Mc-1	2	Removal of skin, meat and periosteum / Formation of striations due to periosteum removal / Periosteum still attached to metaphyses	Further removal of periosteum in small areas / Slight abrasion	Deepening of sulcus	Alteration of striations
2	Mt-1	3	Removal of skin, meat and periosteum / Formation of striations due to periosteum removal / Periosteum still attached to metaphyses	Further removal of periosteum in small areas / Slight abrasion	Pronunciation of sulcus	Pronunciation of striations
3	Mc-2	3	Removal of skin, meat and periosteum / Periosteum still attached to metaphyses	Abrasion		
4	Mt-2	3	Removal of skin, meat and periosteum formation of striations due to periosteum removal / Periosteum still attached to metaphyses	Further removal of periosteum in small areas / Heavy abrasion	Alteration of sulcus	Pronunciation of striations
5	Fem-2-1	3	Removal of skin meat and periosteum / Longitudinal and vertical cut marks due to periosteum removal / Large parts still covered by periosteum and slight exfoliation of periosteum	Removal of periosteum / Heavy abrasion / Outermost bone surface obliterated	Edges rounded / Alteration of breakage patterns	Longitudinal cut marks almost vanished / Vertical cut marks altered and still visible
6	Tib-2-4	2.5	Removal of skin and meat / Scraping off parts of periosteum / Artificial production of longitudinal and vertical cut marks as well as scraping marks	Removal of periosteum / Abrasion in areas already cleaned from periosteum prior to tumbling	Slight rounding of edges	Alteration and obliteration of cut marks / Alteration of scraping marks
7	Mc-3	3	Removal of skin, meat and periosteum / Formation of horizontal scratch during periosteum removal / Periosteum still attached to metaphyses	Slight abrasion	Broadening of sulcus and foramina	Broadening of horizontal scratch / Formation of vertical striation
8	Mt-3	6	Removal of skin, meat and periosteum / Longitudinal striations due to periosteum removal / Fine horizontal and longitudinal scraping marks due to periosteum removal / Fine horizontal and diagonal striations due to periosteum removal / Periosteum still attached to metaphyses	Further removal of periosteum in small areas / Slight abrasion		Broadening of longitudinal striations / Pronunciation of fine horizontal and longitudinal scraping marks / Obliteration of horizontal and diagonal striations, instead fine, oval to round punctures

Table 6.3 (continued)

Tumbling experiment number	Bone	Duration (hours)	Treatment and observed modifications before tumbling	Observed modifications after tumbling		
				Alterations of outer bone tissues	Alterations of bone morphology	Bone-surface modifications
9	Mc-4	3	Removal of skin, meat and periosteum. Longitudinal striation due to periosteum removal. Diagonal striations due to periosteum removal. Horizontal scraping marks due to periosteum removal. Periosteum still attached to metaphyses	Further removal of periosteum in small areas. Slight abrasion		Obliteration of longitudinal and diagonal striations. Pronunciation of diagonal striations. Obliteration of scraping marks and formation of longitudinal striations. Formation of oval to round punctures
10	Mt-4	3	Removal of skin, meat and periosteum. Longitudinal striations due to periosteum removal. Diagonal striations due to periosteum removal. Periosteum still attached to metaphyses	Further removal of periosteum in small areas. Slight abrasion		Broadening of longitudinal striations. Obliteration and smoothing of diagonal striations. Formation of oval to round punctures
11	Mc-5	4	Removal of skin, meat and periosteum. Fine longitudinal striations due to periosteum removal. Periosteum still attached to metaphyses	Further removal of periosteum in small areas. Slight abrasion	Broadening of sulcus	Formation of oval to round punctures. Striations almost vanished
11.1	Mc-5	6	Slight abrasion. Oval to round punctures. Longitudinal striations	Entire bone abraded	Entire bone rounded	Obliteration of punctures. Longitudinal striations almost vanished, only one altered striation remains. Formation of deep diagonal striation
12	Mt-5	4	Removal of skin, meat and periosteum. Fine diagonal striations due to periosteum removal. Fine longitudinal striations due to periosteum removal. Periosteum still attached to metaphyses	Further removal of periosteum in small areas. Slight abrasion	Pronunciation of the sulcus	Pronunciation of striations. Formation of flat striation. Formation of deep striation and punctures
13	Mc-6	3	Removal of skin, meat and periosteum. Periosteum still attached to metaphyses	Further removal of periosteum in small areas. Slight abrasion		Formation of fine striation
14	Mt-6	5	Removal of skin, meat and periosteum. Longitudinal striations due to periosteum removal. Periosteum still attached to metaphyses	Removal of periosteum. Heavy abrasion	Broadening of sulcus	Broadening of striations

Table 6.3 (continued)

Tumbling experiment number	Bone	Duration (hours)	Treatment and observed modifications before tumbling	Observed modifications after tumbling		
				Alterations of outer bone tissues	Alterations of bone morphology	Bone-surface modifications
15	Fem-1-10	16	Edges rounded Heavy abrasion Slight exfoliation Singular striations	Heavy abrasion and polish	Rounding	Obliteration of striations
16	Rad-3-1	16	Removal of skin, meat and periosteum Artificial production of cut mark Meat and periosteum partly attached to the bone	Heavy abrasion and polish	Rounding	Obliteration of cut mark
17	Mt-7	16	Removal of skin, meat and periosteum Vertical striation due to periosteum removal Artificial production of a fine cut mark Periosteum still attached to metaphyses	Removal of periosteum Abrasion	Rounding Perforations	Smoothing of striation Cut mark vanished Formation of round punctures and diagonal striations
18	Hum-2-1	16	Removal of skin and meat Artificial production of a vertical cut mark Periosteum still attached to bone	Periosteum partly removed Abrasion Partly exfoliation to the porous structure	Edges rounded	Alteration of cut mark

we can thoroughly examine the crucial role of the periosteum in striation formation and alteration. The amount of periosteum covering a bone surface is directly responsible for striation formation (e.g., compare Tumbling Experiments 11.1 and 12). The results of Tumbling Experiments 9 and 10 clearly illustrate how the abrasive forces involved in the periosteum-removal processes during tumbling affected existing bone-surface modifications, and occasionally led to the formation of new, fine striations. The formation of clusters of irregular- and round-shaped punctures, a common feature on bone surfaces, appears to have depended on the presence of water (Tumbling Experiment 14). The formation and/or alteration of striations seems to have been a continuous process that took place for the entire duration of periosteum removal during tumbling, and which ceased only once the periosteum had been completely removed. Continuous periosteum reduction led to the disappearance of preexisting bone-surface modifications, while during the final stage of periosteum removal new surface modifications appeared. Only once the periosteum had been completely removed by abrasion could polishing occur. Moreover, once a bone reached this state of abrasion, no additional bone-surface modification could occur.

The surface morphology of each bone was a crucial variable, since it determines the time frame in which the processes described above could take place (e.g., compare Tumbling Experiments 15 and 16). Although a fresh cow bone fragment and a long bone fragment that had been buried for one year were involved in Tumbling Experiments 15 and 16, both displayed considerable abrasion and polishing. The results of the tumbling experiments are summarized in Fig. 6.98a, b.

6.5.2 Results of the Trampling Experiments

The trampling experiments were designed to mimic the effects of trampling on bone surfaces (see also Gaudzinski-Windheuser et al. 2010). A summary of the trampling experiments is given in Table 6.4. In general, the most common bone-surface modifications caused by trampling were abrasion and polishing. The former was controlled by two main factors: the amount of water in the sediment mixture and the shape and structure of the individual bone. The degree of water saturation affected the number of complete versus fragmented mollusks, both before and after trampling. Trampling in muddy conditions resulted in severe damage and destruction of the sediment mixture components (shells and bones).

The importance of the amount of water in the sediment mixture is highlighted by the results of the third dry trampling experiment, which yielded only slight bone abrasion.

During the trampling process, bone and sediment became compact and formed a stable surface that limited movement and helped to prevent further damage. In general, bone-surface polishing increased as more water was mixed with the sediment.

The second factor that controlled the degree of abrasion was the initial bone shape and structure (i.e., species, ontogenetic age, bone preservation), since these factors exert the greatest influence over the nature and timing of the polishing process. Our experiments included fresh and subfossil bovid bones, subfossil hyena skeletal elements, and fresh sheep bones, which all reacted differently to the trampling.

In contrast to bone shape and structure and water saturation, the amount of periosteum covering the bone surface played only a marginal role in surface modification. Moreover, it appeared that the abrasive processes that caused periosteum removal affected bones much more severely during trampling than during tumbling. Trampling Experiment 1 illustrates this quite clearly. Periosteum was scraped from the bovid tibia with a surgical scalpel, resulting in serious periosteum damage, but leaving tiny periosteum scraps attached to the bone. Trampling caused severe bone-surface damage as well as creating a deep flat channel.

Time played a rather marginal role in the degree of bone-surface modification. Its effect on the degree of abrasion depended primarily on the bone shape and structure and water saturation. Contact between bone and sediment during trampling led to the peeling of the external bone surface. This is very clear in Trampling Experiment 2, in which various striations appeared on the diaphysis because of the peeling of the bone surface.

Other surface modifications caused by trampling were highly distinctive and consistent in appearance. They were characterized by a pit mark from which a v-shaped, cross-sectioned striation would emerge. These distinctive marks are extremely similar to those produced by carnivore teeth. Our experiments have determined that striation formation and modification is part of the continuous process of abrasion. Striations that occurred during the first stages of abrasion were sometimes erased, as were any traces that existed prior to trampling (e.g., cut marks). At times, they were replaced by a new generation of striations. The processes that led to the disappearance of striations were selective. Quantitative and qualitative striation formation depended on bone shape and structure, as well as on the degree of water saturation. At a certain stage in the abrasion process, polishing occurred and erased all surface modifications.

One of the side effects of trampling was dry fracturing, which was observed following two hours of trampling in highly saturated sediment (Trampling Experiment 2). It resembles modifications similar to those produced by climatically induced weathering. The appearance of dry fractures did not depend on the trampling duration, and prolonged

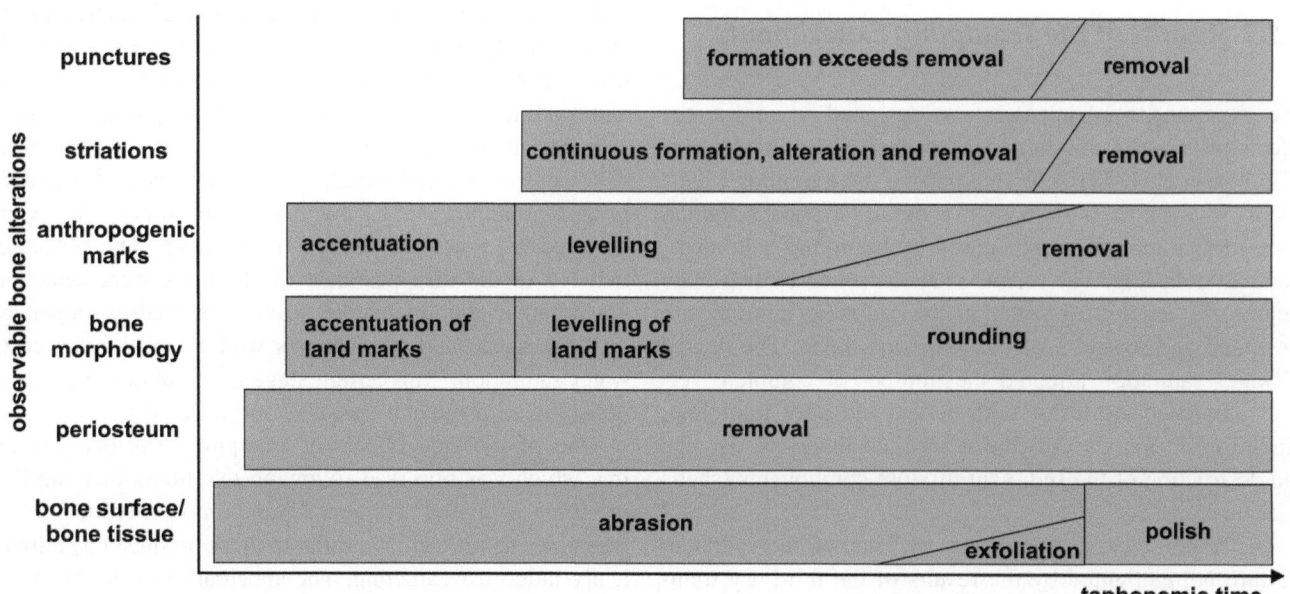

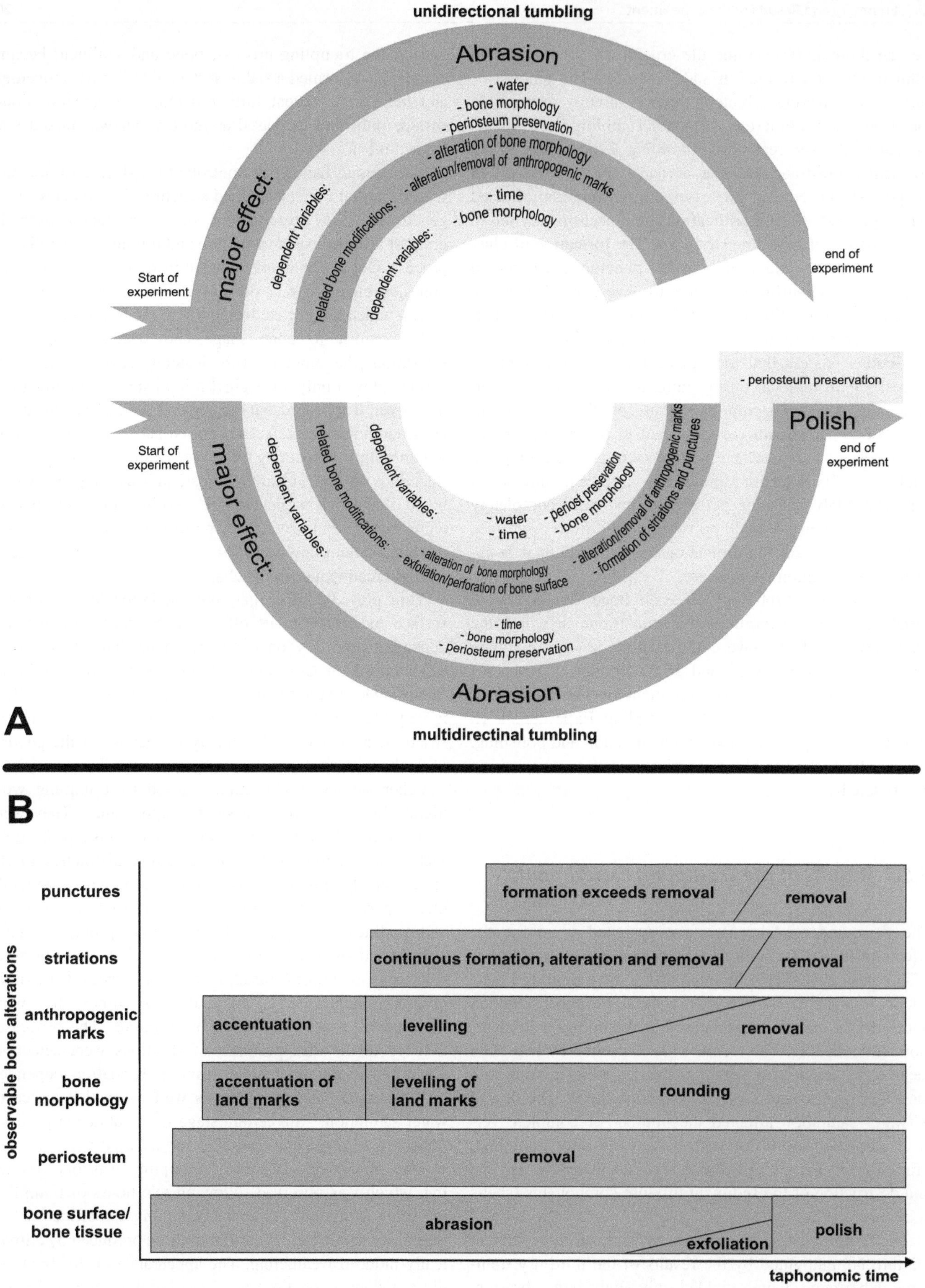

Fig. 6.98

trampling did not change the morphology of these fractures (Trampling Experiment 2.1).

Trampling resulted in the disintegration of several of the hyena ribs into fibrous strands (Trampling Experiment 4), also similar to a modification produced by climatically induced weathering (Fig. 6.99a, b).

6.6 Reconstructing the Taphonomic History at GBY Based on Analysis of the Bone-Surface Modifications

The results of the tumbling and trampling experiments have greatly assisted us in understanding much of the taphonomic history at GBY (see also Gaudzinski-Windheuser et al. 2010). The simulation of uni- and multidirectional water movement in the tumbling experiments resulted in differing degrees of bone-surface abrasion. The amount of water involved, bone shape and structure, and bone and periosteum preservation were all crucial controlling factors in the abrasion process. Time was an additional factor only in multidirectional water-movement experiments.

Preexisting striations were altered during tumbling. The formation of new striations was directly related to periosteum preservation and occurred only rarely. The general scarcity of striations produced by processes other than trampling and the production of differing degrees of abrasion were the most important findings of our experiments which were targeted at problems related to GBY data.

Surface modifications produced during the experiments differ significantly from bone-surface modifications observed on the prehistoric material from GBY. The bone abrasion observed at the site does not correlate with the expected lack of striations resulting from uni- and multidirectional water movement. We conclude that the movement of bones within sediment did not produce the striations observed on the bones at GBY.

Abrasion and polishing are the most common bone-surface modifications to result from the trampling experiments. Although many of the variables remained constant between experiments, it is apparent that bone shape, structure, preservation and water saturation were the dominant controlling factors. In fact, they determined whether or not periosteum preservation and/or trampling duration had an additional impact on the nature and degree of surface modification. Trampling not only regularly produced diagnostic modifications (e.g., pit marks) and, to a certain degree, various striation types, but also altered preexisting modifications. Examination of the GBY assemblages revealed that trampling was the major contributor to bone-surface modifications at the site. Our experiments serve to illustrate the importance of trampling for interpreting the mammalian assemblage at GBY.

As demonstrated in the experiments, bovid bones were much more susceptible to abrasion than bones of smaller mammals. The same was observed in a few cases in the GBY assemblages; here, too, bovid bones differ from other species in their overall higher degree of abrasion, which resulted in the near obliteration of their surface. Some cut-marks remained discernable, although they were mostly erased as well (Fig. 6.100). Other specimens displayed striations along abraded and exfoliated surfaces (Fig. 6.101). Bones in advanced stages of abrasion featured few striations or surfaces where any preexisting striations had been obliterated (Fig. 6.102).

For the large bovid bones in the experimental assemblages, advanced abrasion resulted in striation removal.

Fig. 6.98 Tumbling experiments. **A** Summary of the major effects and influencing variables. Abrasion, followed by the polishing of the bone surfaces, represent the major effects of multi- and unidirectional water movement. Variables influencing the abrasion progress differ for these two water movement types. Whereas water amount, bone morphology, and periosteum preservation are crucial variables for abrasion in multidirectional water movement, time, bone morphology, and periosteum preservation influence the abrasion progress in multidirectional water movement. During the abrasional process, changes in bone morphology and the removal of anthropogenically induced traces can be documented in unidirectional water movement. Time and bone morphology form crucial variables in such a case. Multidirectional water movement results in a variety of bone modifications; i.e., the alteration of bone morphology, the exfoliation and perforation of bone surfaces, the removal of anthropogenically induced marks, and the formation of striations and punctures. The crucial variables here are periosteum preservation, bone morphology, water amount and time. Polishing succeeds abrasion, and in this context, periosteum preservation forms a crucial variable in multidirectional water movement. Additional modifications to bone-surface modifications have not occurred. **B** Model of the taphonomic process for the tumbling experiments that mimic the effects of shoreline conditions. Observable bone modifications of the taphonomic process are given on the ordinate. The progress of the taphonomic process is given on an ordinal scale on the abscissa as taphonomic time (after Lyman 1994: 358). The abrasion process characterizes the taphonomic history of bone modifications in shoreline environments. This process in time is characterized by a distinct sequence of observable bone modifications. The process begins with the removal of the periosteum, followed by the accentuation of bone landmarks and anthropogenically induced traces. Along the process of abrasion periosteum removal leads to the leveling of these accentuated landmarks and traces. Simultaneously, a process begins of continuous formation, modification, striation removal and, later, punctures. During the course of this stage, anthropogenic marks are removed and the rounding of bones occurs, leading to a reduction of bone morphology in time. Disintegration of the bone structure also starts and minor exfoliation of the bone surfaces can be observed. The abrasional process is terminated with the complete polishing of the bone

Table 6.4 Results of the trampling experiments. Observed modifications after trampling are summarized as modification of bone tissue, bone morphology, and bone surface. (For identification of the bone specimens involved in the experiments, see Table 6.1)

Trampling experiment number	Bone	Duration (hours)	Treatment and observed modifications before trampling	Observed modifications after trampling		
				Alterations of outer bone tissues	Alterations of bone morphology	Bone-surface modifications
1	Mc-1	1.75	Slight abrasion Fine longitudinal striations	Abrasion	Rounding Flattening of sulcus	Formation of striation
	Mt-1		Slight abrasion Fine longitudinal striations	Abrasion	Rounding Alteration of foramen	Formation of striation Pronunciation of striation Obliteration of striations
	Fem-2-2		Removal of skin, meat and periosteum Horizontal, longitudinal and vertical cut marks due to periosteum removal Large parts still covered by periosteum	Removal of periosteum Abrasion Partly exfoliation to the porous structure	Rounding of edges	Formation of striation Obliteration of cut marks
	Tib-2-5		Removal of skin, meat and periosteum Large parts still covered by periosteum	Removal of periosteum in larger areas Abrasion Partly exfoliation to the porous structure	Rounding of edges	Formation of striation
2	Cost-1	2	Slight abrasion and exfoliation	Heavy abrasion Dry fractures	Destruction of complete bone portions	Formation of pit marks and striations
	Rad-1		Slight abrasion and exfoliation Fine diagonal striations	Heavy abrasion Dry fractures Porous structure partly exposed	Smoothing and reduction Levelling of land marks	Formation horizontal striations Formation of horizontal to diagonal striations together with round pit marks
2.1	Rad-1	2	Heavy abrasion Dry fractures Pit marks and striations	Increased abrasion Additional dry fractures		Obliteration of striations Formation of striations
3	Cost-2	1	Slight weathering Exfoliations	Slight abrasion, polish and exfoliation	Fracture	
	MtIII-1		Slight weathering Exfoliations	Slight abrasion, polish and exfoliation	Minor levelling of landmarks	Formation of pit marks and striations
	Pel-1		Slight weathering Exfoliations	Slight abrasion, polish and exfoliation	Fracture	
	Tib-3		Slight weathering Exfoliations	Slight abrasion, polish and exfoliation	Minor levelling of landmarks	Formation of pit marks
	Ul-1		Slight weathering Exfoliations	Slight abrasion, polish and exfoliation	Minor levelling of landmarks	Formation of pit marks

Table 6.4 (continued)

Trampling experiment number	Bone	Duration (hours)	Treatment and observed modifications before trampling	Observed modifications after trampling		
				Alterations of outer bone tissues	Alterations of bone morphology	Bone-surface modifications
4	Cost-3	2	Slight weathering / Exfoliations	Heavy abrasion	Destruction of complete bone portions / Compression and disintegration	Formation of horizontal striations and pit marks
	MtIV-1		Slight weathering / Exfoliations	Slight abrasion		Formation of horizontal striations and pit marks
	Pel-2		Slight weathering / Exfoliations	Heavy abrasion / Exfoliations	Fracture / Smoothing and reduction / Levelling of landmarks	Formation of horizontal striations and pit marks
	Rad-2		Slight weathering / Exfoliations	Severe abrasion / Dry fractures	Smoothing and reduction / Levelling of landmarks / Rounding	Formation of horizontal striations and pit marks
	Tib-4		Slight weathering / Exfoliations	Severe abrasion / Dry fractures	Smoothing and reduction / Levelling of landmarks / Rounding	Formation of horizontal striations and pit marks
5	Cost-4	2	Slight weathering / Exfoliations / Shallow scratches	Polish	Destruction of complete bone portions / Compression and disintegration	Formation of pit marks and perforations
	Cost-5		Slight weathering / Exfoliations / Shallow scratches	Polish	Destruction of complete bone portions / Compression and disintegration	Formation of pit marks and perforations
	MtII-1		Slight weathering / Exfoliations / Shallow scratches	Slight abrasion / Polish	Pronunciation of landmarks	Formation of pit marks
	Ul-2		Slight weathering / Exfoliations / Shallow scratches	Heavy abrasion / Polish / Exfoliations	Reduction / Rounding / Levelling of landmarks	Obliteration of exfoliations and scratches
6	Hum-1-1	2	Partly removal of skin, meat and periosteum / Large parts still covered by periosteum	Removal of periosteum / Polish	Rounding and smoothing / Alteration of breakage patterns	Formation of striations / Smoothing and polish of striations
	Mc-7		Partly removal of skin, meat and periosteum / Fine longitudinal striations due to periosteum removal / Large parts still covered by periosteum	Removal of periosteum / Slight abrasion	Pronunciation of landmarks	Obliteration of striations / Formation of horizontal striations and pit marks
	Mc-8		Partly removal of skin, meat and periosteum / Large parts still covered by periosteum	Removal of periosteum / Slight abrasion	Pronunciation of landmarks	Formation of horizontal striations and pit marks

Table 6.4 (continued)

Trampling experiment number	Bone	Duration (hours)	Treatment and observed modifications before trampling	Observed modifications after trampling		
				Alterations of outer bone tissues	Alterations of bone morphology	Bone-surface modifications
	Rad-3-2		Partly removal of skin, meat and periosteum Large parts still covered by periosteum	Removal of periosteum Fracturing Polish	Rounding and smoothing Alteration of breakage patterns	
7	Mc-9	2	Partly removal of skin, meat and periosteum Longtiudinal striations Large parts still covered by periosteum	Removal of periosteum Abrasion	Pronunciation of sulcus	Formation of oval to round punctures Smoothing of striations
	Mc-10		Partly removal of skin, meat and periosteum Longtiudinal striations Large parts still covered by periosteum	Removal of periosteum Abrasion		Formation of pit marks Pronunciation and obliteration of striations
	Rad-4-1		Partly removal of skin, meat and periosteum Large parts still covered by periosteum	Removal of periosteum Polish		Formation of pit marks and oval to round punctures Formation of fine horizontal, longitudinal striations, and puncture marks Formation of deep and broad striation
	Tib-5		Partly removal of skin, meat and periosteum Large parts still covered by periosteum	Removal of periosteum Abrasion		Formation of horizontal striations and pit marks

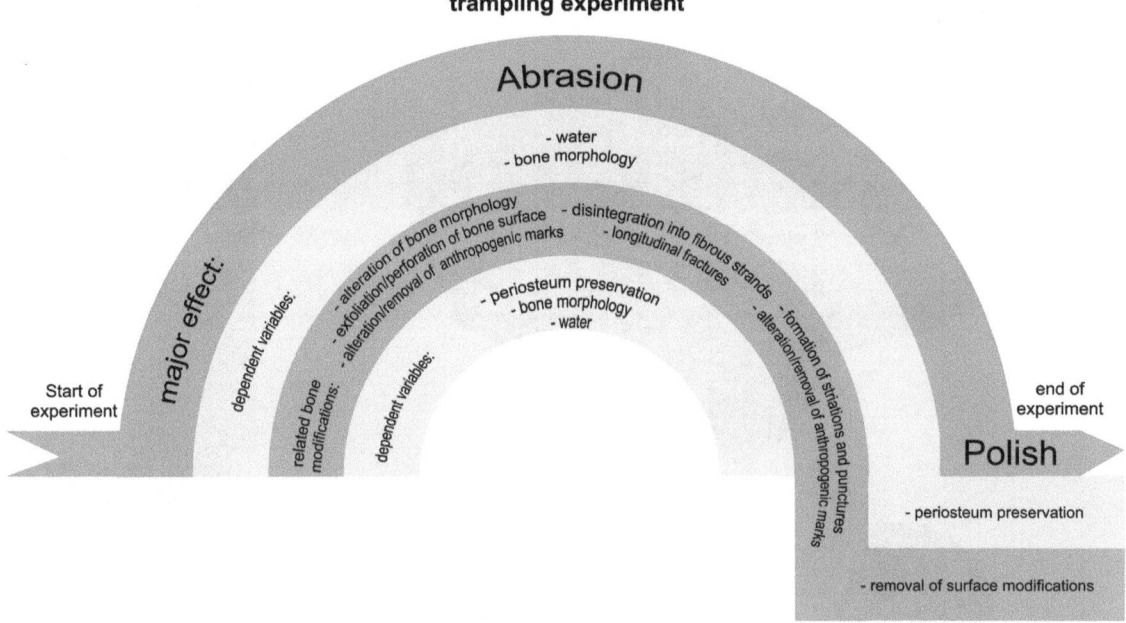

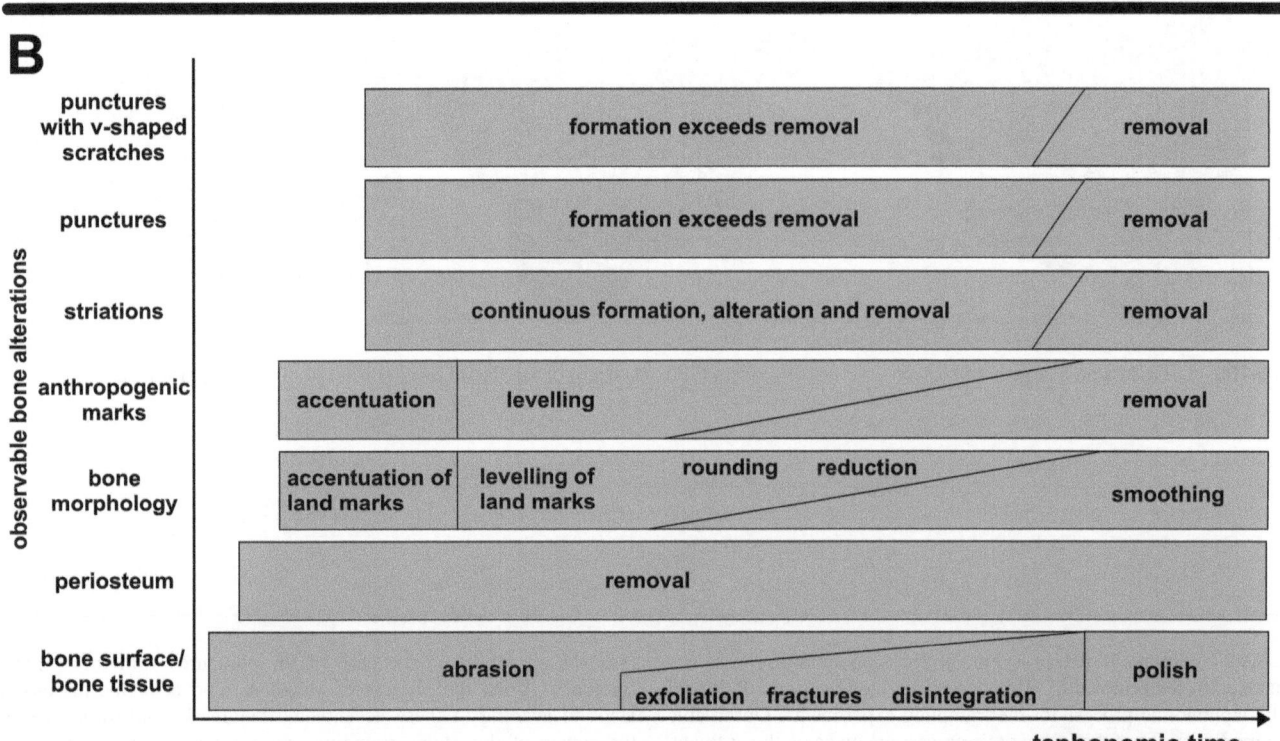

Fig. 6.99 A Model of the processes involved in trampling-induced bone modifications. Abrasion, followed by the polishing of bone surfaces and bones, represent the major effects of trampling. Variables influencing the abrasion progress are water amount and bone morphology. Trampling results in a variety of bone modifications; i.e., the modification of bone morphology, exfoliation/perforation of bone surfaces, disintegration of the bones into fibrous strands, longitudinal bone fractures, alteration/removal of anthropogenically induced traces, formation of striations, and formation of punctures with v-shaped scratches. The crucial variables here are periosteum preservation, bone morphology, and water amount. Polishing succeeds abrasion, and in this context the periosteum preservation forms a crucial variable in multidirectional

Fig. 6.100   Cut marks on a large bovid calcaneum (no. 896) from GBY: (I) detail of the lateral face; (II) detail of the posterior face

Fig. 6.99 (continued) water movement. The taphonomic process terminates with the removal of all bone-surface modifications. **B** Model of the taphonomic process for the trampling experiments. Observable bone modifications of the taphonomic process are given on the ordinate. The progress of the taphonomic process is given on an ordinal scale on the abscissa as taphonomic time (after Lyman 1994: 358). The abrasion process is characterized by a distinct sequence of observable bone modifications. The process begins with the removal of the periosteum, followed by the accentuation of bone landmarks and anthropogenically induced traces. Simultaneously, striations, punctures, and punctures with v-shaped scratches are formed. Their formation exceeds their

removal, especially in the case of the punctures with the v-shaped scratches. Along the process of abrasion removal of the periosteum leads to leveling of these accentuated landmarks and traces. During the course of this stage, rounding and exfoliation occur, and a process begins by which anthropogenically induced marks can be removed. With taphonomic time, reduction of bone morphology begins, accompanied by longitudinal fractures, finally resulting in the disintegration of the bone structure. From now on, bone polishing replaces abrasion, which results in the obliteration of observerable traces and the smoothing of the bone surfaces

Fig. 6.101 Striation on a large bovid metatarsus (no. 1,185) from GBY showing surface abrasion. *Arrow* indicates the area of the bone that was magnified

Bone-surface modifications such as cut marks could only survive in areas protected from abrasive forces (Fig. 6.103). The trampling experiments revealed that both striation-formation and erasure on the bovid bones occurred within a relatively short time period as part of the continuous abrasion process (and in the early stages), and was directly related to preexisting surface preservation.

GBY *Dama* bones are only slightly abraded and, in fact, the majority of the bones displayed no traces of abrasion. As a result, we were able to observe all bone-surface modification attributable to trampling. Numerous bones exhibited fresh-looking cut marks that were not accompanied by any other surface modification (Fig. 6.104), and some

still featured traces of periosteum removal (Fig. 6.105). In the majority of cases, however, distinct cut marks appeared together with a combination of other striation types, and only rarely could we determine their possible order of appearance (Fig. 6.106). A variety of bones displaying such traces are also characterized by partial abrasion of the bone surface (Fig. 6.107). In addition, pit marks sometimes superimposed the striations (Fig. 6.108), and comprise the final stages in the destruction of existing surface modifications. Only a minority of the bones at GBY display abrasion resulting from trampling, which resulted in the complete obliteration of preexisting surface modifications, such as cut marks (Fig. 6.109). In general, the modifications observed on the

Fig. 6.102 A large bovid metatarsus (no. 544) from GBY in an advanced stage of abrasion. An identical detail of the bone is seen at different degrees of magnification. A *rectangle* indicates the area of the bone that was magnified

GBY *Dama* bones occurred within the interval between the formation and destruction of trampling-induced surface modifications. Only a small number of bones exhibited heavy abrasion or the complete obliteration of preexisting bone-surface modifications.

Although preservation of bone-surface modifications differs between bovids and cervids, we cannot assume that each taxon underwent a different taphonomic history. Our experiments indicate that bovid bones are more susceptible to abrasion than bones of smaller species. Certain differences in bone-surface modifications between GBY bovids and cervids may be explained by their specific bone structure. However,

one should take into consideration the extensive difference between sample sizes of the two genera (Table 4.1). We suggest that the same degree of trampling-induced abrasion created different bone-surface modifications between the two taxa (Fig. 6.110). Furthermore, the results of the experiments indicate that the bones were still in fresh condition when trampling ceased, since neither fibrous bone destruction nor dry fractures were observed.

A few *Dama* bones display a degree of abrasion that exceeded that of the bovids or cervids. The *Dama* bones exhibited an advanced stage of abrasion preceding the appearance of polish and could thus be a result of the same

Fig. 6.103 Long, vertical cut mark inside the anterior sulcus of a large bovid metatarsus (no. 544) from GBY. The scale bar of the microscope images is 1 mm

taphonomic processes. As the spatial distribution of these bones did not differ from the remaining faunal elements, both small-scale changes in the matrix and localized differences in water saturation can be excluded as explanations. Therefore, these bones must have been subjected to a more prolonged abrasive process than the other *Dama* and bovid bones. The most likely explanation here lies in the intrusion of these elements from the effects of a shoreline environment.

Our experiments also shed new light on striation formation. Their formation and destruction was a continuous process that was always connected to morphological changes in the preexisting striations, and is well illustrated by overlapping marks (Fig. 6.111). Surface modifications produced during the experiments were comparable to many of the striations that characterized certain *Dama* bones at GBY. In a detailed analysis of the experimental bone surfaces continuous morphological changes were noted, since various clusters of striations on one bone often exhibited varying destruction stages. The striation variety we observed on the experimental bones was also noted in the GBY material. We therefore suggest that the obliteration of striations

was the result of trampling. Some striations, even when overlapping, display minimal evidence of trampling-induced abrasion. Different striation types have been observed at GBY, a pattern that does not correlate with the expected results of our trampling experiments, since the later stages of abrasion involving striation obliteration were missing from the GBY bones.

Cut marks resulting from hominin butchery were found on many bones at GBY. The fact that all the cut marks displayed comparable degrees of surface alteration reinforces the contemporaneousness of the striations, and points to the butchery of the respective animals.

According to our reconstruction of the taphonomic history at GBY, hominins and animals were the most dominant agents responsible for the striations. Our experiment results demonstrate that the GBY bones displayed and preserved a complex variety of surface modification reflective of an entire contemporaneous butchering sequence. The taphonomic processes that produced these modifications were highly dependent on the specific bone structure of differently sized animals. We conclude that tumbling and, especially, trampling processes seriously blurred or destroyed numerous

Fig. 6.104 Fresh-looking cut marks on a *Dama* scapula (no. 2,074) from GBY

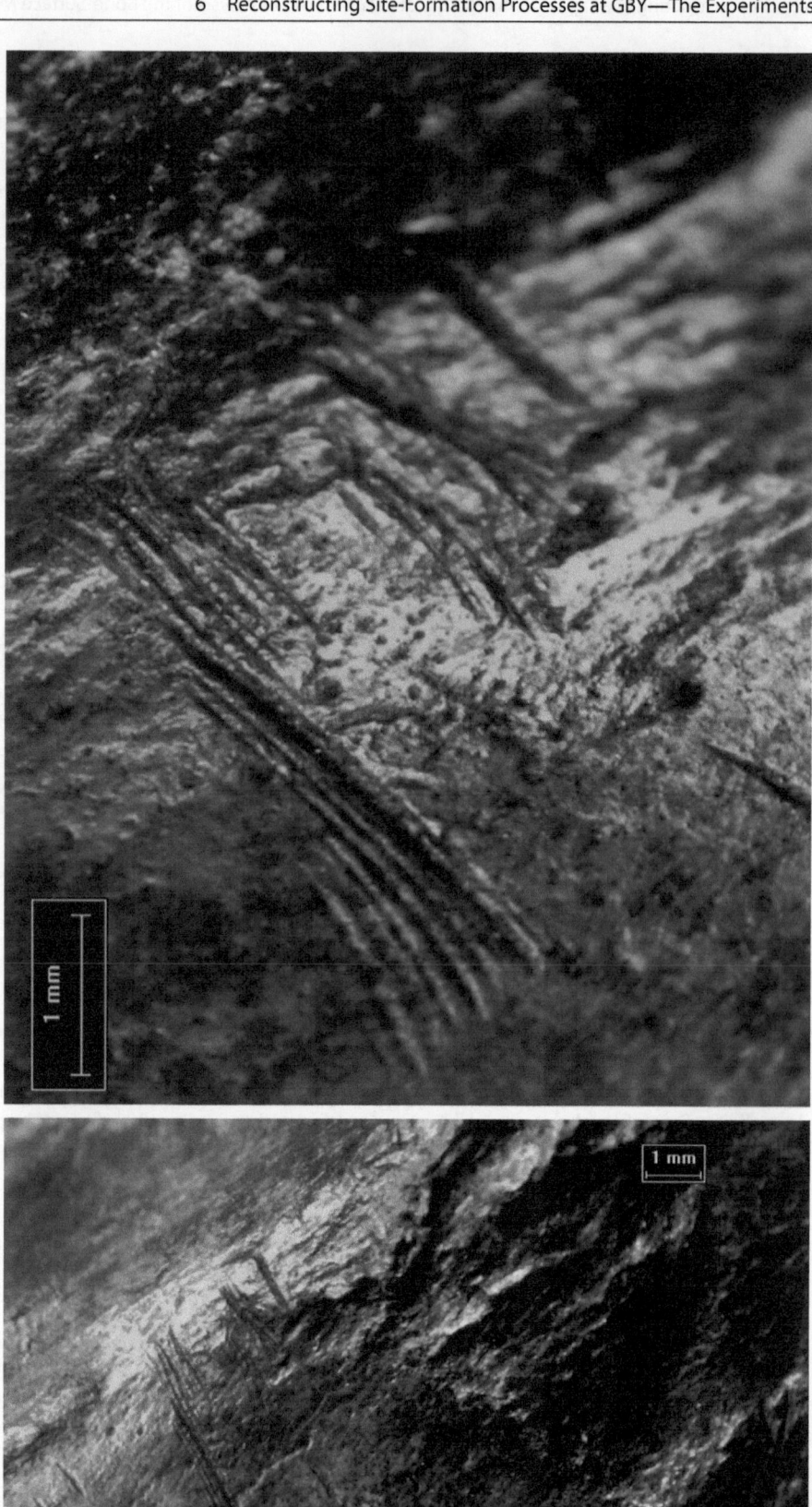

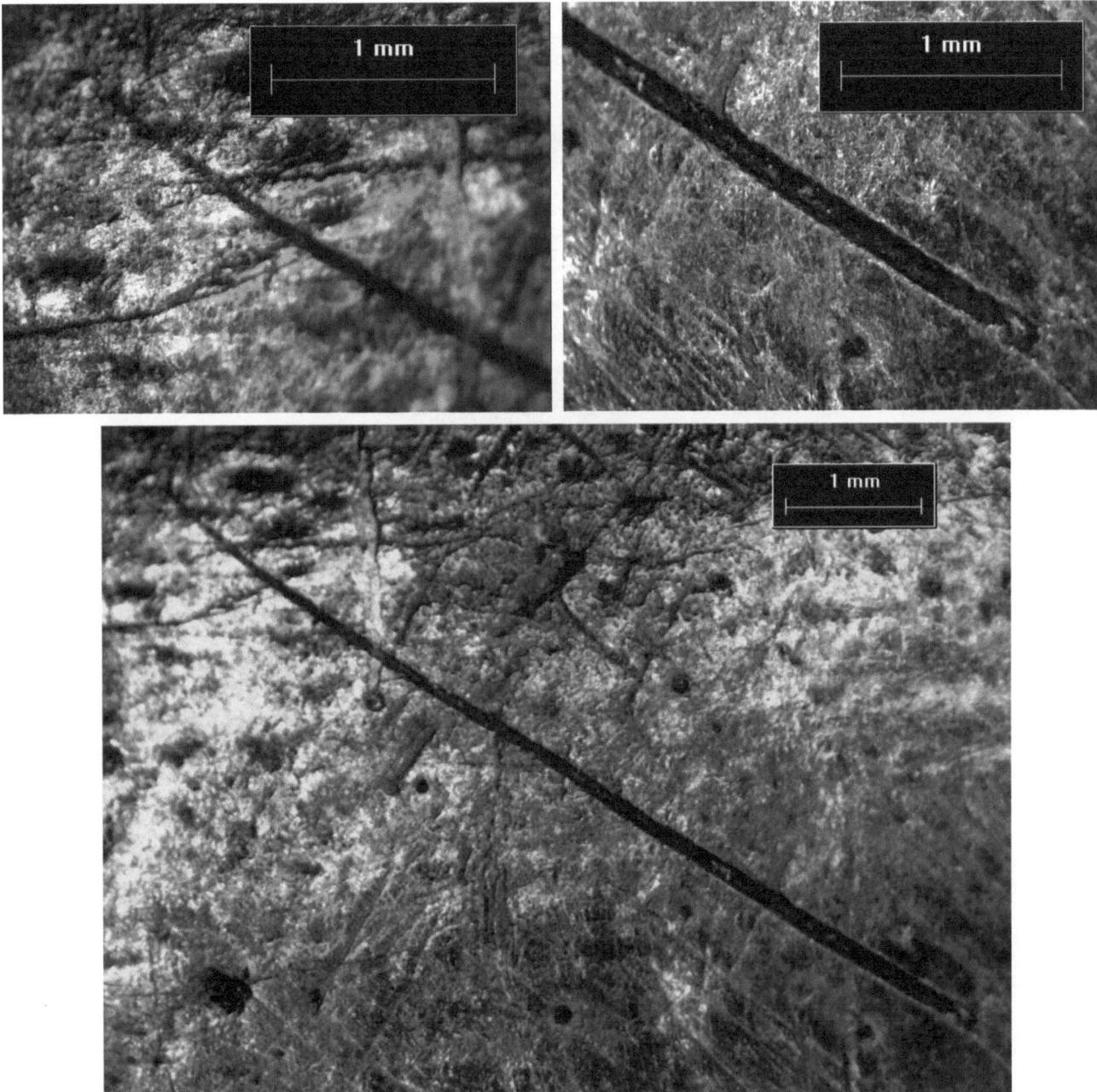

Fig. 6.105 Mark resulting from periosteum removal on a long bone fragment (no. 1,507) from GBY. *Top row*: magnified start- and endpoint of the mark

preexisting butchering traces on bones of animals larger than *Dama*. Thus, the surface appearance of these bones no longer reflects the entire sequence of hominin subsistence strategies.

The implications of these taphonomic processes are highlighted when considering the assumed contemporaneousness of *Dama* and large bovid bones at GBY. The experiments demonstrated that bovid bones are more susceptible to the reproduced abrasive process than are *Dama* bones. Since the bovid bones display both cut- and percussion marks, it seems highly plausible that the quality and quantity of these traces were altered by the above-mentioned processes. As a consequence, only a limited reconstruction of the bovid butchering sequence was possible.

The observation that bovid bones are more susceptible to abrasion than *Dama* bones is highly important for our understanding of past hominin behavior. It seems very unlikely that hominins would have repeatedly transported complete

Fig. 6.106 Cut marks and striations on a *Dama* humerus (no. 883) from GBY

Dama and bovid carcasses obtained by scavenging to the lake margins. Equally unlikely is the assumption that the natural deaths of several *Dama* and bovids occurred at the lake margins during the short time interval that is clearly represented here. The most plausible explanation for the accumulation and alteration of these bones is a hunting scenario; specifically, that individual animals approaching the water were killed by hominins and processed in situ. Rather than assuming a scenario of constant mobility, we suggest that the occupants of GBY considered the site's location as strategically important and attractive, and thus chose to frequently occupy it. That GBY represents such a hot spot of hominin activity becomes obvious when the entire archaeological record is considered, such as plant use, use of fire, extensive stone tool production, etc.

Under these circumstances, the uniquely high level of hominin-induced bone modifications reflects the intensity of activities occurring at the site. It becomes apparent that the very specific hominin behavioral patterns seen at GBY most likely had a much longer ancestry. For example, the nature and intensity of *Dama* carcass processing, wherein hominins exploited small quantities of marrow by splitting phalanges, suggests a cumulative and detailed knowledge of butchery practices.

Fig. 6.107 Striations and partial abrasion of the bone surface of a *Dama* tibia (no. 863) from GBY. The same locations of the bone surface is seen from different directions and at different degrees of magnification

Fig. 6.108 Pit marks among striations shown at different degrees of magnification on **a** a *Dama* tibia (no. 1,616) from GBY and **b** on two locations (I, II) of a long bone shaft fragment (no. 12,829) from GBY

Fig. 6.109 *Dama* bones from GBY. Differences in bone surface preservation ranging from pristine (*above*) to heavily abraded (*below*)

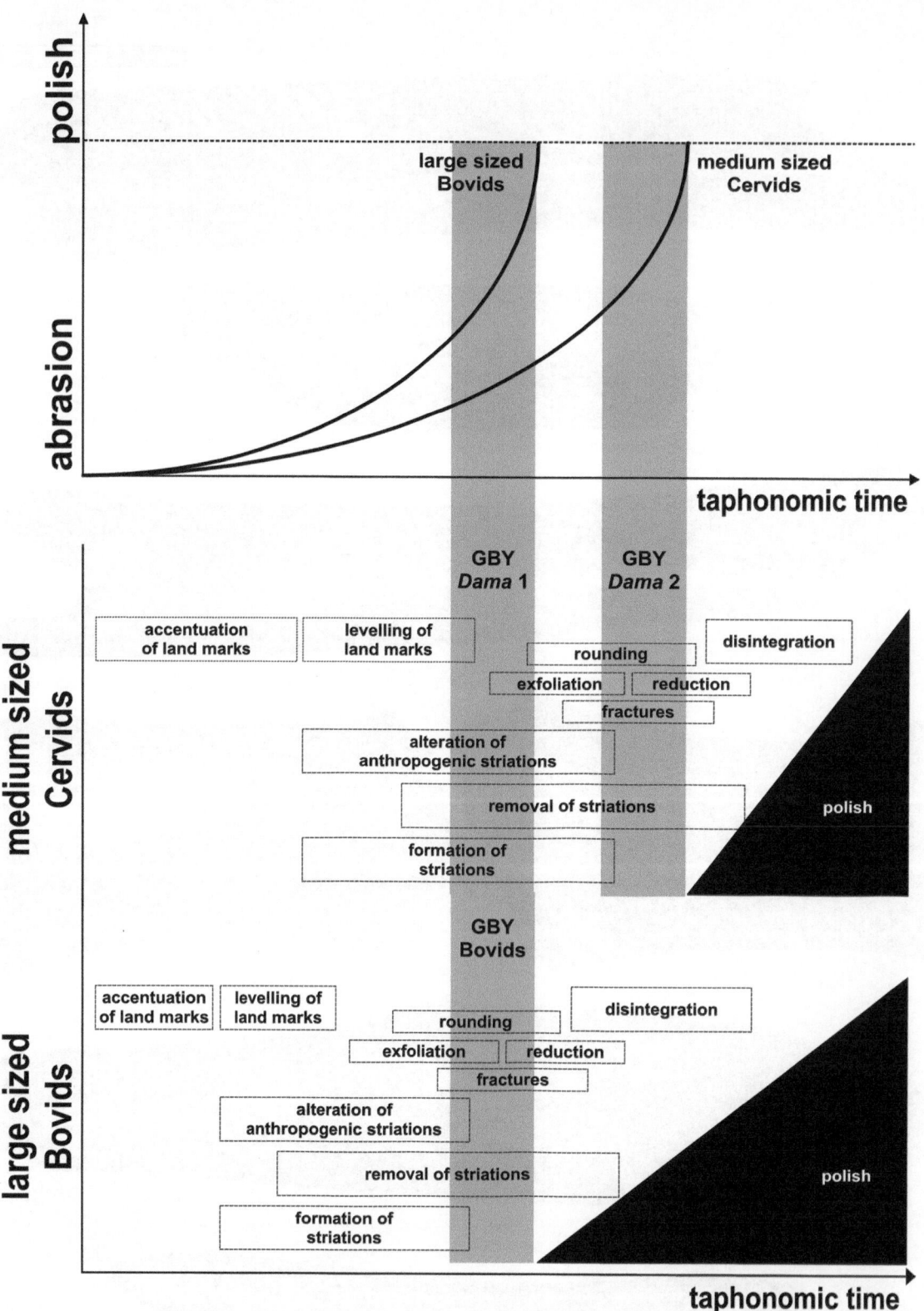

Fig. 6.110 Model of the abrasional process for bones from large bovids and medium cervids from GBY against the background of results obtained by actualistic studies. *Above*: summary of the results of the abrasional process in the trampling experiments on the bones of different-sized taxa. Bones of large-sized bovids undergo this process more rapidly than those of medium-sized cervids. The progress of the taphonomic process is given on an ordinal scale on the abscissa as taphonomic time (after Lyman 1994: 358). *Below*: the progress of the abrasional process for large-sized bovids and medium-sized cervids as

is indicated by the results of the actualistic studies. The *grey bars* illustrate the position of the cervid and bovid bones within the abrasional process. Although differences in bone surface preservation have been observed at GBY, these bones may reflect a corresponding stage within the process itself (*Dama 1*, Bovids), whereas part of the assemblage (*Dama 2*) must have been subjected to a more prolonged abrasive process. The progress of the taphonomic process is given on an ordinal scale on the abscissa as taphonomic time (after Lyman 1994: 358)

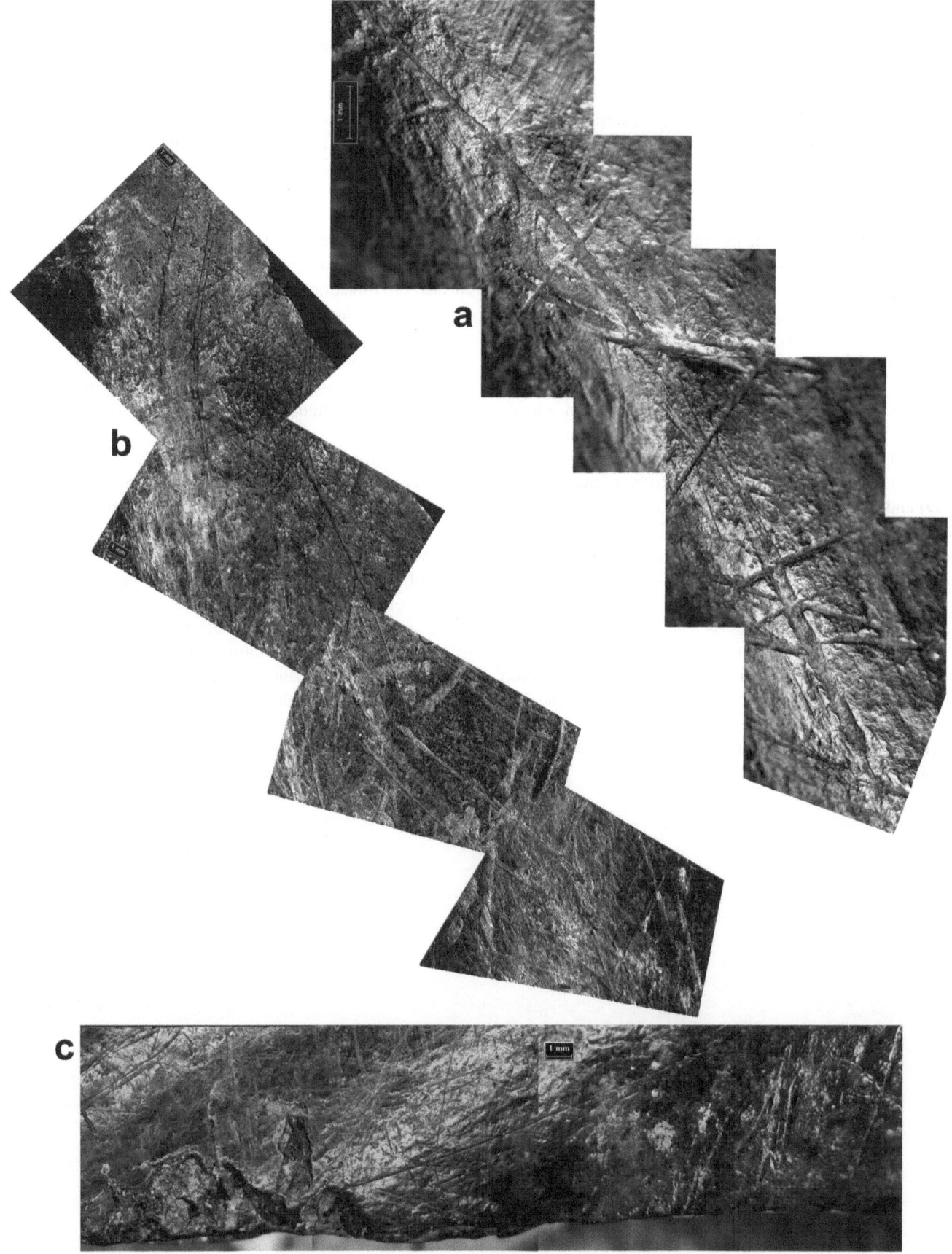

Fig. 6.111 Examples of homogeneity in superimposed striations on **a** a femur (no. 1,015) from GBY, **b** a scapula (no. 2,232) from GBY, and **c** a humerus (no. 881) from GBY

6.7 The Implications of the Experiments for Taphonomic Research

As detailed in the previous chapter, our studies were designed to mimic the individual environmental conditions at GBY. We succeeded in outlining the chronology of biostratonomic factors and processes at the site and, for the first time, were able to obtain insight into the additive sequencing of a taphonomic process.

Our observations have major implications for taphonomic studies, as they show that interpretations based on separated single taphonomic processes may be misleading. Against a biostratonomic background, bone-surface modifications must be considered as a process in which bone surfaces are continuously altered. This implies that interpretations of biotically and abiotically induced modifications according to diagnostic templates, but without their integration into the complete biostratonomic sequence, are highly ambiguous.

Recent studies on isolated taphonomic features—e.g., the frequency of both cut- (Domínguez-Rodrigo and Yarverda 2009) and percussion-marks (Galàn et al. 2009) on bones—illustrate the high variability of these traces, which hampers their verification and quantification and leads to equifinality in their interpretation. These studies acknowledge the modifying qualities of other subsequent and superimposing biostratonomic processes. As the majority of these experimental studies were not site-based, these additional processes should not be taken into account. As a result, interpretations of significant correlations between the various variables involved in these studies, which are expressed in clearly discriminable patterns, are still confronted by equifinality.

A consequent expected problem concerns the question of how diagnostic taphonomic patterns can be with regard to specific processes and agents. Studies of climatically induced weathering and abrasion usually outline that the degree of bone destruction correlates with the time between the death of an individual and its final burial. As a consequence, homogeneity in bone-surface preservation is considered an indication for the assumption that homogeneous environmental conditions prevailed and/or the homogeneity of an entire bone assemblage. Our experiments included a very short abrasive process that produced certain diagnostic bone-surface modifications that are indistinguishable from bone-surface modifications produced, in turn, from long-term, climatically induced weathering. We also demonstrated that the process of trampling is characterised by the formation, modification, and erosion of traces sharing the same morphology as both hominin-induced cut marks and carnivore-induced tooth marks.

Against this background, bone-surface modifications at GBY bear their charateristics only at the beginning of the abrasive process, together with initial striation formation, and again later in the process, before abrasion causes morphological changes in the bones. Moreover, ontogenesis and individual body size are variables that influence the velocity of the abrasive process's impact.

As a consequence, strict determination of the time of cessation of the biostratonomic process is mandatory for diagnosing and interpreting bone-surface modifications (Gaudzinski-Windheuser et al. 2010). In this context, imposing generalized interpretations on fossil assemblages deprives them of their individual history and deprives us of their potential scientific value.

Chapter 7

A Reconstruction of the Taphonomic History of GBY

Abstract This chapter aims to reconstruct the taphonomic processes that influenced the assemblages of Layers V-5 and V-6, as well as to identify the taphonomic role of particular agents that may have influenced the fossil bones and the stone artifacts deposited in these layers, and considers these results vis-à-vis other data, both from the site and experimental.

The chapter begins with a survey of the biogeographical origin and the paleoecology of the species identified in the two layers (mammals, birds, and chelonians), enabling habitat reconstruction. Taphonomic analysis of the bone assemblages (see Chapters 5 and 6) has demonstrated that hominins were the major agent responsible for the accumulation of the medium- to large-sized mammalian fauna. In this context, the habitat reconstruction takes into consideration prey availability, flight distance, hunting method, handling time and return rate (O'Connell et al. 1988, 1990; Kelly 1995; O'Connell et al. 2002), as well as gender and social interaction (Hawkes et al. 2001), all which shed light on hominin prey choice and prey-sampling strategies. Subsequently, a comprehensive taphonomic history of the bone assemblages will provide insight into the character of hominin prey exploitation and butchering strategies reflected in the different assemblages.

Bone taphonomy may also provide valuable information for the interpretation of a site's general character and genesis. In the case of GBY, interpretation of skeletal-element representation, and particularly the results of the bone-striation formation experiments, contributes to the reconstruction of a broader picture.

We believe that this study yielded information that will help identify hominin behavior at GBY more accurately. Subsequently, the taphonomy of the faunal remains and the lithic assemblages, as well as the assemblages' spatial organization, will lead towards a more detailed reconstruction of the site-formation processes in Area C.

7.1 Biogeographical Origin of the Faunal Assemblages

The diverse faunal assemblages at GBY consist of various mammalian taxa (see Chapter 4) dominated by *Dama* sp. (Fig. 7.1). The rest of the vertebrates include micromammals, birds, turtles, reptiles and amphibians (Table 7.1), and fish (Chapter 4 and references therein). The invertebrates are rich and varied, and include both mollusks and crustaceans (as above).

The GBY fauna is contemporaneous with the Middle Pleistocene Galerian (Italian terminology) or Cromerian (British terminology) faunas (Goren-Inbar et al. 2000), whose mammalian biochron is known from numerous Eurasian localities (Palombo et al. 2003; Palombo 2005; Palombo and Sardella 2007).

Levantine assemblages were found to have a unique composition that differs from both European and African assemblages, thus requiring special consideration. The distinct biogeographical signal of the Levant, suggests that the separation of sub-Saharan Africa and the Levant took place earlier than the Early-Middle Pleistocene boundary (O'Regan et al. 2005). If so, then the GBY assemblages already feature the characteristics of a Levantine fauna.

Much of the fauna that prevailed in the Dead Sea Rift, a segment of the Great African rift system, is a mixture of African (e.g., bovids, equids, hippopotamus) and Euro-Asian (e.g., cervids, *Ursus*, *Sus*, *Vulpes*) origins. The Levantine Corridor served as an optimal migration route for many species and, indeed, enabled their survival throughout the Pleistocene.

Several genera were already established in the area by the Early Pleistocene, such as the *Hippopotamus* of African origin, which was present in the area since the Pliocene (Hooijer 1958), and the genus *Gazella*, known in the Mediterranean Basin since the Miocene (Tchernov 1988).

R. Rabinovich et al., *The Acheulian Site of Gesher Benot Ya'aqov Volume III: Mammalian Taphonomy. The Assemblages of Layers V-5 and V-6*, Vertebrate Paleobiology and Paleoanthropology, DOI 10.1007/978-94-007-2159-3_7, © Springer Science+Business Media B.V. 2012

Fig. 7.1 General faunal group
distribution

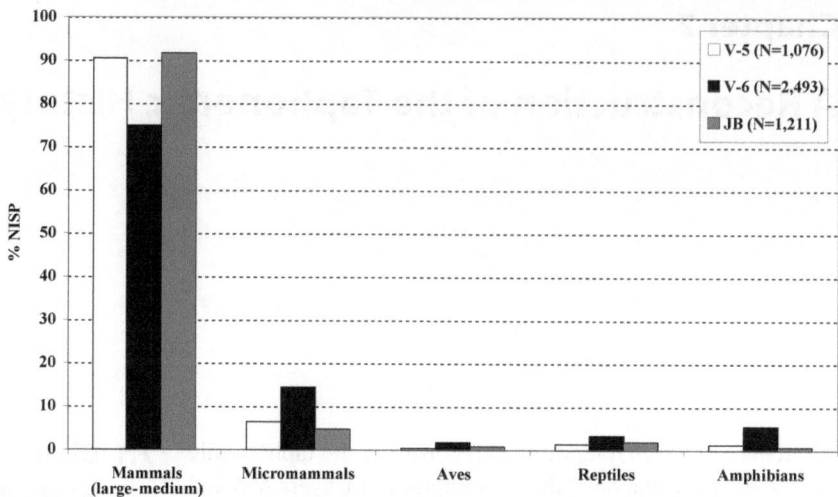

Table 7.1 Various animal taxa from Area C and the JB

	V-5 NISP	V-6 NISP	JB NISP
Mammals (medium and large)	974	1,868	1,111
Micromammals[a]	53	431	50
Birds[b]	4	44	10
Reptiles[c]	14	79	19
Amphibians[c]	9	124	8
Total	1,076	2,493	1,211

[a]Tchernov n.d.; Rabinovich and Biton (2011)
[b]Simmons (2004)
[c]Rabinovich and Biton (2011), Hartman (2004)

Indeed, in a recent analysis of the Bovidae from GBY (Martínez-Navarro and Rabinovich 2011), a mixture of species of Holarctic and Ethiopian origins was observed. More specifically it was suggested that the *Bos* sp. at GBY was probably "a new immigrant evolved in eastern Africa, *Bos buiaensis*, which dispersed into the North during the Early-Middle Pleistocene boundary", while the species Bovini gen. et sp. indet. (cf. *Bison* sp.) represents an Eurasian immigrant into the Levantine Corridor (as above).

The coexistence at GBY of three cervid sizes—Megalocerine, *Cervus elaphus* and *Dama* sp. (large, medium, and small), both browsers and grazers—in the same environment is also known from other Mediterranean sites; e.g., the cave of Yarimburgaz in Turkey (Stiner et al. 1996) and the cave of Petralona in Greece (Tsoukala 1991; Darlas 1995).

The site of 'Ubeidiya represents a broad archive for our knowledge of species occurrence in the Levant 1.4 million years ago. Since 'Ubeidiya's assemblages originated in large excavations that exposed different depositional environments and hence different taphonomic histories, a detailed comparison of species between GBY and 'Ubeidiya is problematic. Nevertheless, we may conclude that several GBY species

appeared earlier at 'Ubeidiya (see Chapter 4), suggesting a continuation of certain faunal Levantine "land marks." This is particularly expressed in the following medium- to large-sized mammals found at GBY, listed in order of descending abundance: cervids, bovids, equids, hippos, and others. The comparison of the GBY bovids with those of 'Ubeidiya indicates that all the bovid genera from GBY have also been recorded at 'Ubeidiya (Geraads 1986; Martínez-Navarro et al. in preparation). In the past, the absence of a Megalocerine at GBY was used to illustrate differences between the older faunas of 'Ubeidiya and Latamne and the younger one of GBY. The taxonomic determination of Megalocerine remains at GBY, which are similar in size to the 'Ubeidiya specimens, may point to a continuity of this species in the Levant (but see Belmaker 2009).

The composition of the community of medium- to large-sized mammals at GBY suggests that *Megaloceros* (Stuart and Lister 2001), *Palaeoloxodon antiquus*, and *Pelorovis/Bos* served as important chrono-stratigraphic markers of the Early-Middle Pleistocene in the Levant. These large species (large deer, large bovid, and elephant) disappeared from the southern Levant by the end of the Middle Pleistocene, but in Europe were known also during the Middle and Late Pleistocene. Lister (2004) proposes that the late appearance of *Palaeoloxodon* ca. 0.8–0.6 Ma, and its competition for woodland habitats with the *Mammuthus trogontherii*, triggered the extinction of the latter.

Martínez-Navarro and others (2007; Martínez-Navarro and Rabinovich 2011) favor the view that the dispersal of the *Bos* lineage from Africa, related to the latest Pleistocene *B. buiaensis* from Buia (Eritrea), was contemporaneous with Acheulean sites outside Africa, such as 'Ubeidiya and GBY, and later colonized the Eurasian continent, evolving into the Middle Pleistocene *B. primigenius*. The dispersal of this species is parallel with that of the Acheulian culture into the northern continent. Other species such as the carnivores

Crocuta crocuta, and later *Panthera leo* and *Panthera pardus*, and the megaherbivore *Palaeoloxodon antiquus* are also part of the same faunal dispersal out of Africa. Martínez-Navarro (2010) termed this faunal turnover "the *Crocuta crocuta* event." Therefore, in the context of Early-Middle Pleistocene faunal turnover and human dispersals, GBY, dated 0.7–0.8 Ma, represents a key site (as above).

The micromammalian species from Area C and the JB (Goren-Inbar et al. 2000: table 1) reflect Afro-Eurasian (*Lepus capensis*), Palearctic (*Mimomys* cf. *ostramosensis*), Southern Levantine (*Microtus guentheri*), and Levantine (*Mus macedonicus, Spalax ehrenbergi, Hystrix* cf. *indica*) biogeographic origins. *Lepus capensis, Procavia syriaca* and *Mimomys* cf. *ostramosensis* have their first appearance in the GBY sequence. Tchernov (in Goren-Inbar et al. 2000) viewed the *Mimomys* cf. *ostramosensis* as a stage in the *M. savini-M. ostramosensis* chronocline, which is missing from the European record and "hence may have taken place in the southern Levant about 1.0–0.7 Ma," (Goren-Inbar et al. 2000: 946).

The biogeographical origin of the birds at GBY is complex, as most species at the site are migratory. These are generally of Palearctic and Holoarctic biogeographic origin. In her analysis of the GBY birds, Simmons (2004) suggested the antiquity of the migratory route of northern birds through the Dead Sea Rift to Africa, via the Great African Rift system. Simmons further proposed that the changes between the avifauna of GBY and that of 'Ubeidiya (the latter include birds of Palearctic, Holoarctic, Mediterranean, and Ethiopian origin; Tchernov 1962, 1980) may have resulted from a climatic condition in which arctic birds were found throughout the year, or, alternatively, the short-term seasonal presence of birds at GBY (Simmons 2004).

The biogeographic origin of the herpetofauna species identified at GBY is Palearctic (Rabinovich and Biton 2011). We can thus conclude that all faunal components present in Area C and the JB reflect the specific location of the site along the Levantine Corridor.

7.2 Paleoecological Reconstruction of GBY Faunal Assemblages

Faunal paleoecological reconstructions are based on the unique niche requirements of the particular species. For a fine tuning of the reconstruction of climatic changes, micromammals are commonly used (i.e., Tchernov 1981, 1988; Fernández-Jalvo 1995; Cuenca-Bescós et al. 2009, 2010), but medium- and large-sized mammals were also revealed to be an important source for paleoclimate reconstruction (i.e., Vrba 1975; Reed 1998; Hernández Fernández and Vrba 2006). Numerous methods exist for faunal paleoecological reconstruction, utilizing the presence and/or absence of species, their relative frequency, body size, diet, locomotion, energetic requirements, etc. Most methods apply complex statistical procedures suitable for large data sets. Ideally, comparisons with assemblages originating in faunal localities with known taphonomic histories (Soligo and Andrews 2005) permit a more complete reconstruction of the paleoenvironment in terms of species distribution and body-size representation (i.e., on Untermassfeld, see Kahlke 2007; on localities in Italy, see Palombo et al. 2003).

Since the assemblages under examination are small, we limit the discussion of the GBY mammalian fauna to paleoecological reconstruction based on the species' presence/absence frequency and niche requirements. It should be noted that the application of some of the methods for niche reconstruction is problematic. The presence/absence of certain species in a fossil assemblage does not necessarily reflect paleoecological conditions, as additional factors—e.g., recovery bias or the presence of micro-habitats—may be responsible for a species' presence and abundance. For example, the scarcity of elephant skeletal-elements at the site may be related to the size of the exposed/excavated area. In the southern Levant, foraging patches are relatively small, due to the influence of the natural surroundings on the herd, the home range, and the population size (Horowitz 2003; Davies and Lister 2007), all which lower this species density/availability.

In the above sections we have tried to show what were the major taphonomic forces involved in the accumulation of the faunal assemblages at GBY. The hominin impact was quite prominent on the formation of the depositional environment, resulting in possible limited species representation, since the fauna accumulated by mainly a single agent over a relatively short period of time. The near absence of large carnivores, few medium-sized carnivores, and the overabundance of *Dama* might also be a result of the same accumulator agent. As was noted above, the palaeoenvironmental conditions at GBY, based on several proxies, were Mediterranean (see Chapter 2). We can only assume what was the presumed habitat of the animal species based on recent animal-habitat associations. Fauna presently found in the Mediterranean does not include many of the species discovered at the site, especially of the medium- to large-sized range. One reason is the eradication of many species in the nineteenth century due to the introduction of firearms, followed by overgrazing and human overpopulating (Mendelssohn and Yom-Tov 1999).

Nevertheless, the community of medium- and large-sized mammals in the fossil bone assemblages of Area C and the JB indicate a rich and diverse environment during the Middle Pleistocene. The abundance of various species within the different layers is consistent and suggests that the fossil remains generally accumulated under similar conditions (Fig. 7.1).

Under the above-mentioned constrains of habitat association, it seems that the mammalian species at GBY point to the existence of both a grassland and woodland habitat, with the latter occupied by elephants, cervids, and some bovids, and the former by equids and other bovids. As seen by the dominance of cervids in the site's faunal assemblages, woodland species prevailed within the community, though this may have been the result of the hominin selection from the environment.

The recovered artiodactyl species from the site are typically Mediterranean: *Gazella* lives in different habitats, but not in dense forest; *Dama* occupies woodlands; *Bos* prefers open parkland, swamps, and river valleys; and *Sus* prefers dense thickets, forest, and riverine habitats (Mendelssohn and Yom-Tov 1999).

Paleoloxodon antiquus was found in a variety of environments from wooded to more open ones (Davies and Lister 2007). As in the case of the modern African ass (a possible descendant of *E.* cf. *africanus*), density, distribution continuity, forage biomass, water sources (Moehlman 2002), and large tracts of land (Saltz 2002) are key factors in sustaining a viable population.

Carnivores occupy a variety of ecological settings due to their various locomotion skills and foraging methods, and the various food types on which they feed (Gittleman 1989). Their scarcity in Early Paleolithic open-air assemblages is known from Evron (Tchernov et al. 1994) and Holon (Monchot and Horwitz 2007), in contrast to 'Ubeidiya, where various small-, medium-, and large-sized carnivores were found (Ballesio 1986; Gaudzinski 2004a, b; Gaudzinski-Windheuser 2005; Belmaker 2009; Martínez-Navarro et al. 2009).

We wish to note that the above-mentioned habitats are a suggested reconstruction, as modern ethology indicates that animals move between habitats and change their diet according to seasonal and/or availability resource constraints. Clearly, the presence of a permanent water body, an ecological focal point, was not only a preferred habitat of hippopotamus, but also for most of the biomass in the vicinity of GBY (Table 7.2).

Rodríguez (1999; Rodríguez et al. 2004), applied a quantitative approach that includes numerous recent faunas, advocating the use of ecological categories that incorporate data on trophic habits, locomotion abilities, microhabitat, and body size (Rodríguez 2004). Following examination of mammalian communities of the Early-Middle Pleistocene sites from Atapuerca (Burgos, Spain), Rodríguez observed a great similarity in community structure throughout the entire period, although not in species composition (as above). He subsequently suggested a new theoretical model in which a limited number of community structures are possible in a particular environment and their exact configuration depends

Table 7.2 Paleoecology of the mammalian species from Area C and the JB

Aquatic habitat (lake, swamp, river valley)	Open parkland	Woodland	Grassland
Hippopotamus amphibius	*Palaeoloxodon antiquus*	*Ursus* sp.	*Gazella* cf. *gazella*
	Bos sp.		*Equus* sp.
Sus scrofa	*Sus scrofa*	*Sus scrofa*	
		Cervidae sp.	
		Megaloceros sp.	
		Cervus elaphus	
		Dama sp.	
	Bovini gen. et sp. indet.	Bovini gen. et sp. indet.	

upon historical and biogeographical factors. It is important to note that the comparison of the various Atapuerca sites was performed regardless of their different taphonomic histories.

The total number of elements of each class (mammals, micromamamls, birds, reptiles and amphibians) can provide a rough estimation of the environment as reflected by the presence-abundance of the animal groups represented in the studied assemblages. A significant similarity in the frequency of animals from Layers V-5 and V-6 ($r = 0.988$, $p = 0.002$) and the JB (JB-V-5; $r = 1$, $p = 0.000$; JB-V-6: $r = 0.986$, $p = 0.002$) can be outlined. Thus, based on the various animal groups, the assemblages reflect similar paleoenvironmental conditions.

Birds are a good paleoenvironmental marker, and the main factor that determines the presence or absence of an avian population is the availability of their relatively strict habitats (Ashkenazi 2004). Sixteen bird taxa were defined at GBY, most aquatic and shore birds, and only a small percentage of grassland species (Simmons 2004: fig. 2). Moreover, *Anhinga rufa* can serve as an indicator for the paleo-lake's water depth, as it requires water deeper than 1–2 m in order to fish. Its absence from the Area C and JB assemblages may indicate a retreat in the lake's water level.

The freshwater turtle (*Mauremys caspica*) from the GBY assemblages (Hartman 2004) points to the existence of a freshwater lakeshore and marshland environment at the site. Based on the anomaly of one of the neural bony plate in both fossil and recent turtle populations, Hartman (2004) suggests continuity between the species represented in the Early-Middle Pleistocene and the extant population, demonstrating environmental stability. A water source is essential for most of the avifauna species, the turtles, and the amphibians (Fig. 7.1).

Current research on the GBY faunal sequence located above the Brunhes-Matuyama boundary has shown a continuity of the medium- and large-sized mammalian communities, although marked differences in species abundance exist (Rabinovich and Biton 2011). Area C and the JB represent the youngest archaeological occupations along this sequence. The medium- and large-sized mammalian faunal assemblages that originated in the earlier parts of the sequence are quite small, and we relate this differences to the various hominin tasks that were performed at the site (as above; Sharon et al. 2011). Yet, we are also aware of minor fluctuations that are reflected in the presence-absence and abundance of more sensitive species. For example, certain fluctuations did occur among the amphibians, apparently related to the fact that the aquatic environment was more diverse at the lower part of the GBY sequence than in the upper part of Area C and the JB.

7.3 Bone Taphonomy and Subsistence Strategies

Analysis of the large mammal taphonomy at GBY revealed that hominins were the main agent responsible for the bone accumulation within the different layers. However, differences in sediment composition in Layers V-5 and V-6 have been noted: the former comprises a molluscan packstone indicative of beach deposition (Feibel 2001, 2004), whereas the latter is made of organic mud, indicative of an offshore environment that was exposed due to a brief drop in lake level, which, in turn, exposed the mud and enabled the accumulation of artifacts and bones. These differences at GBY raise the question of whether various differences in the taphonomic agents were responsible for bone accumulation. Against this background, the results of our taphonomic analysis will be considered in a broader context.

Reconstruction of the paleoecological setting of GBY, based on the bone assemblages of the Layers V-5 and V-6, testified to the absence of marked environmental changes. This is also supported by the representation of medium- to large-sized mammal species in the three assemblages, as species composition has proven to be similar in all samples (Table 4.1, Fig. 5.37) (significantly positive correlations: Layer V-5–Layer V-6: $r = 0.888$, $p = 0.000$; Layer V-5–JB: $r = 0.768$, $p = 0.001$; Layer V-6–JB: $r = 0.751$, $p = 0.002$). In addition, body-size group distribution is also markedly similar among the assemblages (Layer V-5–Layer V-6: $r = 0.986$, $p = 0.000$; Layer V-5–JB: $r = 0.812$, $p = 0.050$; Layer V-6–JB: $r = 0.824$, $p = 0.044$), thus implying that species composition underwent similar taphonomic processes that were responsible for the similarities in body-size group distribution. The latter comprise bones that were broken due to hominin exploitation, after which they underwent trampling and weathering during a short accumulation period.

The waterlogged nature of the GBY deposits had a major effect on bone preservation. *Dama*, the most common species of all three assemblages, typically reflects bone preservation of medium-sized mammals. The assemblages are dominated by slightly weathered bones that are either blackish in color, with mollusks fragments, or of a dull bluish-gray metallic color. Layer V-5 has more cases of bones with adhered mollusk fragments, clearly a result of its particular sedimentary characteristics.

Striations appear on most of the identified mammal bones and also on the unidentified ones. Striation frequency varies for individual species, as well as for bones attributed to different body-size groups. Striations are more common on bones of *Dama*, *Gazella*, and their corresponding body-size groups that exhibit relatively numerous cut marks and evidence of marrow extraction. The same phenomenon is true for the highly fragmented bones ascribed to body-size groups, which are characterized by slightly higher striation frequencies. Striation frequency per layer is similar among the assemblages, thus indicating that the agent responsible for striation formation is similar in each layer.

The frequency of *Dama* skeletal elements is similar in the three assemblages, with those from the JB somewhat less fragmented, as is reflected by comparison of various skeletal counts. This was also suggested by analysis of the bone fragmentation, where fragments assigned to BSGD from the JB tend to be larger than those from this body size group from the other layers. BSGC and BSGE display similar skeletal-element distributions, indicating somewhat similar taphonomic histories. The general picture emerging from the GBY mammalian skeletal-element representation is that of a range of several large-sized species that were not affected by density mediated factors or by winnowing.

This study has not addressed the abundance of deer antler fragments at the site, which are probably *Dama*, but does raise the possibility that the fragments are the result of the antlers' use as soft hammers (Sharon and Goren-Inbar 1999; Goren-Inbar and Sharon 2006; Goren-Inbar 2011b).

The taphonomic history of biotic modifications on mammalian bones from GBY involves the influence of animals. Damage includes tooth scratches, gnaw marks, and traces of digestion. Animal modifications occurred on less than 2% of the bones from Layer V-5 and the JB, with a total of 2.84% in Layer V-6. Among the several agents identified as responsible for these traces are rodents, small- (e.g., fox) and medium-sized (e.g., wolf) carnivores, and large-sized carnivores (e.g., hyena). Small- and medium-sized carnivores are the major animal agents that left their mark on the Area C and JB fauna.

The animal agent responsible for the damage is important when reconstructing the role of animals in the assemblages' taphonomic history. Clearly, rodent gnawing and tooth scratches can occur at any time—when hominins are in the vicinity or after they had left. Rodent gnawing was observed in numerous open-air sites, such as Holon (Monchot and Horwitz 2007) and Quneitra (Rabinovich 1990).

Carnivore modifications help provide better insight into their role in bone accumulation and damage. The issue of carnivore access to the carcass is of great importance, as carnivores compete with hominins (Capaldo 1997; Domínguez-Rodrigo 1997; Selvaggio 1998). Most carnivore marks at GBY are located on long-bone shafts, and most probably occurred after hominins broke them. While large carnivores left their mark in the form of tooth scratches on three *Dama* bones from Layer V-6, there are no signs of large carnivores in the Layer V-5 assemblage.

In the near absence of carnivores at the site, the information concerning their possible presence is based on species that are known from earlier (e.g., 'Ubeidiya) and later assemblages (numerous sites from the Middle Pleistocene) where a community of large felids, hyenids (two species), and numerous small- and middle-sized carnivores were present (Ballesio 1986; Dayan 1989; Rabinovich 2002; Gaudzinski 2004a, b; Gaudzinski-Windheuser 2005).

Compared with known accumulations of carnivores (Kruuk 1972, 1976; Binford 1981; Kerbis-Peterhans and Horwitz 1992; Lam 1992; Blumenschine and Marean 1993; Faith 2007) and mixed assemblages of both carnivores and hominins (Stiner 1994; Fosse 1999; Rabinovich 2002), there is evidence that carnivores, particularly hyenas, had only a marginal impact on the formation and modification of the Area C bone assemblage. This is emphasized by the frequency of carnivore species in each assemblage, which is less than one percent; by the few surface modifications caused by small- and particularly large-sized carnivores; and by the rare occurrence of tooth marks and numerous long-bone shaft fragments.

The presence of both low and high meat and marrow skeletal elements in the GBY equids, together with a large carnivore tooth mark on the mandible of a young foal (see Chapter 5, no. 2,033), may indicate carnivore ravaging of this specific animal. Since the mandible with cut marks resulting from disarticulation and tongue removal co-occur with elements of high meat and marrow value, we assume that the carnivores were ravaging hominin leftovers.

We could not find any significant correlation between *Dama* skeletal elements and their relative nutritional value (MGUI—meat, marrow, and grease). According to economical models (i.e., Binford 1981; Lyman 1994), no correlation is expected when the majority of the medium-sized fauna represented complete carcasses. In contrast, only selected parts of the large-sized animals were present; however, a negative statistically relevant correlation was not observed.

An influential component affecting skeletal-element representation is the character of the site. In the case of a kill site, the best parts are taken away, so that only parts of lower nutritional values are left and those of higher nutritional values are transported to the campsite (Schlepp Effect). This model, among others, is supported by studies of recent hunter-gatherers but does not necessarily mirror past hominin behavior.

The *Dama* butchery sequence of GBY reflects the butchering of complete carcasses. The hominin-induced bone damage butchering comprises both cut- and percussion marks. Since *Dama* is the most common species in the studied assemblages, we analyzed the exploitation patterns of this species in detail. On the basis of the data presented in the previous chapters, we suggest a reconstruction of *Dama* carcass-processing, which began with the skinning of the animal and ended with marrow extraction. The high frequency of hominin-induced damage marks on the *Dama* skeletal elements suggests a step-by-step sequence of butchery processes, which were repeated time and again in the same pattern. The presence of most of the major bones of the *Dama* skeleton and of all hominin-induced damage types in a small area leads us to conclude that the entire sequence of carcass-processing took place in situ (Rabinovich et al. 2008a).

Due to their good state of preservation, the GBY bones constitute the earliest Eurasian evidence of hominin-induced damage patterns that permits the identification, evaluation, and comparison of hominin-induced damage patterns on *Dama* bones. Qualitative and quantitative analyses of these patterns observed on the assemblages of Layers V-5 and V-6 indicate that in situ *Dama* carcass processing occurred on a regular basis. From the recurrent nature of this processing, we conclude that the occupants regularly hunted medium-sized mammals. The repetition and complexity of the observed damage patterns indicate an in-depth knowledge of the anatomy of fallow deer.

The *Dama* butchery sequence comprises a complete set of activities, from dismemberment to marrow extraction, and reflects a high intensity of carcass exploitation. In order to consume the marrow, the bones were fractured at mid-shaft or by breaking off the articulation, resulting in numerous fragments, as is seen in the assemblages.

The technique of splitting phalanges, present at GBY and also observed at sites associated with Neanderthals and modern humans (Bar-Oz and Munro 2007), is considered an intensity measure of carcass exploitation (Binford 1978; Munro and Bar-Oz 2005). In the absence of evidence that suggests a shortage of meat resources, such behavior can also be interpreted in terms of a group habit (snacking phalanges) representing ad-hoc consumption regardless of utility or benefit.

Apart from *Dama* and *Dama*-sized taxa, numerous bones representing seven taxa, belonging to species of five body-size groups, also display traces of hominin interference. Their analysis supports and emphasizes the results obtained by the *Dama* bone analysis in that all stages of the butchering sequence, except the initial carcass skinning, have been documented for these taxa. The evidence illustrates highly flexible hominin interaction with a variety of subsistence resources. As for the larger animals, it appears that meat strips of high nutritional value can be transported unattached to bone without leaving a telltale sign in the archeological record. Thus, the elephants and the hippopotamus that are represented by few bones in the assemblages may not accurately reflect their role in the hominin diet.

In most Eurasian Early-Middle Pleistocene assemblages the medium- to large-sized mammals outnumber the remaining species, reflecting species availability and hominin capability to deal with prey: to hunt, process, and transport. Body size dictates prey density, so that the highest ranked taxa, in terms of edible parts (the megafauna), were taken whenever possible, but were most likely never plentiful enough to be the major prey species for hominin subsistence (Byers and Ugan 2005).

Even though elephant, being the largest animal (<1,000 kg), can theoretically provide most of the hominins' meat-fat nutritional needs, there are no indications from the southern Levant that this was a largely exploited animal, or even the most abundant animal in the Paleolithic record (i.e., Rabinovich 2002; Stiner 2005; Chazan and Horwitz 2007; for European data, see Gaudzinski et al. 2005).

The abundance of *Hippopotamus* teeth is also taphonomically significant, due to the fact that either the rest of the skeleton was outside the excavated area (a similar explanation was suggested by this study for the skeletal representation of elephants or another large animal), or that the mode of the hippopotamus' introduction into the site was selective. Alternatively, they could have been a part of the natural background fauna. Examination of both the hippopotamus bones and those of BSGB slightly increases the diversification of the skeletal-element representation at the site, but it remains minimal and does not accept any of the above hypothesis.

The taphonomic agent responsible for the accumulation of the micromammals and birds has yet to be identified.

The freshwater turtle (*Mauremys caspica*) represents a population that lived and died on the site (Hartman 2004). The amphibians, mainly the anurans, could have been predated by an agent such as aquatic and near-shore birds (details in Rabinovich and Biton 2011). Indeed, the aquatic and near-shore avifauna of the present Hula Valley are important predators of anurans (Ashkenazi and Dimentman 1998); several of these species are present in the GBY assemblage, namely *Podiceps cristatus*, *Mergus serrator*, *Anhinga rufa* and *Fulica atra* (Simmons 2004).

Analysis and identification of fossil freshwater crabs from Layers V-5 and V-6 indicate the minimal impact of taphonomic processes on the paleo-Lake Hula shores. This is illustrated by the crabs' elements, which are, though fragile, extremely well-preserved and include complete mandibles and pincers. Occasionally fitting pairs of pincers were found close to each other. Once again, this analysis indicates only negligible post-depositional disturbances at the site (Ashkenazi et al. 2005). Moreover, it should be noted that the existence of numerous crab fragments in the assemblages, mainly Layer V-6, may represent the remains of an ad-hoc hominin foraging strategy at the site (Ashkenazi et al. 2005), including the occasional consumption of large specimens, as was found in the Late Pleistocene site of Eynan (Ashkenazi in Valla et al. 2007).

In addition, we gain complementary taphonomic knowledge about the impact of currents and wave action at the margins of the paleo-Lake Hula from the ongoing malaecological study of the gastropod that created the bulk of the Layer V-5 sediments—*Viviparus apamaea*. It has recently become evident that the burial of this coquina was relatively rapid and that both its formation and exposure to the elements did not destroy the mollusks.

The background species (birds, reptiles, and amphibians) thus provide additional information on the site's paleoenvironment, and may have been occasionally part of the hominin diet.

The medium- and large-sized mammal assemblages of GBY may further contribute to the understanding of site-formation processes and to certain hominin behavioral aspects. One avenue of research is to examine evidence for seasonality. Kahlke and Gaudzinski (2005) have shown that species ageing and mortality profiles in an assemblage should be interpreted on the basis of each species' specific life history and its preferred habitat.

The age composition of the GBY *Dama*, as well as the presence of shed and unshed antlers, suggest either winter or winter to spring occupation. This reconstruction is based on the fact that present-day *Dama* birthing occurs during March to May in the Mediterranean zone, and antlers are shed during February and March. The presence of young specimens, with and without antlers, implies that the site was occupied during more than a single season. This hypothesis is further supported by the young age of some of the equid, hippopotamus, and wild boar specimens. Thus, based on the ageing of the GBY mammal species, we are inclined to propose an autumn-winter-spring occupation. This is also underlined by the analysis of bird paleoecology, which proposes a winter occupation for the site (Simmons 2004), as Anatidae (ducks, geese, and swans), which represent most of the birds from Area C and the JB due to their migratory nature, are likely to be found in the Levant between November and February.

In a much broader view, evidence of hominin interference with fauna is observed as early as 2.5 million years ago by the presence of sporadic cut marks (de Heinzelin et al. 1999;

Domínguez-Rodrigo et al. 2005), and the systematic butchery of many taxa, as early as 1.75 Ma (Bunn and Kroll 1986; Bunn 2001; Domínguez-Rodrigo 2002; for a different view, see Lupo and O'Connell 2002).

The earliest Levantine site, 'Ubeidiya (1.4 Ma) (Tchernov 1992; Bar-Yosef and Goren-Inbar 1993), yields only limited evidence of hominin bone modification (Gaudzinski 2004a, b; Gaudzinski-Windheuser 2005), contrasting the patterns observed at GBY. Although cut marks appear in both sites, they are much more frequent at the latter. Percussion marks, indicative of marrow extraction, are abundant at GBY and entirely absent at 'Ubeidiya. A possible explanation may stems from differences in subsistence strategies (Gaudzinski 2004a, b; Gaudzinski-Windheuser 2005), access to animals (Belmaker 2006), or behavioral modes.

In the southern Levant, only the Lower Paleolithic site of GBY offers insight into an entire butchery sequence of single taxon individuals of such antiquity. This sequence of carcass exploitation has been observed in later records of hominin behavior, which were found to resemble those of modern hunter-gatherers. The records derive from Acheulian Holon (Horwitz and Monchot 2002; Monchot and Horwitz 2007), the Acheulo-Yabrudian Qesem Cave (Lemorini et al. 2006; Stiner et al. 2009), the Mousterian Misliya Cave (Yeshurun et al. 2007), the Neanderthal Kebara Cave (Speth and Tchernov 2001; Speth and Clark 2006), the Middle to Epipaleoltihic Hayonim Cave (Stiner 2005) Upper Paleolithic Hayonim Cave (Rabinovich et al. 1997), and a number of Epipaleolithic sites (Munro and Bar-Oz 2005).

Butchery marks and impact fractures occur in Lower Paleolithic Europoean sites, such as level TE9 at Sima del Elephante (Carbonell et al. 2008) and level TD6 at Gran Dolina (Díez et al. 1999) at Atapuerca; Isernia, Italy (Anconetani 1999); and Boxgrove, England (Parfitt and Roberts 1999). These marks, indicative of efficient animal exploitation, appear repeatedly throughout the Paleolithic sequence.

According to Lupo, "Foraging models have the best explanatory and predictive potential value when the full suite of subsistence choices available to the forager is considered" (2007: 173). Against this background, GBY provides valuable information for the reconstruction of Pleistocene subsistence strategies, as it provides insight into the exploitation of medium-sized mammal resources. Water birds, crabs, and fish were probably exploited as well. The same is true for plant resources (Melamed 1997; Goren-Inbar et al. 2000, 2001; Melamed 2003; Goren-Inbar et al. 2004), which must also be taken into consideration when the full hominin diet is examined. Thus, a broad diet is proposed, combining foraging and hunting, most probably multi-seasonal. Assuming that the Pleistocene Mediterranean seasonality was similar

to the present, a marked difference would have occurred between the rainy and dry season, and the exploitation of floral components probably resulted in intensive foraging, at which time more nutritional or other favorable elements would be available (i.e., nuts and fruit). The potential of foraging opportunities reflected in the floral and faunal data of GBY likely fulfills the required daily caloric consumption estimated for the paleo-hunter-gatherers.

7.4 Conclusions Drawn from the Experiments

Taphonomic analysis indicates that in-situ *Dama* butchering took place at GBY. However, these results also suggest that a time delay may have existed between these butchering episodes and those of other taxa, which, in turn, suggests that the site's faunal assemblages accumulated over a prolonged period of time (see Chapter 5). These results contradict other evidence that indicates a rather short deposition time at the site. In order to investigate these differences, a series of experiments was undertaken, which would mimic pre- and post-burial processes during the early stages of bone deposition at GBY.

Bone surfaces at the site are covered by cut marks and striations. While the former is generally considered as an indicator for interaction between hominins and animals, the latter, on the other hand, is generally assumed to result from interaction between bone surfaces and their surrounding matrix. This has been shown by a number of studies (Lyman 1994 and references therein), which aimed to provide a diagnostic mean for the definition of cut marks.

Though we still lack detailed knowledge about the ranking of striation-producing processes in the taphonomic chain, we believe that analysis of striations may also provide a key to the understanding of depositional processes at GBY. We understand that taphonomic processes are additive and, therefore, the processes that produced striations must have had considerable effect on the character of the fossil assemblages from the site.

The site of GBY is characterized by a very distinct sediment composition. The bones display good surface preservation with a high number of cut marks and striations. The general environmental setting of the site at a lake margin is significant for most of the early hominin sites we know to date. The results of our studies may, therefore, also be applied to the disentangling of the taphonomic histories of other sites featuring a similar setting.

Tumbling, trampling, scratching, and burial experiments were undertaken in order to provide additional insight into the formation processes of the bone assemblages. The tumbling experiments demonstrated that the dominant taphonomic effect caused by uni- and multidirectional water movement is bone abrasion. Abrasion leads to bone-surface polishing by sediment particles. Part of the abrasion process is characterized by continuous striation formation and destruction. Striations occur initially as a result of periosteum removal. In the later stages of the abrasion process the polish leads to the final destruction of bone-surface modifications.

Unidirectional water movement did not produce striations but significantly altered the morphology of the surface modifications documented prior to tumbling. The amount of water involved, the bone morphology, and, above all, the state of periosteum preservation, served as the controlling factors, whereas time effects did not operate in a simple linear manner.

Multidirectional water movement produced a variety of bone-surface modifications and accelerated their formation. The controlling crucial variables are the state of periosteum preservation and bone morphology, but especially time.

The tumbling experiments only barely produced striations on bone surfaces, and we thus concluded that bone movement within a moderate hydrodynamic setting was not a variable in GBY's taphonomic history.

The trampling experiments produced abrasion and polish on the bone surfaces. The main controlling factors were the degree of water saturation of the sediment and, most importantly, bone morphology (i.e., species, ontogeny, and bone preservation). Trampling in muddy conditions resulted in the severe destruction of the sediment components (i.e., mollusks and bones). Trampling produced diagnostic modifications and, to a certain degree, striations of various morphology, but also led to the alteration of pre-existing modifications. Experiments revealed that both striation formation and erasure occurred within a relatively short time period in the early stages of the continuous abrasion process, and was directly related to pre-existing bone-surface preservation. We conclude that bone movement in the sediments during relatively short trampling events was a major contributor to the taphonomic history of the site.

Results of our experiments have enabled us to reconstruct the effects of the different stages of the taphonomic processes involved in GBY's site-formation history. The results further demonstrate that the degree to which abrasive forces affect bones is highly dependant on specific bone structure that varies by animal species, ontogenetic age, and bone shape and preservation. Therefore, different degrees of bone abrasion of different species within a homogeneous archaeological sample do not necessarily point to different taphonomic histories of the species involved. Only major differences in the degree of abrasion within one species can indicate a longer duration of the depositional sequence. Abrasion can accelerate a bone's disintegration process and thus imitate traces that may otherwise have been caused by climatically induced weathering.

The experiments also illustrate that in the course of abrasion, cut marks undergo severe morphological change and can become indistinguishable from striations. In this light, the origin of striations on bone surfaces cannot be attributed solely to sediment interaction, but can also originate in anthropogenic activities.

To conclude, even though differences in striations and the bone preservation of different species were observed at GBY, it seems most likely that all bones originated from a homogeneous faunal assemblage, which was deposited over a short period of time. The striations associated with cut marks indicate intensive short-term interaction between bones and other sediment components (e.g., mollusks) by trampling. After integrating the experiment results with the taphonomic bone analysis, we argue that the bones were rapidly buried since the abrasion process caused by trampling stopped at a very early stage.

Considering the results of previous studies of the site, the most likely interpretation of this reconstruction is that of a relatively short time period during which different animals in the vicinity of the lake margins were killed by hominins and processed in situ, followed by the rapid burial of their remains.

7.5   Summary of the Paleontological Analyses

Results of the studies of the faunal assemblages' taphonomy provide insight not only into subsistence strategies but also into site-formation processes. Skeletal-element representation of different-sized mammals can reflect hominin decision-making in prey exploitation, as the skeletal elements can mirror carcass transportation patterns. These patterns thus shed light on the character of an archaeological site in terms of hunting and/or base camps, as has been observed by studies of hunter-gatherers (Binford 1978, 1981, 1984; Kelly 1995). Taking into consideration primarily the size of the prey and the bulky nature of its parts in correlation with the benefit return, it could reimburse the effort invested in its hunting, butchering, and transportation. Numerous studies have addressed this issue, combining ethnographic data with foraging models and comparing them with the archaeological data (Brain 1976; Binford 1978; Lee 1979; Bunn et al. 1988; O'Connell et al. 1988; Fowler 1989; Gifford-Gonzales 1989; O'Connell et al. 1990) in order to understand past hominin behavior/attitude concerning animal acquisition and consumption (e.g., summary in Lupo 2006).

The Levant, and particularly the Jordan Valley, lacks any comparative living hunter-gatherers that survive, and only African ethnographic or North American analogies can be used for model building. The lack of such analogies limits our ability to reconstruct past behavior in the Levant, and thus also limits the comparisons to contemporaneous assemblages from archaeological sites from other regions or to Levantine sites from different periods (Rabinovich et al. 2008a, b).

However, it is clear that the main prey—medium-sized mammals—was brought to the site complete, and that the site had a homogeneous and short life history that was characterized by short but highly intensive mobility as well as rapid and undisturbed sedimentation.

Against this background, it can be assumed that a group of hominins frequented the site during a short period of time. They hunted and butchered mainly small- and medium-sized animals in situ. Judging from the skeletal-element representation and evidence of butchering traces, animals weighing up to 50 kg were probably brought to the site complete or were hunted ad hoc while they approached the water. Larger animals were probably brought selectively to the site.

The duration of occupation can also be measured by taking carcass treatment efforts into account. However, investment in carcass treatment depends on several variables, among them environmental and climatic conditions, exploitation intensity (meat, marrow, fat, and stripped meat for future drying—biltong) (Lupo 2006 and references therein), and group size. Social behavioral patterns and local group preferences are additional variables. How can we measure processing time in a Paleolithic context? Perhaps by taking into consideration tool type, details of cut marks, spatial distribution of bones, and a plethora of other aspects of the past inhabitants and their surroundings.

In the following sections we shall describe the taphonomic history of the flint assemblages, the spatial distribution of the artifacts and the fauna.

7.6 Taphonomic History of the Lithic Assemblages

7.6.1 Introduction

Site-formation processes and, in particular, taphonomic aspects, comprise a fundamental component of all archaeological and archaeozoological analysis. The taphonomic study of the GBY mammalian bone assemblage is only one aspect of the multitude of taphonomic agents that have left their imprint on the archaeological finds. While the present study focuses mainly on the GBY fossils and the experiments attempting to mimic their formation, this section discusses additional aspects relating to the taphonomic processes that left their mark on the archaeological horizons at the site; in particular, the lithic assemblages from Layers V-5 and V-6.

While the theoretical and, to a certain degree, experimental foundations of site-formation processes (including aspects of post-depositional processes), were established in the past by several pioneering scholars, their application to the archaeological record is still quite limited. Studies by Isaac (1967, 1983), Petraglia and Potts (1994), Morton (2004), Schiffer (1983, 1987), Schick (1986, 1987), and Tappen (1992), to mention but a few, laid the groundwork for extensive examination of the agents and mechanisms that created the initial internal structure of prehistoric assemblages. This acquired knowledge was applied mainly to Plio-Pleistocene African sites, mostly those featuring fluvial settings. Examination of the Acheulian record reveals only a few cases that were analyzed from this perspective (e.g., Petraglia and Potts 1994; Schick 1992, 2001). However, a closer examination of the scholarly literature supplies important data that directly relates to these taphonomic issues.

At GBY, the lithic and the faunal assemblages from Layers V-5 and V-6 are considered to be associated finds that were used for specific tasks and activities carried out by Acheulian hominins on the lake margins. The presence of numerous percussors, percussion- and cut marks on the bones, as well as the many very thin flint flakes in close vicinity, indicate the integrity and unique characteristics of the various components at the site.

The sedimentological components of Layers V-5 and V-6 differ markedly. The sediment of the former comprises coquina, primarily that of *Viviparus apamaea galileae* and *Bellamya* sp. (Ashkenazi et al. 2010; Spiro et al. 2009) with minimal inclusion of small grain matrix. The presence of an extensive coquina deposit on the shore of a shallow lake, however non-turbulent as it may have been (analogous to the waters of the drained Recent Lake Hula; see Dimentman et al. 1992), clearly leaves its mark (Morton 1995, 2004). The sediment of the latter constitutes black mud with a species composition that includes *Melanopsis*, *Theodoxus* and a variety of others (see Chapter 2; Ashkenazi and Mienes 2005; Ashkenazi et al. 2010). It is in these two layers' different depositional contexts that the diverse assemblages underwent their final formation and burial processes, leaving a distinct taphonomic mark that is reflected in the retrieved assemblages. This section will examine the extent of these differences and will demonstrate to what extent the lithic artifacts, their original composition, and possibly their spatial arrangement, had been taphonomically influenced.

The following sections will evaluate the possible impact of different depositional environments, along with examining the taphonomic histories of the various components (bones and stone artifacts) of Layers V-5 and V-6. The history of site-formation processes and post-depositional processes at the site, as imprinted in the lithic artifacts, is presented below through a selection of attributes that illustrate aspects of the flint assemblages' depositional history.

7.6.2 The Lithic Assemblages of Layers V-5 and V-6

Three types of raw material were used for the production of stone artifacts in both layers—basalt, limestone, and flint, although the first two are minimally represented. While most of the Acheulian tools at the site were made of basalt, those from Layers V-5 and V-6 were primarily of flint (Sharon and Goren-Inbar 1999; Goren-Inbar and Sharon 2006; Sharon et al. 2011). Thus, most of the lithic analyses and discussions presented below are based on flint, as its sample sizes are statistically more representative of the original hominin population at the site. Samples are considered valid when they comprise 20 or more artifacts (Tables 7.3 and 7.4).

The condition of the artifacts reflects their "life history," the combination of physical and chemical events that left their imprint on the stones. Therefore, in the following paragraphs the artifacts' size properties and condition (preservation, patination, weathering, dimensions, and breakage patterns) will be used to compare the assemblages from the two layers and to reach a better understanding of the impact of the sediment type and the array of post-depositional processes on the lithic component. We refrained from integrating the lithic assemblage of the JB from the present analysis since it comprises a mixture of the lithics from Layers V-5 and V-6 (see Chapter 2).

The association between animal bones and flint artifacts may be seen in every exposure of each layer along the left bank of the Jordan River. While an explanation for such a concentration of finds from a functional and social perspective remains to be determined, its occurrence at the site is both repeated and homogeneous. This homogeneity is expressed in the techno-typological characteristics of the lithic assemblage, and the association between the damaged hominin bones (displaying cut- and percussion marks) and the various lithic artifacts that may have produced the damage. Together with the seemingly high degree of freshness of both the lithics and fauna, these traits support the notion that the finds were the result of relatively short occupation durations. Here and in future studies we hope to gain further evidence regarding each layer's homogeneity.

Due to their bifacial characteristics, the GBY lithic assemblages have been assigned to the Acheulian Technocomplex. Both handaxes (flint) and cleavers (basalt) occur in very low frequencies in Layers V-5 and V-6 (Goren-Inbar and Sharon 2006; Sharon et al. 2011). Conversely, the flint assemblages of both layers are comprised of flakes, many of which are products of the modification of the flint handaxes, defined as *éclats de taille de bifaces* (Sharon and Goren-Inbar 1999; Goren-Inbar and Sharon 2006).

7.6.2.1 Size

The integrity of the lithic assemblages of Layers V-5 and V-6 is demonstrated by the range of artifact size, from bifaces to microartifacts (between 2 mm and 2 cm). Size sorting (to either end of the scale) would have resulted in an absence of a particular size fraction, but this situation was not encountered in the assemblages. From a taphonomic standpoint, the presence of both large and small items is considered an indication of the assemblages' original state, as demonstrated by lakeshore experiments that were carried out by Morton (1995). Evidence for the assemblages' original composition derives from the abundance of microartifacts (Layer V-5: N = 36,770, Layer V-6: N = 6,585), constituting over 90% of each layer's entire lithic assemblage. Clearly, extensive winnowing would have resulted in much lower frequencies than these (see Tables 6.4 and 7.1).

A comparison of flint flakes and flake tool sizes is presented in Fig. 7.2 and Table 7.5.

Table 7.5 illustrates the similarities between the size of the flakes and flake tools of the two assemblages, and reveals that the mean values of the flint artifacts from Layer V-6 are larger than those from Layer V-5. This tendency is further emphasized in Fig. 7.2, which illustrates the higher frequency of smaller artifacts in Layer V-5 when compared to Layer V-6 (whether the differences are due to the re-working of material originally deposited in the older layer (V-6), or whether the artifacts of each layer simply differ in size, will be addressed below). This size difference is also clearly

Table 7.3 The lithic assemblage from Layer V-5

Category	Flint N	Flint %	Basalt N	Basalt %	Limestone N	Limestone %	Total
Microartifacts	30,875	98.97	5,562	98.52	333	98.81	36,770
Flakes and flake tools	289	0.92	74	1.31	1	0.29	364
Cores and core tools	32	0.10	8	0.14	3	0.89	43
Handaxes	0	0.00	0	0.00		0.00	0.00
Cleavers		0.00	1	0.01	0	0.00	1
Total	31,196	99.99	5,645	99.98	337	99.99	37,178

Table 7.4 The lithic assemblage from Layer V-6

Category	Flint N	Flint %	Basalt N	Basalt %	Limestone N	Limestone %	Total
Microartifacts	4,507	94.05	1,982	97.73	96	79.3	6,585
Flakes and flake tools	275	5.73	46	2.26	1	0.82	322
Cores and core tools	9	0.18	0.00	0.00	19	15.70	28
Handaxes	1	0.02	0	0.00	0	0.00	1
Cleavers	0	0.00	0	0.00	5	4.13	5
Total	4,792	99.98	2,028	99.99	121	99.95	6,941

Fig. 7.2 Length (mm) distribution of flint flakes and flake tools from Layers V-5 and V-6

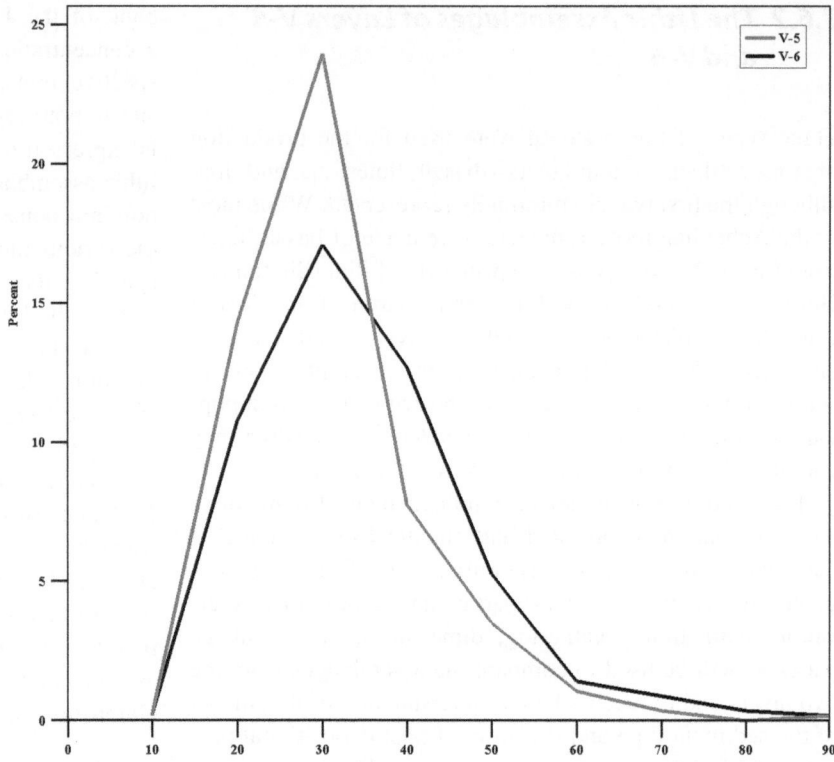

Table 7.5 Size statistics of flint flakes, flake tools, cores and other waste products

Attribute	Layer V-5			Layer V-6		
	N	Mean	Std. deviation	N	Mean	Std. deviation
Flakes and flake tools						
Length	291	26.60	10.07	277	30.07	12.33
Maximal length	291	29.79	10.20	277	34.51	11.93
Width	291	21.42	8.39	277	25.60	9.40
Thickness	291	7.29	2.77	277	7.04	5.61
Cores and other waste products						
Length	30	27.67	11.05	9	41.11	12.49
Maximal length	30	30.50	11.19	9	43.0	14.21
Width	30	20.63	8.26	9	33.67	15.81
Thickness	30	12.27	5.49	9	17.67	7.10

reflected by flint cores and other waste products. Although the sample sizes of the latter are much smaller than other artifacts, Table 7.5 clearly indicates that this component in Layer V-6 is significantly larger than in Layer V-5.

7.6.2.2 Breakage

It is evident that the occupation surface of both Layer V-5 and V-6 was exposed to atmospheric conditions for a period of time until the one (Layer V-5) sealed the other (Layer V-6) and Layer V-4 sealed Layer V-5. Although current

sedimentological methods preclude a precise estimation of the duration of each layer's exposure, after considering the preservation state of the flint artifacts, one should rule out prolonged exposure. Among the means of evaluating the role of taphonomic processes in a lithic assemblage is estimating exposure duration to atmospheric conditions through evaluation of breakage patterns, assuming that lengthy trampling results in a higher frequency of broken artifacts.

Figure 7.3 illustrates the similarity in the breakage pattern of flakes and flake tools in Layers V-5 and V-6. The only marked difference between the two is the higher frequency of fragments in the former, which is also associated with the higher frequency of smaller items, as shown in Fig. 7.2. Despite the many studies and experiments undertaken on prehistoric flint assemblages, it is still impossible to associate particular breakage patterns with taphonomic events. The abundance of fragments in Layer V-5 may have resulted either by knapping (an original situation derived from instantaneous breakage) or by a possible taphonomic agent.

7.6.2.3 Preservation

The preservation state of basalt and limestone differs; Tables 7.6. and 7.7. present the distribution in each assemblage of raw materials for flakes and flake tools and cores and core tools. No basalt or limestone core or core tool has been classified as "fresh," an example of the different traits of the raw materials and the differing weathering processes that

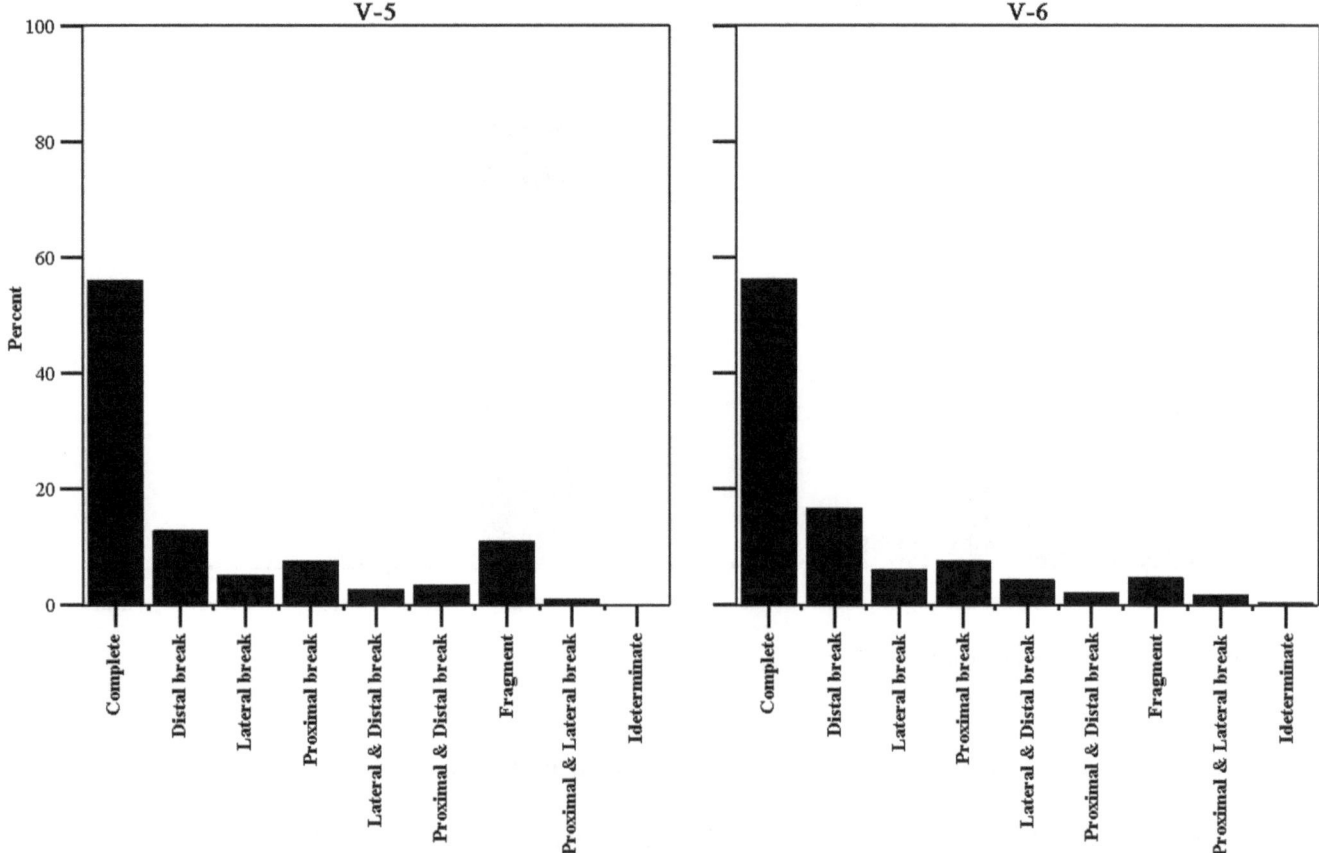

Fig. 7.3 Breakage pattern of flakes and flake tools from Layers V-5 and V-6

Table 7.6 State of preservation of cores and other waste products

Raw material	Layer		Fresh	Slightly abraded	Abraded	Total
			State of Preservation			
Flint	V-5	N (%)	13 (43.3)	11 (36.7)	6 (20.0)	30 (100.0)
	V-6	N (%)	7 (77.8)	0 (.0)	2 (22.2)	9 (100.0)
	Total	N (%)	20 (51.3)	11 (28.2)	8 (20.5)	39 (100.0)
Limestone	V-5	N (%)		1 (33.3)	2 (66.7)	3 (100.0)
	Total	N (%)		1 (33.3)	2 (66.7)	3 (100.0)
Basalt	V-5	N (%)		7 (87.5)	1 (12.5)	8 (100.0)
	V-6	N (%)		12 (63.2)	7 (36.8)	19 (100.0)
	Total	N (%)		19 (70.4)	8 (29.6)	27 (100.0)

Table 7.7 State of preservation of flakes and flake tools according to raw material and layer

Raw material	Layer		Fresh	Slightly abraded	Abraded	Rolled	Chemically weathered	Total
			State of Preservation					
Flint	V-5	N (%)	228 (78.1)	44 (15.1)	17 (5.8)	3 (1.0)	0 (.0)	292 (100.0)
	V-6	N (%)	252 (91.0)	20 (7.2)	4 (1.4)	0 (.0)	1 (.4)	277 (100.0)
	Total	N (%)	480 (84.4)	64 (11.2)	21 (3.7)	3 (.5)	1 (.2)	569 (100.0)
Limestone	V-5	N (%)	0 (.0)		1 (100.0)			1 (100.0)
	V-6	N (%)	1 (100.0)		0 (.0)			1 (100.0)
	Total	N (%)	1 (50.0)		1 (50.0)			2 (100.0)
Basalt	V-5	N (%)	4 (5.3)	14 (18.7)	52 (69.3)	3 (4.0)	2 (2.7)	75 (100.0)
	V-6	N (%)	4 (8.7)	22 (47.8)	19 (41.3)	0 (.0)	1 (2.2)	46 (100.0)
	Total	N (%)	8 (6.6)	36 (29.8)	71 (58.7)	3 (2.5)	3 (2.5)	121 (100.0)

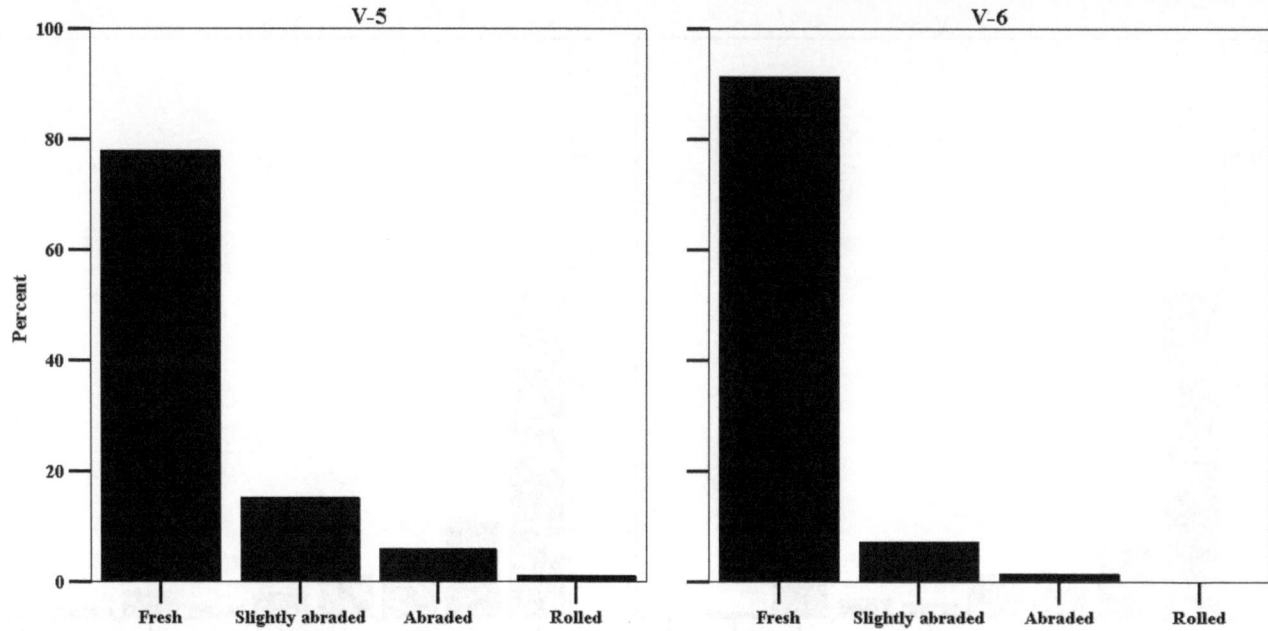

Fig. 7.4 Preservation of flint flakes and flake tools

they underwent as a result of prolonged burial and, to a certain extent, post-depositional processes. The basalt artifacts at the site sometimes were completely decayed (the rock was transformed into clay), while all the limestone artifacts—from each sequence of the site—are weathered, with not a single fresh item found (Goren-Inbar et al. in preparation). Since the number of basalt and limestone artifacts in Layers V-5 and V-6 is very small, flint should be the focus of any discussion concerning the preservation state of raw materials from both layers. Figure 7.4 presents the preservation state of flint flakes and flake tools. It reveals that the flint material from Layer V-6 is in optimal preservation condition, as is illustrated by the higher frequency of fresh items and the lower frequency of slightly abraded items than those of the flint assemblage from Layer V-5. However, the presence of rolled items in each layer is negligible. In short, it is evident that the lithics from Layer V-6 underwent extremely rapid burial, while those from Layer V-5 were somewhat more influenced by environmental condition due to longer exposure to atmospheric conditions.

7.6.2.4 Patination

Most of the flint artifacts from Layers V-5 and V-6 are a dark color (Delage 2007), which, when patinated, attain a lighter shade as a result of biogenic activity (Friedman et al. 1995). This patina type differs markedly from that observed on artifacts deposited in the black muds of Layer V-6, which attain their color by absorbing dark anaerobic organic-rich sediment particles. Exposure to atmospheric conditions leads to biogenic activity and is expressed in a rapid change of color. Therefore, trampling episodes of lithic assemblages, which were exposed to atmospheric conditions, could result in the formation of a double patina.

Figure 7.5 presents the different generations of patination seen on the flint flakes and flake tools. Patina distribution from both layers is very similar and clearly demonstrates rapid burial. Items from Layer V-5 display a higher frequency of double patina than those from Layer V-6, indicating somewhat longer exposure to atmospheric conditions.

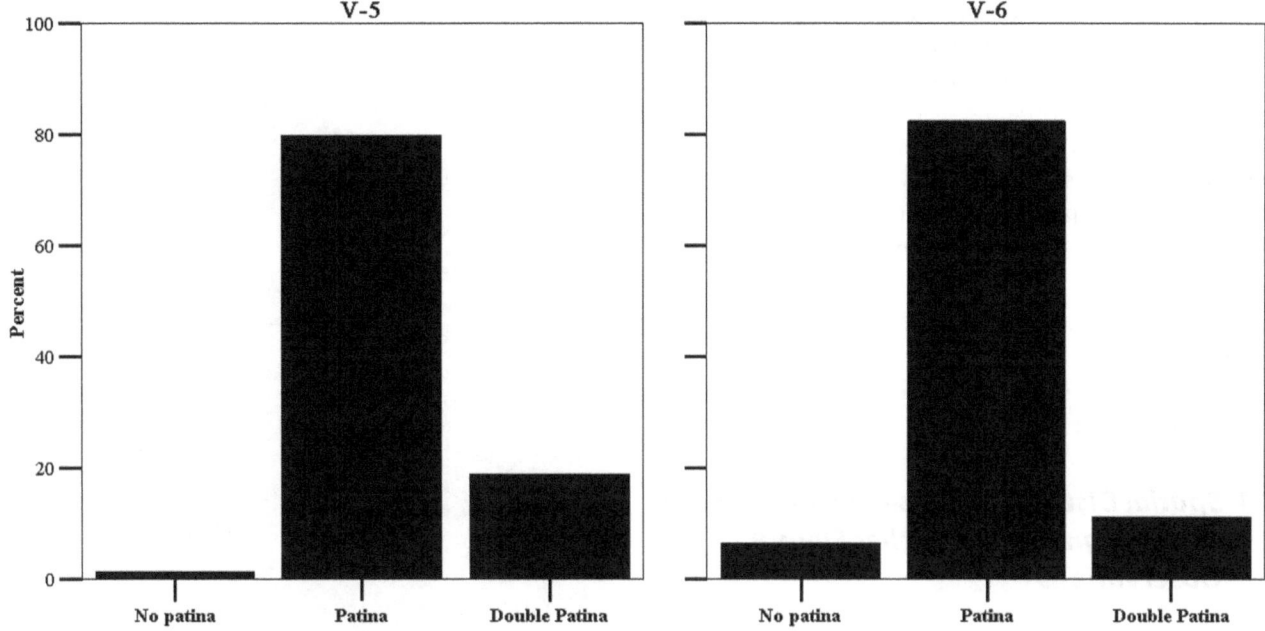

Fig. 7.5 Frequency breakdown of patination in flakes and flakes tools from Layers V-5 and V-6

7.6.2.5 Signs of Utilization

The "signs of utilization" attribute is derived from an observation that does not classify edge modification (edge damage) as retouch. Signs of utilization or edge damage are neither continuous nor homogeneous in size, morphology, and spacing along the edge. Frequency of the marks on flint flakes and flake tool samples is higher in Layer V-6 (Table 7.8). Since, in terms of patination and breakage, this layer is better preserved than Layer V-5, we conclude that the higher frequency of signs of utilization do not reflect trampling—a very common explanation for the presence of this attribute—but a variety of activities carried out with the help of these artifacts, which took place on the archaeological horizon during initial stages of deposition. Such activities included the use of the flint artifacts on bone, wood, and other materials.

Table 7.8 Frequencies of signs of utilization on flakes and flake tools

Layer	Signs of utilization		Total flakes and flake tools examined
	N	%	180
V-5	78	43.3	91
V-6	57	62.6	271

7.6.3 Summary of the Lithic Assemblages' Taphonomic Analysis

The lithic assemblages of Layers V-5 and V-6 are very similar regarding the impact of the various factors attributed to post-depositional processes, although the former comprises a coarser sedimentological environment. This difference is expressed by the higher frequencies of small macroartifacts, fragmented flakes, and double-patinated objects in Layer V-5.

It is most likely the coarser sedimentological environment of Layer V-5 that is responsible for the observed differences between the two layers. In discussing the coquina of Layer V-5, Feibel states that "[a] rise in lake level accompanied by storm concentration of offshore mollusks buried the archaeological accumulation, and in the process incorporated some of the material into the coquina. This was a short-term and discrete … event burying the archaeological level in a transgression as is reflected in the unbroken and unabraded character of the bulk of the molluscan material" (Feibel 2001: 137, 139). Indeed, there are minimal individual cases (a basalt cleaver and a long bone) that are positioned vertically to the sedimentation plane, proving that a different agent was responsible for their incorporation, but is clearly expressed minimally. Feibel's reconstruction should be viewed as a possible scenario as it suits, to a certain extent, the findings from Layer V-6, as expressed in the presence of larger artifacts, including cores and waste products of limestone and basalt (see Table 7.5, Fig. 7.2).

Indeed, the finds of Layer V-5 are very similar to those of Layer V-6, and we could theoretically suggest that they originated in the same assemblage, thus supporting the assumption that they were incorporated from Layer V-6. However, this similarity cannot explain the abundance of finds, including lithic artifacts of all sizes, which characterize Layer V-5. Therefore, it seems that the mechanism that created the assemblage of Layer V-5 is somewhat more complex than that of Layer V-6, and that the agents that produced and/or influenced the assemblages of Layer V-5 comprise a slightly different configuration.

Additional insight into the complexity of the post-depositional processes at the site may be gained through an examination of the spatial organization of the artifacts and bones of each layer, as presented below.

7.7 Taphonomy and Aspects of Spatial Distribution

7.7.1 Spatial Distribution Based on Conjoining Bone Fragments and Other Faunal Observations

In order to more precisely define the character of hominin activities at GBY, the taphonomic study was extended to include an examination of bone fractures by bone conjoins. We refitted *Dama* long bones, unidentified bones of the BSGD, as well as elephant and equid bones. In some cases, conjoins lend additional insight into the process of bone breakage for marrow consumption by hominins.

For refitting, bone fragments were sorted according to anatomical element (humerus, radius-ulna, femur, tibia, and metapodial bones). Unidentified fragments assigned to the *Dama*-sized group were grouped by grid squares (mimicking the excavation areas). In addition, fragments bearing signs of surface modifications were classified according to modification type (i.e., cut marks, percussion marks, striations).

In the first stage, we looked for conjoins between the unidentified bones in each excavated unit, and then across squares. The fragments, which may belong to the same bone, were positioned in their anatomical location and element category, according to shape, size, color, and morphological characteristics. Bone conjoins were not expected to be refitted because of post-depositional weathering, but weathering was expected to be more pronounced on breaks that predate fossilization, e.g., those made by hominins on fresh bones. Even in situ weathering caused by sediment particles may obliterate the fracture edge.

Several conjoins were identified. Most were fragments of unidentified *Dama*-sized long bones, which were exposed in close proximity to each other. Several long bone fragments (nos. 12,580, 12,581 and 12,582) were refitted (Fig. 7.6), to form part of a *Dama*-sized femur shaft. Microscopic analyses of some of the conjoined edges revealed that sediment particles are still adhered to the bones, indicating that the breakage is an old one. In addition to the conjoined sets of *Dama* and *Dama*-sized bones, skeletal elements of elephants,

Fig. 7.6 Conjoinable pieces of a femur shaft of a *Dama*-sized animal (nos. 12,580, 12,581, and 12,582)

Fig. 7.7 The elephant hyoid fragment conjoins (nos. 1,652, 7,899, 12,901)

equids and BSGA bones were refitted. Three bone fragments were refitted to an elephant hyoid (no. 1,652) (Fig. 7.7) and to rib fragments (no. 762) from the JB belonging to a BSGA animal. In addition, two fragments equid pelvis fragments (no. 905) from Layer V-6 were refitted.

The GBY avian bones are excellently preserved, and in general, these bones weather in a manner similar to that of mammalian bones (Behrensmeyer et al. 2003). We undertook a refitting project for these bones as well. The number of near complete bones or sizable fragments (about half the original size) relative to the entire assemblage of bird bones is considerably larger than that of the mammalian bones. The surface appearance, a shiny, mottled brownish-black surface, is also similar to the mammalian bones, and a small number have a dull, slightly rough brown surface. The splinters' edges are sharp, and in some cases may have resulted from bone exposure that led to cracking and breakage. Striations were observed on the surface of the bird bones, similar to those on the *Dama* bones. Most of the conjoins were fresh

Table 7.9 Conjoined skeletal elements of mammals and birds from Area C and the JB

| Body element | Layer | Species/Body-size group | Coordinates | | Elevation |
			X	Y	
Hyoid (no. 1,652)	V-5	Elephant	181.99	130.92	60.36/60.35
(no. 7,899)	V-5		182.66	130.92	60.30
(no. 12,901)	V-5		182.72	130.73	60.36/60.34
Tibia (no. 11,985)	V-5	BSGD	182.8	130.6	60.43
(no. 11,906)	V-5		182.78	130.68	60.32
Femur distal (no. 12,582)	V-6	BSGD	181	130	60.2/60.13–59.89/59.86
	V-6		181	130	60.2/60.13–59.89/59.86
Long bone shaft (no. 12,589)	V-6	BSGD	185	129	60.18–59.99/59.93
(no. 12,598)	V-6		185	129	60.18–59.99/59.93
Old break: long bone shaft (no. 13,936)	V-6	BSGD	184.4	129.3	60.06–60.05
Rib shaft (no. 762)	JB	BSGA	199.10	124.68	60.00
(no. 773)	JB		197.01	124.94	59.87
Pelvis (no. 905)	JB	Equid	205.60	122.92	60.18/60.11
	JB		205.00	122.00	60.00
Probably the same element: atlas (no. 1,033)	JB	BSGE	194.1	125.6	59.87/59.84
(no. 10,160)	JB		194.1	125.5	59.91/59.90
Bird bones					
Frontal (no. 5)	V-6	*Cygnus* sp.	183	130d	60.37/60.31 60.27/60.18
	V-6		183	130d	60.37/60.31 60.27/60.18
Shaft (no. 1)	V-6	Unidentified Aves	184	129d	60.32/60.32 60.02/59.96
(3 pieces)	V-6		185	129a	60.12
Shaft (no. 7)	V-6	Unidentified Aves	184.34	129.5	60.03/60.02
	V-6		184.33	129.52	60.03/60.02
Shaft (no. 16)	V-6	Unidentified Aves	182.21	130.25	59.9
(2 pieces)	V-6		182.21	130.25	59.9
Shaft (no. 191)	V-6	Unidentified Aves	181	130a	60.2/60.13 59.89/59.76
	V-6		181	130a	60.2/60.13 59.89/59.76

and closely located. The frontal part of *Cygnus* sp. (no. 5) exhibits an old break, which was found together, probably broken in situ following trampling or due to minor sediment changes (i.e., dryness, compaction, water absorption) (Table 7.9).

Examination of the data summarized in Table 7.9 reveals that the maximal distance between the refitted bones is ca. 70–100 cm along the north–south axis, although most are only a few centimeters apart. The movement along the east–west axis is minimal, if at all. Several elements were conjoined from pieces found after sorting from the general bags (for excavation methodology, see Chapter 2).

Although the number of the refitted bone sets is small, it sheds some light on the dynamics of spatial distribution at the site. The refits reinforce and fine-tune the previous observations concerning certain taphonomic aspects at the site, indicating that only minor post-depositional weathering may have occurred either in situ or as a result of a slight movement. Together with the results of bone-marrow breakage by hominins, the refitting demonstrates the post-depositional splitting of some of the bones; the splinters remained in situ due to the surrounding matrix.

Embryos and faecal pellets within female viviparids were found in Layer V-5 and in the interface with Layer V-6 (Ashkenazi et al. 2010; Mienis and Ashkenazi 2011). Delicate ostracod valves were found inside the shells, near their aperture. Many were found articulated and whole. Also found inside the shells were articulated embryonic specimens of the bivalve *Unio terminalis*. These unique finds survived due to the minimal disturbance during the deposition and creation of the sediment plug in the aperture. This state of preservation suggests that either death occurred as a fast event (Mienis and Ashkenazi 2011), or due to a gentle rise of water level after the deposition of Layer V-6 (Feibel 2004).

7.7.2 The Spatial Organization of Stone Artifacts

7.7.2.1 Stone Artifact Refitting

Refitting is considered to be one of the best indicators of minimal taphonomic disturbances. Refitting attempts were therefore carried out in the three assemblages, but failed to yield positive results (Davidzon n.d.). This may possibly be due to the very small excavated surface and volume from both Layer V-5 and V-6 (see Chapter 2), although this is mere conjecture that has yet to be examined in the rest of the site.

7.7.2.2 Microartifact Spatial Distribution

Although stone microartifacts (basalt, flint, and limestone) were mentioned earlier in this chapter as part of a discussion

of all artifact sizes present at the site, and particularly in respect to their small fraction, their abundance and spatial organization are highly relevant to this section. In order to avoid a repetition, we have chosen not to repeat the above points, but clearly all information should be considered together.

The spatial distribution of the microartifacts supplies evidence for the original configuration of the finds in both layers. There are three types of raw materials at the site within the size fraction of microartifacts: basalt, flint, and limestone. Their spatial distribution in every GBY archaeological horizon comprises different patterns. Examination of these patterns in Layers V-5 and V-6 (Fig. 7.8) reveals a distinct clustering pattern of each particular raw material, indicating its individuality.

7.7.2.3 Microartifacts and Fire

Further indication for each assemblage's integrity is derived from burned flint microartifacts. We have previously demonstrated that the burned flint artifacts are clustered (Goren-Inbar et al. 2004; Alperson-Afil and Goren-Inbar 2006; Alperson-Afil 2008; Alperson-Afil et al. 2009; Alperson-Afil and Goren-Inbar 2010) and not randomly arranged (Alperson-Afil 2007). Furthermore, these clusters are differently spatially located in each layer, thus supporting the notion that each layer represents an individual entity (Fig. 7.9).

We interpret the clusters as Acheulian hearth residue (of phantom hearths), from which the ash, charcoal, and other lightweight components have disappeared and only the heavier flint microartifacts kept them in place. However, the concentrations of the burned flint microartifacts imply only minimal post-depositional taphonomic disturbances.

7.7.2.4 Basalt Microartifacts

While the most abundant raw material of microartifacts in Layers V-5 and V-6 is flint, both limestone and basalt are present in small quantities as well (see Tables 6.4 and 7.1). As noted above, the spatial clustering of the burned flint microartifacts reflects the presence of phantom hearths. Although basalt is much less common than flint (Layer V-6: N = 1,982), an examination of its microartifact distribution in Layer V-6 reveals that it also forms clusters. When basalt cleavers are added to the plotted microartifacts, several are found to be located in the immediate vicinity of the highest concentrations of the basalt microartifacts (Fig. 7.8). Such a spatial configuration supports previous observations that post-depositional processes did not disturb the arrangement of these components, and that these are, in fact, the original clustering.

A) Layer V-5

Flint Microartifacs (N=30,315)   Basalt Microartifacs (N=5,562)   Limestone Microartifacs (N=333)

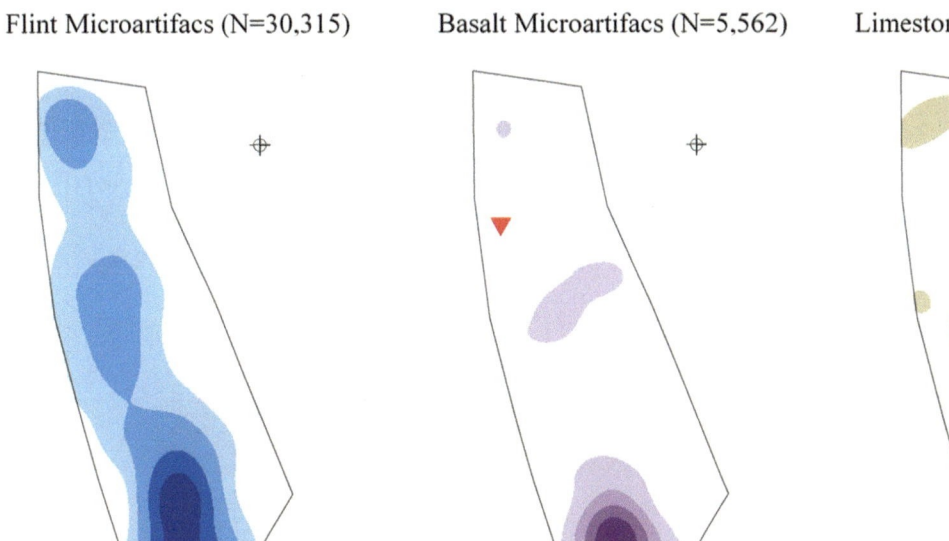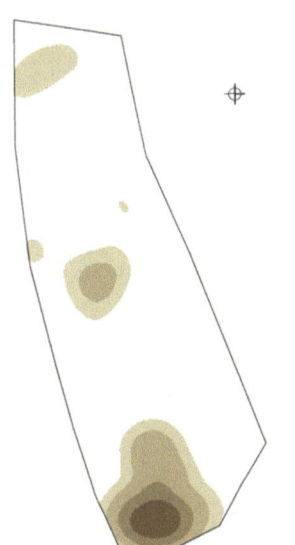

B) Layer V-6

Flint Microartifacs (N=4,424)   Basalt Microartifacs (N=1,982)   Limestone Microartifacs (N=96)

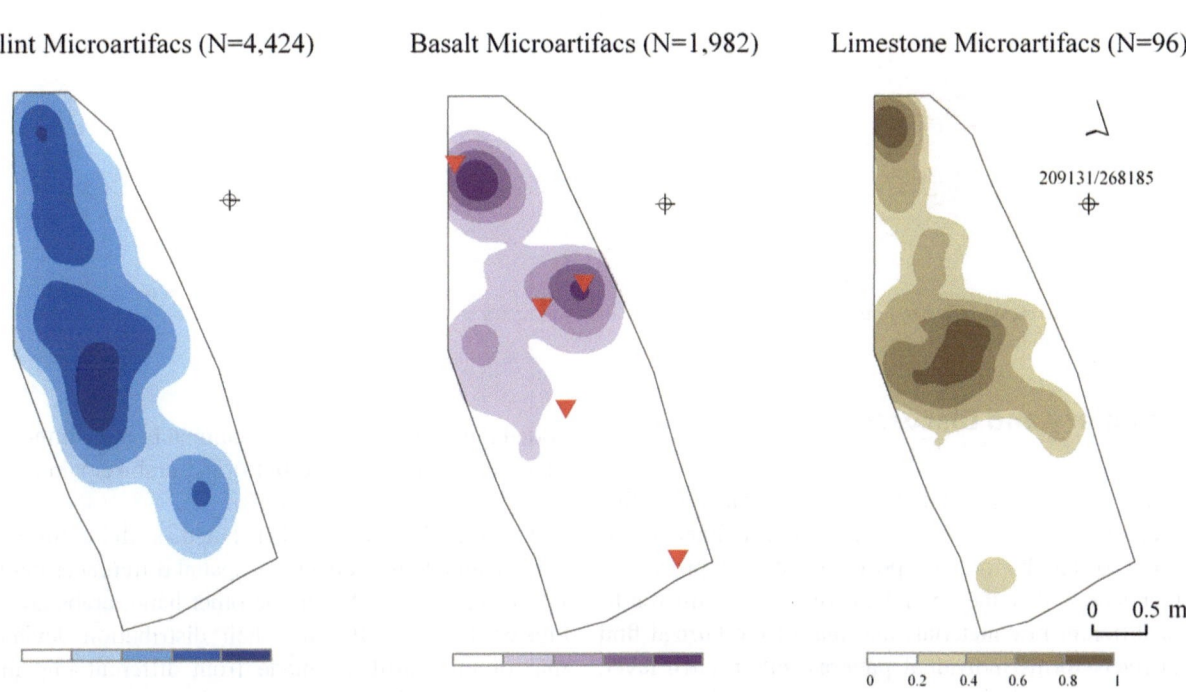

Fig. 7.8 Spatial location of basalt, flint, and limestone microartifacts and cleavers in Layers V-5 and V-6

Layer V-5 (N=402)   Layer V-6 (N=332)

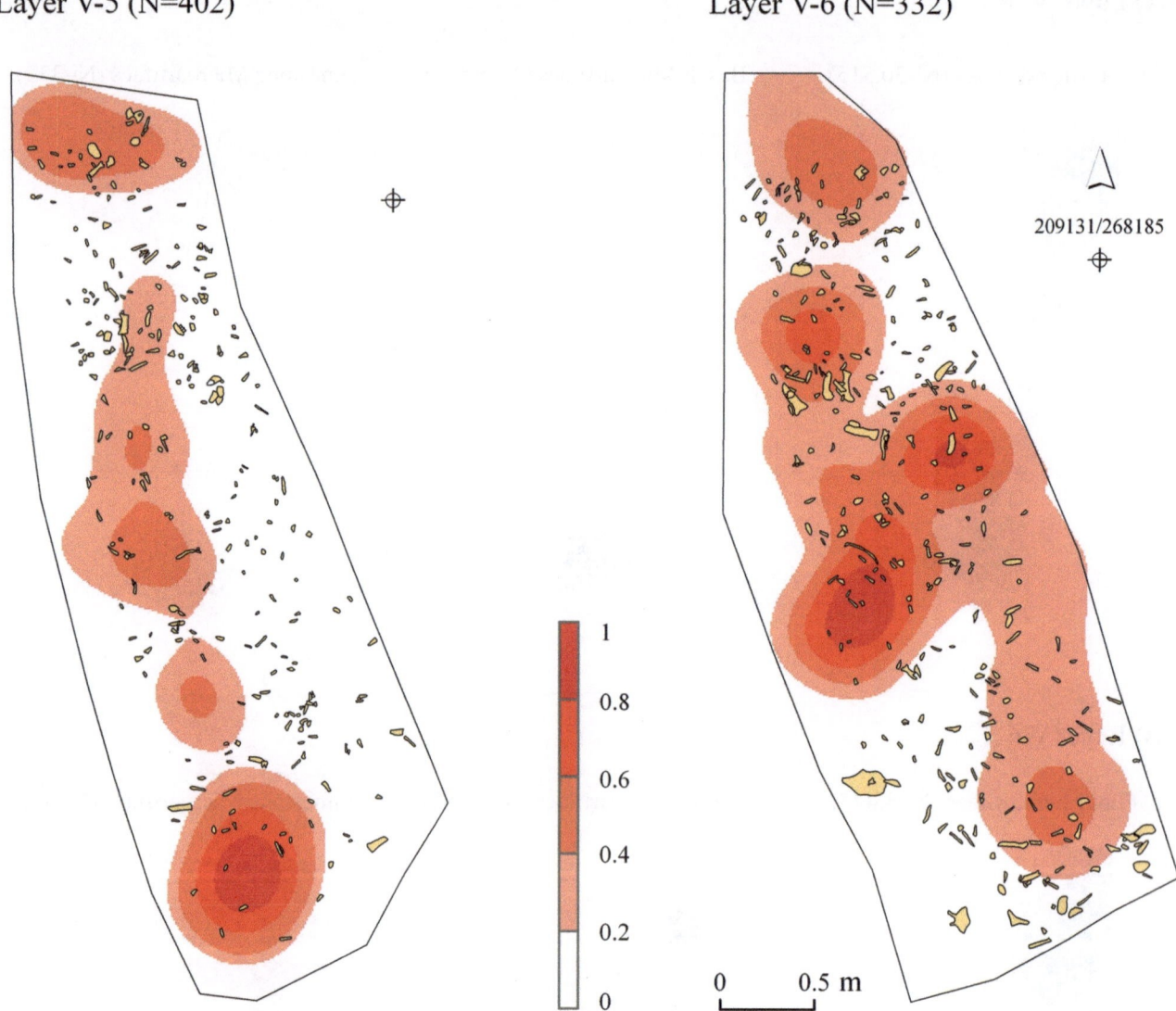

Fig. 7.9 Spatial distribution of burned microartifacts and mammal bones in Layers V-5 and V-6

7.8 Summary and Discussion

The lithic evidence presented above demonstrates that the assemblages of the two layers were neither transported nor sorted by pre- or post-depositional processes. It further shows that the spatial distribution of microartifacts of different raw materials and that of the burned flint microartifacts forms individual patterns within each layer. In addition, the spatial organization of the different raw materials and those of the burned flint is similar in the layers themselves: the microartifacts concentrations in Layer V-5 are located in the southern corner of the excavated area and those of Layer V-6 in the central space of the excavated area. In order to rule out the possibility that the similar distribution pattern within layers is a result of taphonomic agents, we added the spatial location of mammal bones (limited to items that were mapped in the field) and crab remains (the entire collection) (Figs. 7.9 and 7.10).

Mammal bones were distributed on the entire excavated surface and do not reflect any spatial differences between the two layers (Fig. 7.9). On the other hand, crabs are spatially clustered (Fig. 7.10) and their distribution deviates from that of microartifacts made from different raw materials, and those of the burned items in each layer. Had taphonomic agents seriously affected the archaeological record, we would have expected a different spatial distribution of mammal bones, crab remains, and other classes of finds sorted by their specific weight and morphology. The distinct individual spatial distributions—the clustering of burned flint, microartifacts of different raw materials, mammal

Layer V-5 (N=206)   Layer V-6 (N=2,686)

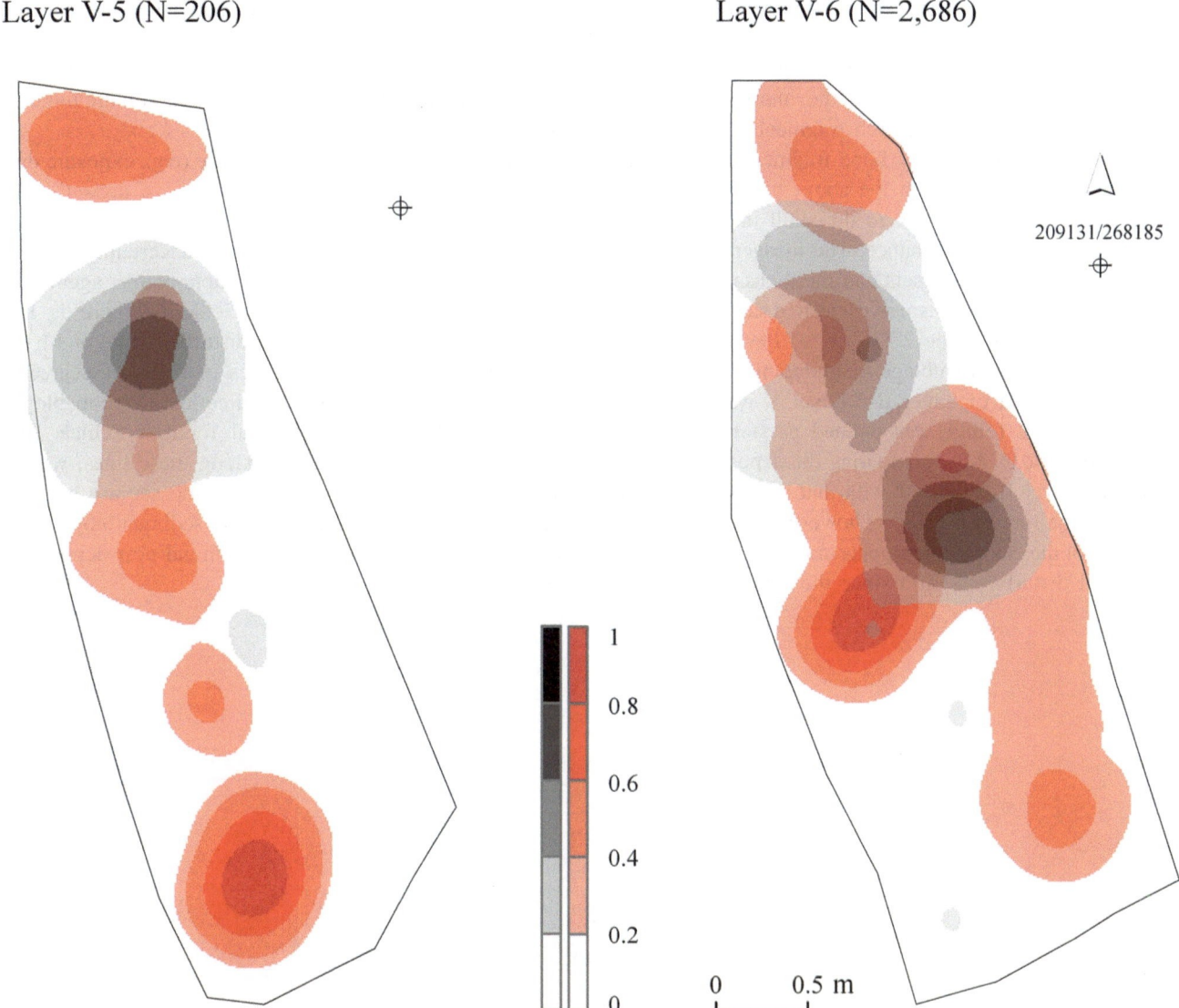

Fig. 7.10 Spatial distribution of burned microartifacts and crab remains in Layers V-5 and V-6

bones and crab remains—supports the notion of an in situ preservation of the original occupation surfaces.

The small size of the excavation area in Area C is clearly a drawback and imposes difficulties on the interpretation and the evaluation of the spatial analysis described above. Nevertheless, there are indications that the importance of the specific spatial distribution results is further supported by other archaeological horizons at GBY, which are not presented in this study and whose excavated surfaces are much larger. Another synergetic study focusing on Layer II-6 Level 2 analyzed different finds categories (crabs, fish, mammal bones, lithics and burned flint microartifacts, and organic remains) whose spatial distributions are of great interest and which seem to feature meaningful associations interpreted as indications of their primary depositional location, as well

as lack evidence of taphonomic disturbances (Alperson-Afil et al. 2009).

Supporting evidence for a negligible influence of post-depositional processes is encountered frequently in other layers at the site. Excavation of Layer II-6 Level 1 revealed associations between stone artifacts, fossil fauna, and wood. The most impressive case is that of the butchered elephant (*Paleoloxodon antiquus*), which furnished evidence relevant to the taphonomic issue (Goren-Inbar et al. 1994). A nearly complete elephant skull (missing mainly the occipital) was associated with the many small skull fragments that were recovered from its immediate proximity. Many of these are extremely light and small and look similar to the "inner lining of the braincase and sinuses of an elephant skull, and are very likely to belong to it" (Goren-Inbar et al. 1994: 100).

In addition, the following were also identified as elephant elements and found in the close vicinity of the elephant skull: molar fragments, tusk fragments, larger cranium fragments, two fragments of the internal surface of the braincase, two zygomatic arch fragments, a left occipital condyle, two premaxillary bone fragments, and a limb bone fragment. Together with the above conjoined pieces, the fragments close proximity to the skull indicates minimal post-depositional disturbance. The meager evidence of more than six refitted bone sets and the thousands of small skull fragments from a very small excavated surface (approximately 14.5 m^2 out of the 27.75 m^2 of Level 1 in Layer II-6) emphasizes this assemblage's taphonomic integrity. The sedimentary context of Layer II-6 Level 1, where the elephant bones were found, was defined as storm-generated beach (Feibel 2001). After comparing the depositional environment of Layer II-6 Level 1 with that of Layer V-6, it is evident that the former is coarser; however, the close spatial association of the elephant fragments indicates minimal post-depositional disturbance. And since the lithic assemblages of Layers V-5 and V-6 resemble each other, we conclude that even within the different depositional environments at the site (storm beaches, fine-grained muds, and coquina), the influence of post-depositional processes on the lithic and bone assemblages is minimal.

As the extent of the taphonomic role (e.g., exposure duration) is extremely difficult to evaluate, we presently lack any means, apart from the above data, to enable a more detailed reconstruction. This kind of uncertainty, of measuring the precise extent of the taphonomic agents, has been theoretically described by Schick: "There is most certainly no such thing as an 'undisturbed' site, and it will be useful to begin evaluating sites in terms of degrees of disturbance and the nature or direction of probable site modification by post-occupational forces" (Schick 1987: 105). If we take a step forward along these lines, we may conclude that the differences that we do see are primarily a result of short-exposure duration, sedimentological differences between the layers, and minimal post-depositional involvement.

Chapter 8

Summary and Conclusions

Abstract Taphonomic analysis of the fossil animal bones from two archaeological layers of Gesher Benot Ya'aqov (GBY) has enabled a comprehensive interpretation of the archaeological remains and sheds light on site-formation processes. We were able to reconstruct the environmental nature of the early Middle Pleistocene layers and the ecological background of the fauna, and to estimate their duration of accumulation, the possible occupation seasons, and the major agent responsible for the accumulation. Various aspects of hominin behavioral patterns could also be reconstructed, especially modes of animal exploitation and other subsistence traits. By studying and interpreting the various bone modifications found at the site, we were able to reconstruct aspects of site-formation processes. Combined with the study of the stone artifacts and other types of classes of evidence, a broad scenario as to the character of the site can be drawn and the site function discussed.

Taphonomic analysis of three assemblages of fossil animal bones from two archaeological layers (Layers V-5 and V-6) excavated at the site of GBY has enabled a comprehensive interpretation of the archaeological remains and sheds light on certain site-formation processes. We were able to reconstruct the environmental nature of the early Middle Pleistocene layers and the ecological background of the fauna, and to roughly estimate their duration of accumulation, the possible occupation seasons, and the major agent responsible for the accumulation. Various aspects of hominin behavioral patterns could also be reconstructed, most importantly for this volume, modes of animal exploitation and other subsistence traits. By studying and interpreting the various bone modifications found at the site, we were able to reconstruct aspects of site-formation processes. Combined with the study of the stone artifacts and other types of archaeological evidence, a broad scenario as to the character of the site can be drawn and the site function discussed.

Taphonomic analysis of faunal assemblages is a powerful tool in interpreting archaeological sites (see Chapters 1

and 3). Every site is a unique entity within its geographic setting, and in relation to its techno-cultural affiliations and particular formation history. As taphonomic processes are very complex and unique in nature, their interpretation should be undertaken on an individual basis. Actualistic studies (see Chapter 6) that, to a certain extent, replicate a site's paleoenvironmental and sedimentological setting serve as an important key in refining taphonomic-based interpretations. However, these interpretations will remain vague so long as the finds' archaeological origins, their excavation and recovery methods, analyses, and results of multidisciplinary studies (stone tools, botanical remains, etc.) are not taken into account. Hence, the taphonomic analyses described in this volume are an example of a synergetic approach towards investigating an archaeological site. Clearly, under any circumstances in Paleolithic research, only a small fraction of the finds is recoverable, and we were thus frequently left with many unresolved questions.

The bones of medium- and large-sized mammalian taxa that were analyzed in this monograph were found together in waterlogged sediments, associated with many other taxa, such as micromammals, birds, turtles, fish, crabs and mollusks, all spatially associated with macro-botanical remains in each excavated layer. These finds were also found to be associated with rich lithic assemblages comprising flint, basalt, and limestone.

Throughout this study we have demonstrated that the assemblages associated with the coquina deposit (Layer V-5) seem to be slightly more modified, either due to processes that formed the layer or to post-depositional ones. These two alternatives relate to two different scenarios concerning the formation mode of the assemblages. While the integrity and primary nature of Layer V-6 assemblages are not disputable, the situation in Layer V-5 is slightly different. Two explanations are suggested for its formation. The first views it as a distinct, discrete, autochthonous assemblage, which attained its taphonomic signature from the coquina in which it is embedded. The second views it as a derived, para-autochthonous, assemblage, whose origin lies in Layer V-6 and which later formed following an accretion event—while being deposited by waves that carved out the finds from the clayey matrix of Layer V-6 and integrated/deposited it in the coquina sediments of Layer V-5 (Feibel 2001).

These two models are temporally different: one assumes the formation of a single event while the other assumes a diachronic one. Yet, the extensive degree of the assemblages' similarities demonstrate that no matter which of these alternatives is closer to the Pleistocene scenario, both layers and their rich assemblages were sealed rapidly and were not exposed afterwards to additional taphonomic processes apart from those that occurred under the surface. A third possibility is that both scenarios took place, and that the sum of events that created the two archaeological records originated under different depositional mechanisms.

The lithic assemblages of Layers V-5 and V-6 do not reflect evidence of sorting, winnowing, and accumulation triggered by the presence of obstacles, nor any other major diagnostic disturbances. In contrast, the presence of a wide range of sizes, from microartifacts to cores and bifaces supports our conclusion of minimal post-depositional disturbances.

The analyses of stone tools demonstrate that the artifacts in each layer essentially represent an undisturbed assemblage that is well preserved and indicates rapid burial processes (see Chapter 7). It further indicates that several stages of stone tool production are present, including the smallest fractions of microartifacts, which form an excellent measure for the assemblages' integrity and/or undisturbed nature.

While the bulk of the lithics were made of flint, bifacial tools are scarce and were made primarily of basalt and are cleavers. In contrast with the paucity of the bifaces, it is evident that a substantial part of the flint flakes is a product of the final modification stages of flint bifaces, and that the initial stages of their reduction sequences are entirely missing (Sharon and Goren-Inbar 1999; Goren-Inbar and Sharon 2006).

The abundance of sharp-edged flint artifacts is viewed as the agent that caused many different modification signs on bones, primarily in the form of cut marks, but also striations. Hominin bone breakage activities, which are very abundant in both assemblages, were carried out for marrow extraction and were most probably achieved by the application of larger artifacts (e.g., basalt manuports).

The excavation of the two layers yielded a large number of mammal bones in very good condition and whose bone surfaces are mostly intact. These are the richest fossil assemblages of the entire GBY record and include 15 species of medium- and large-sized mammals.

Analysis of the taxonomic composition of the fauna and flora has enabled the reconstruction of the paleo-habitats at the site. The fauna from the paleo-Lake Hula margins is both African and Eurasian in origin, and the identified taxa

demonstrate the presence of a rich and diverse environment during the Middle Pleistocene. The mammalian species indicate the existence of both grasslands and woodlands habitats at GBY. The woodlands were occupied by elephants, cervids and some bovids, while the open grasslands were occupied by equids, other bovid species, and additional taxa. The background species (birds, reptiles, and amphibians) provide additional information regarding the site's paleoenvironment, and may have occasionally been part of the hominin diet.

Such a habitat reconstruction is strongly supported by the preservation of the organic material found in association with the fossil bones in the form of wood fragments, fruits, and seeds. The vegetation is also documented in the pollen record (van Zeist and Bottema 2008), and in the wood fragments of trees species such as *Cedrus*, *Ficus carica*, *Fraxinus syriaca*, *Juniperus*, *Olea europea*, *Quercus ithaburensis*, *Salix* and *Ulmus*, all typical Mediterranean flora (Goren-Inbar et al. 2002).

As we have seen, the site's faunal assemblage is dominated by *Dama*, which is represented by almost all the skeletal elements. The overall picture emerging from GBY's mammalian skeletal-element representation is that of several medium- and large-sized species that were not affected by density mediated factors or winnowing. Furthermore, the various taxa distribution and the quantitative representation of their elements are similar in both assemblages.

Carnivores are very rare, and analysis has indicated that carnivore bone modification is minimal, thus supplying additional evidence for refuting a hypothesis that they could have created the assemblages. In a few cases, carnivores left their signs on the bones after hominins already left their signs of consumption. On the one hand, the study indicates contemporaneity of the accumulated bones and, on the other hand, very rapid burial. The latter is deduced by the experiments, which have shown that bone abrasion processes that are caused by trampling stopped at a very early stage. The presence, though minimal, of refitted *Dama*- and *Dama*-sized bones, elephant, equid, and BSGA bones further supports these conclusions.

We maintain that hominins were the biotic taphonomic agents who left a heavy impact on the bone assemblages: seven taxa and the bones attributed to five body-size groups display traces of hominin interference. Analysis of these anthropogenic-induced marks (cut marks, hack marks, percussion marks, and split phalanges) has shown that complete *Dama* carcasses were brought to the site and entirely exploited for meat and marrow. The same butchery sequence is also documented on bones of other medium- and large-sized mammals, but whose details are less known due to the size of the sample. Thus, a large range of species of different body sizes was exploited by hominins for meat and marrow. Based on the animals' age and the *Dama* antlers (shed and unshed), we assume that occupation on the lake margins took place from autumn to spring in both layers.

Striations appear on most of the mammal bones, and their frequency varies in individual species and in different body-size groups. They occur mainly on *Dama* and *Gazella* bones, and on bones of their corresponding body-size groups, coinciding with the abundance of cut marks and damage caused by marrow extraction. Striation frequency per layer is similar among the assemblages, thus suggesting that the agent responsible for striation formation was similar in each layer.

The high density of bones and the small excavated volumes in Area C hampered the identification of particular and/or significant large-scale spatial-distribution patterning. The spatial distribution of mammal bones did not yield any discrete patterning, most probably due to the large density in a very small volume. Moreover, stone artifacts larger than 2 cm have yet to be spatially analyzed, but some spatial patterning has emerged from the spatial clustering of microartifacts, in particular those which are interpreted as depicting the location of phantom hearths (Alperson-Afil 2007). In addition, evidence of discrete spatial-patterning of river crab clustering, as well as that of basalt cleavers found in the immediate vicinity of basalt microartifacts, demonstrates that some original spatial patterning was retained and that despite the small fractions and susceptibility of the microartifacts and crab remains, their configuration was not influenced by the natural agents of a typical lake margin environment.

Taphonomic analysis of mammal bones, crabs, and lithic artifacts, along with, in particular, analysis of the spatial organization of distinct concentrations of different raw materials, burned microartifacts, and conjoinable pieces, leads us to conclude that the assemblages were minimally affected by post-depositional processes. The synergy of this multidisciplinary research also led to the conclusion that the accumulation of archaeological remains in Layers V-5 and V-6 does not reflect a scenario of individual, unrelated occupations. In short, it seems that the evidence presented in this study allows us a glimpse of a very small time frame approximately 750 ka.

How, then, can we interpret this evidence in terms of hominin behavior? In attempting to interpret the archaeological record presented in this volume, we unveiled complex patterns of functional and subsistence strategies, which are documented in the occupations' characteristics and considered to be, among others, a signature of hominin mobility. Results of experiments concerning the interpretation of bone-surface modifications indicate that the site clearly represents a central spot of hominin activities. Hominins revisited the margins of the paleo-Lake Hula over the course of several seasons. They arrived at this particular locality equipped with various sets of artifacts, as illustrated by the presence of three raw material types—basalt, flint, and limestone—used for tool production, none of which are available autochthonically. The need to transport the lithics resulted in a complex mobility pattern, which is evident in both layers: basalt bifaces (cleavers) were brought to the site as nearly finished products and were minimally curated, while flint handaxes that produced the bulk of the flakes and flake tools used in the carcasses exploitation are represented by a single biface (Sharon and Goren-Inbar 1999; Goren-Inbar and Sharon 2006) and were obviously heavily exploited during site occupation. The fact that the flint handaxes that produced most of the flake material are documented by only a single case leads us to conclude that hominins took these items with them when they abandoned the site.

One of the most discernible activities characterizing the data of GBY concerns the hunting of various animals and/or the handling of animal carcasses in a systematic way and their thorough exploitation. Bone taphonomy suggests a hunting scenario with either simultaneous killing of several *Dama* individuals approaching the edge of the lake or that of bringing several complete *Dama* carcasses to the vicinity of the lake margin. Additional elements of larger animals and/or complete larger mammals may also have been brought into the site. As complete butchering sequences were also demonstrated for larger species, it seems clear that hominins interacted with complete carcasses, which makes a scenario of a regular scavenging mode of behavior highly unlikely. Evidently, the transportation of medium-sized animals, or portions of hunted animals, should be viewed as an additional aspect of hominin mobility at the site.

The presence of several large-sized mammals at the site supports the notion that these animals may have been driven to or brought to the site and not simply hunted on the spot while drinking. Animal ethology coupled with evidence from the site (Layer II-6 Level 1) supports the suggestion that large-sized species may have been killed on the spot as well as transported to the site. In the same level, giant basalt cores (Madsen and Goren-Inbar 2004; Goren-Inbar et al. 2011) were transported to the site. Their reduction process produced an extensive inventory of bifacial tools, including many bifaces and waste products together with a large array of other artifact types. Along with the phantom hearths (Alperson-Afil 2007), all are associated with evidence of an elephant butchery scene (Goren-Inbar et al. 1994). It is difficult to assume that elephants routinely drank water at this particular spot, which was intensively occupied and characterized by hominin activity, as elephants are known to withdraw from the presence of humans unless attracted by the availability of edible crops.

Indication of further aspects of the hominin subsistence strategies can be outlined from the systematic exploitation of nuts and other edible plant organs. The presence of several edible fruit trees—including *Euryale ferox, Quercus,*

Vitis sylvestris, *Trapa natans*, and *Rubus* (Melamed 2003)—furnishes additional support for the diverse subsistence options that existed at the site. Some of the nuts were submerged and bank vegetation plants and thus available along the lake margins, while others were most probably gathered and brought to the lake margin occupations (Goren-Inbar et al. 2001). The high dietary value of the latter is well known and has been consumed along the entire course of history, as has been shown archaeologically, historically, and ethnographically.

The high bone density in Layers V-5 and V-6, along with their high taxonomic diversity and the results of actualistic studies, which point to extensive traffic at the site during a relatively short period of time, lead us to assume that a large group of hominins was involved in the assemblages' formation. In such a scenario, a large hominin group of mixed ages and gender gather at a particular spot in the landscape to divide the labor, share food, and exploit the environment on a seasonal basis, similar to known hunter-gatherer societies.

This and other GBY-related research has illustrated a variety of aspects that reflect a behavioral modernity of the GBY Acheulian hominins. Movement of artifacts and prey, spatial organization, long-term planning and insight, integration of environmental and landscape knowledge, foraging, exploitation of plant resources during the different seasons and a variety of ecological niches, all testify to the hominin ability to adapt and utilize a mosaic of resources and to affect the natural landscape some 750 ka. These results have tremendous consequences for our perception of the abilities of our Middle Pleistocene ancestors, and for interpreting the archaeological evidence.

Future research of the GBY data will unveil the nature of other Acheulian occupations at this particular hot spot of hominin activity. Since we have previously demonstrated that these occupations differ drastically from one another, only their thorough study, combining the results obtained from the present analysis, will contribute to and improve and enrich the interpretation and reconstruction of hominin behavior both at GBY and throughout the Acheulian Old World.

References

Abbazzi, L., & Masini, F. (1996). *Megaceroides solilhacus* and other deer from the middle Pleistocene site of Isernia La Pineta (Molise, Italy). *Bollettino della Società Paleontologica Italiana, 35*(2), 213–227.

Aiello, L. C., & Wheeler, P. (1995). The expensive tissue hypothesis: The brain and the digestive system in human and primate evolution. *Current Anthropology, 36*, 199–221.

Alperson-Afil, N. (2007). *Ancient flames: Controlled use of fire at the Acheulian site of Gesher Benot Ya'aqov, Israel.* Ph.D. dissertation, The Hebrew University of Jerusalem, Jerusalem.

Alperson-Afil, N. (2008). Continual fire-making by hominins at Gesher Benot Ya'aqov, Israel. *Quaternary Science Reviews, 27*, 1733–1739.

Alperson-Afil, N., & Goren-Inbar, N. (2006). Out of Africa and into Eurasia through controlled use of fire: Evidence from Gesher Benot Ya'aqov, Israel. *Archaeology Anthropology & Ethnology of Eurasia, 4*(28), 63–78.

Alperson-Afil, N., & Goren-Inbar, N. (2010). *The Achulian site of Gesher Benot Ya'aqov: Ancient flames and controlled use of fire.* Dordrecht: Springer.

Alperson-Afil, N., Sharon, G., Zohar, I., Biton, R., Melamed, Y., Kislev, M. E., et al. (2009). Spatial organization of hominin activities at Gesher Benot Ya'aqov, Israel. *Science, 326*(18), 1672–1680.

Anconetani, P. (1999). An experimental approach to intentional bone fracture: A case study from the Middle Pleistocene site of Isernia La Pineta. In S. Gaudzinski & E. Turner (Eds.), *The role of early humans in the accumulation of European lower and middle palaeolithic bone assemblages* (pp. 121–138). Monographien des Römisch-Germanischen Zentralmuseums, Vol. 42. Bonn/Mainz: Verlag des Römisch-Germanischen Zentralmuseums.

Andrews, P., & Cook, J. (1985). Natural modifications to bones in a temperate setting. *Man, 20*, 675–691.

Ashkenazi, S. (2004). Wetland drainage in the Levant (Lake Hula, Amik Gölö, and el-Azraq Oasis): Impact on avian fauna. In N. Goren-Inbar & J. D. Speth (Eds.), *Human paleoecology in the Levantine Corridor* (pp. 167–190). Oxford: Oxbow Books.

Ashkenazi, S., & Dimentman, C. (1998). Foraging, nesting and roosting habitats of the avian fauna of the Agmon wetland, northern Israel. *Wetlands Ecology and Management, 6*(2–3), 169–187.

Ashkenazi, S., & Mienis, H. K. (2005). *The taxonomy of the mollusc assemblage of the Pleistocene site of Gesher Benot Ya'aqov (GBY).* Irene Sala-Levi CARE Archaeological Foundation, Final report of 2004–2005.

Ashkenazi, S., Motro, U., Goren-Inbar, N., Bitton, R., & Rabinovich, R. (2005). New morphometric parameters for assessment of body size and population structure in freshwater fossil crab assemblage from the Pleistocene site of Gesher Benot Ya'aqov (GBY), Israel. *Journal of Archaeological Science, 32*, 675–689.

Ashkenazi, S., Klass, K., Mienis, K. H., Spiro, B., & Abel, R. (2010). Fossil embryos and adult Viviparidae from the early–middle Pleistocene of Gesher Benot Ya'aqov, Israel: Ecology, longevity and fecundity. *Lethaia, 43*, 116–127.

Ballesio, R. (1986). Les carnivores du gisement d'Oubeidiyeh (Israël). In E. Tchernov & C. Guérin, (Eds.), *Les Mammifères du Pléistocène Inférieur de la Vallée du Jourdain à Oubeidiyeh* (pp. 63–91). Mémoires et Travaux du Centre de Recherche Français de Jérusalem, Paris: Association Paléorient. Vol. 5.

Bar-Oz, G., & Munro, D. N. (2007). Gazelle bone marrow yields and Epipalaeolithic carcass exploitation strategies in the southern Levant. *Journal of Archaeological Science, 34*, 946–956.

Bar-Yosef, O., & Goren-Inbar, N. (1993). *The lithic assemblage of 'Ubeidiya.* Qedem Vol. 34, Jerusalem: Institute of Archaeology, Hebrew University.

Bar-Yosef, O., Vandermeersch, B., Arensburg, B., Belfer-Cohen, A., Goldberg, P., Laville, H., et al. (1992). The excavations in Kebara Cave, Mt. Carmel. *Current Anthropology, 33*, 497–549.

Bate, D. M. A. (1927). On the animal remains obtained from the Mugharet-el-Zuttiyeh in 1925 and 1926. In F. Turville-Petre (Ed.), *Research in prehisotric Galilee 1925–1926* (pp. 27–52). Jerusalem: British School of Archaeology.

Bate, D. M. A. (1937). Paleontology: The fossil fauna of the Wady-el-Mughara Caves. In D. A. E. Garrod & D. M. A. Bate (Eds.), *The stone age of mount carmel* (pp. 139–231). Oxford: Clarendon Press.

Beden, M. (1986). Le Mammoth d'Oubeidiyeh. In E. Tchernov, & C. Guérin (Eds.), *Les Mammifères du Pléistocène Inférieur de la Vallée du Jourdain à Oubeidiyeh* (pp. 213–234). Mémoires et Travaux du Centre de Recherche Français de Jérusalem, Paris: Association Paléorient. Vol. 5.

Behrensmeyer, A. K. (1975a). The taphonomy and paleoecology of Plio-Pleistocene vertebrate assemblages east of Lake Rudolf, Kenya. *Bulletin Museum of Comparative Zoology, 146*, 473–578.

Behrensmeyer, A. K. (1975b). Taphonomy and paleoecology in the hominid fossil record. *Yearbook of Physical Anthropology, 19*, 36–50.

Behrensmeyer, A. K. (1978). Taphonomic and ecological information from bone weathering. *Paleobiology, 4*, 150–162.

Behrensmeyer, A. K. (1988). Vertebrate preservation in fluvial channels. *Palaeogeography, Palaeoclimatology, Palaeocology, 63*, 183–199.

Behrensmeyer, A. K., & Boaz, D. E. D. (1980). The recent bones of Amboseli National Park, Kenya, in relation to East African paleoecology. In A. K. Behrensmeyer & A. P. Hill (Eds.), *Fossils in the making* (pp. 72–92). Chicago: University of Chicago Press.

Behrensmeyer, A. K., & Hill, A. P. (1980). *Fossils in the making.* Chicago: University of Chicago Press.

Behrensmeyer, A. K., & Kidwell, S. M. (1985). Taphonomy's contribution to paleobiology. *Paleobiology, 11*, 105–119.

Behrensmeyer, A. K., Gordon, K. D., & Yanagi, G. T. (1986). Trampling as a cause of bone surface damage and pseudo-cutmarks. *Nature, 319*, 768–771.

Behrensmeyer, A. K., Stayton, C. T., & Chapman, R. E. (2003). Taphonomy and ecology of modern avifaunal remains from Amboseli Park, Kenya. *Paleobiology, 29*(1), 52–70.

Belfer-Cohen, A., & Bar-Yosef, O. (1981). The Aurignacian at Hayonim cave. *Paléorient, 7*(2), 19–42.

Belitzky, S. (2002). The structure and morphotectonics of the Gesher Benot Ya'aqov area, Northern Dead Sea Rift, Israel. *Quaternary Research, 58*, 372–380.

Belmaker, M. (2006). *Community structure through time: 'Ubeidiya, a lower Pleistocene site as a case study*. Ph.D. dissertation, The Hebrew University of Jerusalem, Jerusalem.

Belmaker, M. (2009). Hominin adaptivity and patterns of faunal turnover in the early to middle Pleistocene transition in the Levant. In M. Camps, & P. Chauhan (Eds.), *Sourcebook of Paleolithic Transitions* (pp. 211–227). New York: Springer.

Bermudez de Castro, J. M., Carbonell, E., Cáceres, I., Diez, J. C., Fernández-Jalvo, Y., Mosquera, M., et al. (1999). The TD6 (Aurora stratum) hominid site. Final remarks and new questions. *Journal of Human Evolution, 37*, 695–700.

Binford, L. R. (1978). *Nunamiut ethnoarchaeology*. New York: Academic Press.

Binford, L. R. (1981). *Bones: Ancient man and modern myths*. New York: Academic Press.

Binford, L. R. (1984). Butchering, sharing, and the archaeological record. *Journal of Anthropological Archaeology, 3*, 235–257.

Blasco, R., Rosell, J., Férnandez Peris, J., Cáceres, I., & María Vergès, J. (2008). A new element of trampling: An experimental application on the level XII faunal record of Bolomor Cave (Valencia, Spain). *Journal of Archaeological Science, 35*, 1605–1618.

Blumenschine, R. J. (1986). *Early hominid scavenging opportunities*. BAR International Series, Vol. 283. Oxford: Archaeopress.

Blumenschine, R. J. (1988). An experimental model of the timing of hominid and carnivore influence on archaeological bone assemblage. *Journal of Archaeological Science, 15*, 483–502.

Blumenschine, R. J. (1995). Percussion marks, tooth marks, and timing of hominid and carnivore access to long bones at FLK *Zinjanthropus* Olduvai Gorge, Tanzania. *Journal of Human Evolution, 29*, 21–51.

Blumenschine, R. J., & Madrigal, T. C. (1993). Variability in long bone marrow yields of East African ungulates and its zooarchaeological implications. *Journal of Archaeological Science, 20*, 555–587.

Blumenschine, R. J., & Marean, C. W. (1993). A carnivore's view of archaeological bone assemblages. In J. Hudson (Ed.), *From bones to behavior: Ethnoarchaeological and experimental contributions to the interpretations of faunal remains* (pp. 271–300). Carbondale, IL: Southern Illinois University.

Blumenschine, R. J., & Selvaggio, M. M. (1988). Percussion marks on bone surfaces as a new diagnostic of hominid behaviour. *Nature, 333*, 763–765.

Blumenschine, R. J., & Selvaggio, M. M. (1991). On the marks of marrow bone processing by hammerstones and hyaenas: Their anatomical patterning and archaeological implications. In J. D. Clark (Ed.), *Cultural beginnings: Approaches to understanding early hominid life-ways in the African savanna* (pp. 17–32). Bonn/Mainz: Verlag des Römisch-Germanischen Zentralmuseums.

Bocherens, H., Billiou, D., Patou-Mathis, M., Otte, M., Bonjean, D., Toussaint, M., et al. (1999). Palaeoenvironmental and palaeodietary implications of isotopic biogeochemistry of late interglacial Neandertal and mammal bones in Scladina Cave (Belgium). *Journal of Archaeological Science, 26*(6), 599–607.

Bouchud, J. (1987). *La Faune du Gisement Natoufien de Mallaha (Eynan), Israël*. Mémoires et Travaux du Centre de Recherche Français de Jérusalem. Paris: Association Paléorient, Vol. 4.

Brain, C. K. (1967a). Bone weathering and the problem of bone pseudo-tools. *South African Journal of Science, 63*, 97–99.

Brain, C. K. (1967b). Hottentot food remains and their bearing on the interpretation of fossil bone assemblages. *Scientific Papers of the Namib Desert Research Station, 32*, 1–7.

Brain, C. K. (1976). Some principles in the interpretation of bone accumulations associated with man. In G. Isaac & E. McCown (Eds.), *Human Origins* (pp. 97–116). Menlo Park, CA: W.A Benjamin.

Brain, C. K. (1981). *The Hunters or the hunted? An introduction to African cave taphonomy*. Chicago: University of Chicago Press.

Brantingham, P. J. (1998). Hominid-carnivore coevolution and invasion of the predatory guild. *Journal of Antropological Archaeology, 17*, 327–353.

Brown, W. A. B., & Chapman, N. G. (1990). The dentition of fallow deer (*Dama dama*): A scoring scheme to assess age from wear of the permanent molariform teeth. *Journal of the Zoological Society of London, 221*, 659–682.

Brown, W. A. B., & Chapman, N. G. (1991). Age assessment of fallow deer (*Dama dama*): From a scoring scheme based on radiographs of developing permanent molariform teeth. *Journal of the Zoological Society of London, 224*, 367–379.

Bukhsianidze, M., & Vekua, A. (2006). *Capra dalii* nov. sp. (Caprinae, Bovidae, Mammalia) at the limit of Plio-Pleistocene from Dmanisi (Georgia). In R.-D. Kahlke, L. C. Maul, & P. P. A. Mazza (Eds.), *Late Neogene and Quaternary biodiversity and evolution: Regional developments and interregional correlations* (pp. 159–171), Vol. I. Courier Forschungsinstitut Senckenberg, Stutgartt: E. Schweizerbart'sche Verlagsbuchhandlurg, 256.

Bunn, H. T. (1997). The bone assemblages from the excavated sites. In G. L. L. Isaac (Ed.), *Koobi Fora research project* (pp. 402–458). Vol. 5: Plio-Pleistocene Archaeology. Oxford: Clarendon Press.

Bunn, H. T. (2001). Hunting, power, scavenging, and butchering in the Hadza foragers and Plio-Pleistocene Homo. In C. B. Stanford & H. T. Bunn (Eds.), *Meat-eating and human evolution* (pp. 199–218). Oxford: Oxford University Press.

Bunn, H. T., & Ezzo, J. A. (1993). Hunting and scavenging by Plio-Pleistocene hominids: Nutritional constraints, archaeological patterns, and behavioural implications. *Journal of Archaeological Science, 20*, 365–398.

Bunn, H. T., & Kroll, E. M. (1986). Systematic butchery by Plio-Pleistocene hominids at Olduvai Gorge, Tanzania. *Current Anthropology, 27*, 431–452.

Bunn, H. T., Bartram, L. E., & Kroll, E. M. (1988). Variability in bone assemblage formation from Hadza hunting, scavenging, and carcass processing. *Journal of Anthropological Archaeology, 7*, 412–457.

Byers, D. A., & Ugan, A. (2005). Should we expect large game specialization in the late Pleistocene? An optimal foraging perspective on early Paleoindian prey choice. *Journal of Archaeological Science, 32*, 1624–1640.

Campana, D. V., & Crabtree, P. J. (1987). Animals. A C language computer program of the analysis of faunal remains and its use in the study of Early Iron age fauna from Dun Ailinne. *Archaeozoologia, 1*(1), 57–68.

Capaldo, S. D. (1997). Experimental determinations of carcass processing by Plio-Pleistocene hominids and carnivores at FLK 22 (Zinjanthropus), Olduvai Gorge, Tanzania. *Journal of Human Evolution, 33*, 555–597.

Carbonell, E., Bermúdez de Castro, J. M., Parés, J. M., Pérez-González, A., Cuenca-Bescós, G., Ollé, A., et al. (2008). The first hominin of Europe. *Nature, 452*, 465–469.

Chaplin, R. E., & White, R. W. G. (1969). The use of tooth eruption and wear, body weight and antler characteristics in the age estimation of male wild and park Fallow deer (*Dama dama*). *Journal of the Zoological Society of London, 157*, 125–132.

Chazan, M., & Horwitz, L. K. (Eds.) (2007). *Holon: A Lower Paleolithic site in Israel*. American School of Prehistoric Research, Bulletin 50, Peabody Museum of Archaeology and Ethnology, Cambridge: Harvard University Press.

Cleghorn, N., & Marean, C. W. (2007). The destruction of skeletal elements by carnivores: The growth of a general model for skeletal element destruction and survival in zooarchaeological assemblages. In T. R. Pickering, K. Schick, & N. Toth, (Eds.), *Breathing life into fossils: Taphonomic Studies in Honor of C.K. (Bob) Brain* (pp. 37–66). Gosport: Stone Age Press.

Coard, R. (1999). One bone, two bones, wet bones, dry bones—transport potentials under experimental conditions. *Journal of Archaeological Science, 26*, 1369–1375.

Coard, R., & Dennell, R. W. (1995). Taphonomy of some articulated skeletal remains: Transport potential in an artificial environment. *Journal of Archaeological Science, 22*, 441–448.

Crader, D. C. (1983). Recent single-carcass bone scatters and the problem of "butchery" sites in the archaeological record. In J. Clutton-Brock & C. Grigson, (Eds.), *Animals and archaeology: Vol. 1. Hunters and their prey* (pp. 107–141). BAR International Series, Vol. 163. Oxford: Archaeopress.

Crégut-Bonnoure, E. (2006). European ovibovini, ovini and caprini (Caprinae, Mammalia) from the Plio-Pleistocene: New interpretations. In R.-D. Kahlke, L. C. Maul, & P. P. A. Mazza (Eds.), *Late Neogene and Quaternary biodiversity and evolution: Regional developments and interregional correlations* (pp. 139–158), Vol. I. *Courier Forschungsinstitut Senckenberg*, Stutgartt: E. Schweizerbart'sche Verlagsbuchhandlung, 256.

Cuenca-Bescós, G., Rofes, J., López-García, J. M., Blain, H.-A., De Marfá, R. J., Galindo-Pellicena, M. A., et al. (2010). Biochronology of Spanish Quaternary small vertebrate faunas. *Quaternary International, 212*, 109–119.

Cuenca-Bescós, G., Straus, L. G., González Morales, M. R., & García Pimienta, J. C. (2009). The reconstruction of past environments through small mammals: From the Mousterian to the Bronze Age in El Mirón Cave (Cantabria, Spain). *Journal of Archaeological Science, 36*, 947–955.

Darlas, A. (1995). The earliest occupation of Europe: The Balkans. In W. Roebroeks & T. van Kolfschoten (Eds.), *The earliest occupation of Europe* (pp. 51–59). Leiden: University of Leiden.

Davidzon, A. (n.d.). Attempts at lithics refiting in layers V-5 and V-6, GBY.

Davies, P., & Lister, A. M. (2007). Palaeoloxodon. In M. Chazan & L. K. Horwitz (Eds.), *The Lower Paleolithic site of Holon, Israel* (pp. 123–131). American School of Prehistoric Research Bulletin 50. Peabody Museum of Archaeology and Ethnology. Cambridge, MA: Harvard University Press.

Davis, S. (1980). A note on the dental and skeletal ontogeny of *Gazella*. *Israel Journal of Zoology, 29*, 129–134.

Davis, S. (1985). A preliminary report of the fauna from Hatoula: A Natufian Khiamina (PPNA) site near Latroun, Israel. In M. Lechevallier & A. Ronen (Eds.), *Le Site Natoufien-Khiamien de Hatoula* (pp. 71–98). Cahiers du CNRF 1. Paris: Association Paléorient.

Dayan, T. (1989). *The succession and community structure of the carnivores of the middle east in time and space*. Ph.D. dissertation, Tel-Aviv University, Tel-Aviv.

De Giuli, C. (1986). Late Villafranchian faunas of Italy: The Selvella local fauna in the southern Chiana Valley – Umbria. *Palaeontographia Italica, 74*, 11–50.

de Heinzelin, J., Clark, D. J., White, T., William, H., Renne, P., Wolde Gabriel, G., et al. (1999). Environmental and behavior of 2.5-million-year-old Bouri Hominids. *Science, 284*(5415), 625–629.

Delage, C. (2007). Chert procurement in the Acheulean of Gesher Benot Ya'aqov (Israel): Preliminary assessments. In J. Hedges & E. Hedges (Eds.), *Chert availability and prehistoric exploitation in the near east* (pp. 240–257). BAR International Series, Vol. 1615. Oxford: Archaeopress.

Delagnes, A., & Roche, H. (2005). Late Pliocene hominid knapping skills: The case of Lokalalei 2C, West Turkana, Kenya. *Journal of Human Evolution, 48*, 435–472.

Delagnes, A., Lenoble, A., Harmand, S., Brugal, J.-P., Prat, S., Tiercelin, J.-J., et al. (2006). Interpreting pachyderm single carcass sites in the African Lower and early Middle Pleistocene record: A multidisciplinary approach to the site of Nadung'a 4 (Kenya). *Journal of Anthropological Archaeology, 25*, 448–465.

Dennell, R., & Roebroeks, W. (1996). The earliest colonization of Europe: The short chronology revisited. *Antiquity, 70*, 535–542.

Denys, C. (2002). Taphonomy and experimentation. *Archaeometry, 44*(3), 469–484.

Dewbury, A. G., & Russell, N. (2007). Relative frequency of butchering cutmarks produced by obsidian and flint: An experimental approach. *Journal of Archaeological Science, 34*(3), 354–357.

Díez, J. C., Fernández-Jalvo, Y., Rosell, J., & Caceres, I. (1999). Zooarchaeology and taphonomy of Aurora Stratum (Gran Dolina, Sierra de Atapuerca, Spain). *Journal of Human Evolution, 37*, 623–652.

Dimentman, C., Bromley, H. J., & Por, F. D. (1992). *Lake Hula, reconstruction of the fauna and hydrobiology of a lost lake*. Jerusalem: The Israel Academy of Science and Humanities.

Domínguez-Rodrigo, M. (1997). Meat-eating by early hominids at the FLK 22 *Zijanthropus* site, Olduvai Gorge (Tanzania): An experimental approach using cut-mark data. *Journal of Human Evolution, 33*, 669–690.

Domínguez-Rodrigo, M. (2002). Hunting and scavenging by early humans: The state of the debate. *Journal of World Prehistory, 16*, 1–54.

Domínguez-Rodrigo, M., & Barba, R. (2006). New estimates of tooth mark and percussion mark frequencies at the FLK Zinj site: The carnivore-hominid-carnivore hypothesis falsified. *Journal of Human Evolution, 50*, 170–194.

Domínguez-Rodrigo, M., & Yarverda, J. (2009). Why are cut mark frequencies in archaeofaunal assemblages so variable? A multivariate analysis. *Journal of Archaeological Science, 36*, 884–894.

Domínguez-Rodrigo, M., Pickering, T. R., Semaw, S., & Rogers, M. J. (2005). Cutmarked bones from Pliocene archaeological sites at Gona, Afar, Ethiopia: Implications for the function of the world's oldest stone tools. *Journal of Human Evolution, 48*, 109–121.

Domínguez-Rodrigo, M., de Juana, S., Galán, A. B., & Rodríguez, M. (2009). A new protocol to differentiate trampling marks from butchery cut marks. *Journal of Archaeological Science, 36*, 2643–2654.

von den Driesch, A. (1976). *A guide to the measurements of animal bones from archaeological sites*. Peabody Museum Bulletin 1. Cambridge, MA: Harvard University Press.

Eisenmann, V. (1986). Les Equidés d'Oubeidiyeh. In E. Tchernov & C. Guérin (Eds.), *Les Mammifères du Pléistocène Inférieur de la Vallée du Jourdain à Oubeidiyeh* (pp. 191–212). Mémoires et Travaux du Centre de Recherche Français de Jérusalem, Paris: Association Paléorient. Vol. 5.

Eisenmann, V. (1992). Systematic and biostratigraphical interpretation of the equids from Qafzhe, Tabun, Shkul and Kebara (Acheuloyabrudian to Upper Paleolithic of Israel). *Archaeozoologica, V*(1), 43–62.

Eisenmann, V. (2006). Pliocene and Pleistocene equids: Palaeontology versus molecular biology. In R.-D. Kahlke, L. C. Maul, & P. P. A. Mazza (Eds.), *Late neogene and quaternary biodiversity and evolution: Regional developments and interregional correlations* (pp. 71–85). Courier Forschungsinstitut Senckenberg, Stutgartt: E. Schweizerbart'sche Verlagsbuchhandlung, 256.

Eisenmann, V. (n.d.). *The equids of Gesher Benot Ya'aqov, Israel*.

Eltringham, S. K. (1999). *The hippos. Poyser natural history series*. London: Academic Press.

Evans, L. A., Mccutcheon, A. L., Dennis, G. R., Mulley, R. C., & Wilson, M. A. (2005). Pore size analysis of fallow deer (Dama dama) antler bone. *Journal of Material Science, 40*, 5733–5739.

Faith, J. T. (2007). Sources of variation in carnivore tooth-mark frequencies in a modern spotted hyena (*Crocuta crocuta*) den assemblage, Amboseli Park, Kenya. *Journal of Archaeological Science, 34*, 1601–1609.

Faith, J. T., & Gordon, A. D. (2007). Skeletal element abundances in archaeofaunal assemblages: Economic utility, sample size, and assessment of carcass transport strategies. *Journal of Archaeological Science, 34*, 872–882.

Feibel, S. C. (2001). Archaeological sediments in lake margin environments. In J. K. Stein & W. R. Farrand (Eds.), *Sediments in archaeological contexts* (pp. 127–148). Salt Lake City, UT: University of Utah Press.

Feibel, S. C. (2004). Quaternary lake margins of the Levant Rift Valley. In N. Goren-Inbar & J. D. Speth (Eds.), *Human paleoecology in the Levantine Corridor* (pp. 21–36). Oxford: Oxbow Books.

Feibel, C. S. (in preparation). *The Acheulian Site of Gesher Benot Ya'aqov Volume V: Sediments and Paleoenvironment.* Dordrecht: Springer.

Feibel, S. C., Goren-Inbar, N., Verosub, K. L., & Saragusti, I. (1998). Gesher Benot Ya'aqov, Israel: New evidence for its stratigraphic and sedimentologic context. *Journal of Human Evolution, 34*, A7.

Fernández-Jalvo, Y. (1995). Small mammal taphonomy at La Trinchera de Atapuerca (Burgos, Spain). A remarkable example of taphonomic criteria used for stratigraphic correlations and palaeoenvironment interpretations. *Palaeogeography, Palaeoclimatology, Palaeoecology, 114*, 167–195.

Fernández-Jalvo, Y., & Andrews, P. (2003). Experimental effects of water abrasion on bone fragments. *Journal of Taphonomy, 1*(3), 145–161.

Fernández-Jalvo, Y., Sánchez-Chillón, B., Andrews, P., Fernández-López, S., & Alcalá Martínez, L. (2002). Morphological taphonomic transformations of fossil bones in continental environments, and repercussions on their chemical composition. *Archaeometry, 44*(3), 353–361.

Fiorillo, R. A. (1989). An experimental study of trampling: Implications for the fossil record. In R. Bonnichsen & H. M. Sorg (Eds.), *Bone modification* (pp. 61–71). Orono, ME: Center for the Study of the first Americans.

Fosse, P. (1999). Cave occupation during Palaeolithic times: Man and/or hyaena? In S. Gaudzinski & E. Turner (Eds.), *The role of early humans in the accumulation of European Lower and Middle Palaeolithic bone assemblages* (pp. 73–88). Monographien Des Römisch-Germanischen Zentralmuseums, Vol. 42. Bonn/Mainz: Verlag des Römisch-Germanischen Zentralmuseums.

Fowler, C. S. (Ed.) (1989). *Willard Z. Park's ethnographic notes on the Northern Paiute of Western Nevada, 1933–1940* (pp. 77–79), Vol. 1. Salt Lake City, UT: University of Utah Press.

Friedman, E., Goren-Inbar, N., Rosenfeld, A., Marder, O., & Burian, F. (1995). Hafting during Mousterian times – further indication. *Mitekufat Ha'even, 26*, 8–31.

Frison, G. C., & Todd, L. C. (1986). *The Colby Mammoth site: Taphonomy and archaeology of a Clovis kill in northern Wyoming.* Albuquerque, NM: University of New Mexico Press.

Galàn, A. B., Rodríguez, M., de Juana, S., & Domínguez-Rodrigo, M. (2009). A new experimental study on percussion marks and notches and their bearing on the interpretation of hammerstone-broken faunal assemblages. *Journal of Archaeological Science, 36*, 776–784.

Gamble, C. (1986). *The Palaeolithic settlement of Europe.* Cambridge: Cambridge University Press.

Garrard, A. N. (1980). *Man-animal-plant relationships during the Upper Pleistocene and early Holocene of the Levant.* Ph.D. dissertation, University of Cambridge, Cambridge.

Gaudzinski, S. (1995). Wallertheim revisited. *Journal of Archaeological Science, 22*, 51–66.

Gaudzinski, S. (1996). On bovid assemblages and their consequences for the knowledge of subsistence patterns in the Middle Palaeolithic. *Proceedings of the Prehistoric Society, 62*, 19–39.

Gaudzinski, S. (2004a). Subsistence patterns of early Pleistocene hominids in the Levant – taphonomic evidence from the 'Ubeidiya formation. *Journal of Archaeological Science, 31*, 65–75.

Gaudzinski, S. (2004b). Early hominid subsistence in the Levant: Taphonomic studies of the Plio-Pleistocene 'Ubeidiya formation (Israel) – Evidence from 'Ubeidiya Layer II-24. In N. Goren-Inbar & J. D. Speth (Eds.), *Human paleoecology in the Levantine Corridor* (pp. 75–87). Oxford: Oxbow Books.

Gaudzinski, S. (2005). Monospecific or species dominated faunal assemblages during the Middle Palaeolithic in Europe. In E. Hovers & S. Kuhn (Eds.), *Transitions before the Transition. Evolution and stability in the Middle Palaeolithic and Middle Stone age* (pp. 137–147). New York: Springer.

Gaudzinski-Windheuser, S. (2005). *Subsistenzstrategien frühpleistozäner Hominiden in Eurasien. Taphonomische Faunenbetrachtungen der Fundstellen der `Ubeidiya Formation (Israel).* Monographien des Römisch-Germanischen Zentralmuseums, Vol 61. Bonn/Mainz: Verlag des Römisch-Germanischen Zentralmuseums.

Gaudzinski, S., & Roebroeks, W. (2000). Adults only: Reindeer hunting at the Middle Palaeolithic site Salzgitter Lebenstedt, Northern Germany. *Journal of Human Evolution, 38*, 497–521.

Gaudzinski, S., Turner, E., Anzidei, A. P., Àlvarez-Fernández, E., Arroyo-Cabrales, J., Cinq-Mars, J., et al. (2005). The use of Proboscidean remains in every-day Palaeolithic life. *Quaternary International, 126–128*, 179–194.

Gaudzinski-Windheuser, S., Kindler, L., Rabinovich, R., & Goren-Inbar, N. (2010). Testing heterogeneity in faunal assemblages from archaeological sites. Tumbling and trampling experiments at the Early-Middle Pleistocene site of Gesher Benot Ya'aqov (Israel). *Journal of Archaeological Science, 37*, 3170–3190.

Geraads, D. (1986). Ruminants pleistocene d'Oubeidiyeh. In E. Tchernov & C. Guérin (Eds.), *Les Mammifères du Pléistocène Inférieur de la Vallée du Jourdain à Oubeidiyeh* (pp. 143–181). Mémoires et Travaux du Centre de Recherche Français de Jérusalem, Paris: Association Paléorient. Vol. 5.

Geraads, D., & Tchernov, E. (1983). Femurs Humains du Pleistocene Moyen de Gesher Benot Ya'aqov (Israel). *L'Anthropologie, 87*, 138–141.

Geraads, D., Guérin, C., & Faure, M. (1986). Les Suidae (Mammalia, Atiodactyla) du gisement pléistocene ancien d'Oubeidiyeh. In E. Tchernov & C. Guérin (Eds.), *Les Mammifères du Pléistocène Inférieur de la Vallée du Jourdain à Oubeidiyeh* (pp. 93–105). Mémoires et Travaux du Centre de Recherche Français de Jérusalem, Paris: Association Paléorient. Vol. 5.

Gifford-Gonzales, D. (1989). Ethnographic analogues for interpreting modified bones: Some cases from East Africa. In R. Bonnichsen & M. H. Sorg (Eds.), *Bone modification* (pp. 179–246). Peopling of the Americas Series. Orno, ME: Center for the Study of the First Americans, University of Maine.

Gifford-Gonzales, D. P., Damrosch, D. B., Damrosch, D. R., Pryor, J., & Thunen, R. L. (1985). The third dimension in site structure: An experiment in trampling and vertical dispersal. *American Antiquity, 50*, 803–818.

Gilead, D. (1968). Gesher Benot Ya'aqov. *Hadashot Archeologiot, 27*, 34–35.

Gittleman, J. L. (1989). *Carnivore behavior, ecology, and evolution.* London: Chapman and Hall.

Goldman, T., & Hovers, E. (2009). Methodological issues in the study of Oldowan raw material selection: Insights from A. L. 894 (Hadar,

Ethiopia). In E. Hovers & D. R. Braun (Eds.), *Multidisciplinary approaches to the Oldowan* (pp. 71–84). Dordrecht: Springer.

Goren-Inbar, N. (2011a). Culture and cognition in the Acheulian industry – A case study from Gesher Benot Ya'aqov. *Philosophical Transactions of the Royal Society B, 366,* 1038–1049.

Goren-Inbar, N. (2011b). Behavioral and cultural origins of Neanderthals: A Levantine perspective. In S. Condemi & G.-C. Weniger (Eds.), *150 years of Neanderthal discoveries. Continuity and discontinuity* (pp. 89–100). Dordrecht: Springer.

Goren-Inbar, N., & Belitzky, S. (1989). Structural position of the Pleistocene Gesher Benot Ya'aqov site in the Dead Sea Rift Zone. *Quaternary Research, 31,* 371–376.

Goren-Inbar, N., & Saragusti, I. (1996). An Acheulian biface assemblage from the site of Gesher Benot Ya'aqov, Israel: Indications of African affinities. *Journal of Field Archaeology, 23,* 15–30.

Goren-Inbar, N., & Sharon, G. (2006). Invisible handaxes and visible Acheulian biface technology at Gesher Benot Ya'aqov, Israel. In N. Goren-Inbar & G. Sharon (Eds.), *Axe age: Acheulian tool-making from quarry to discard* (pp. 111–135). London: Equinox.

Goren-Inbar, N., Belitzky, S., Verosub, K., Werker, E., Kislev, M., Heimann, A., et al. (1992a). New discoveries at the Middle Pleistocene Gesher Benot Ya'aqov Acheulian site. *Quaternary Research, 38,* 117–128.

Goren-Inbar, N., Belitzky, S., Goren, Y., Rabinovich, R., & Saragusti, I. (1992b). Gesher Benot Ya'aqov – the "Bar": An Acheulian assemblage. *Geoarchaeology, 7*(1), 27–40.

Goren-Inbar, N., Lister, A., Werker, E., & Chech, M. (1994). A butchered elephant skull and associated artifacts from the Acheulian site of Gesher Benot Ya'aqov, Israel. *Paléorient, 20*(1), 99–112.

Goren-Inbar, N., Feibel, C. S., Verosub, K. L., Melamed, Y., Kislev, M. E., Tchernov, E., et al. (2000). Pleistocene milestones on the Out-of-Africa corridor at Gesher Benot Ya'aqov, Israel. *Science, 289,* 944–947.

Goren-Inbar, N., Sharon, G., Melamed, Y., & Kislev, M. (2001). Nuts, nut cracking, and pitted stones at Gesher Benot Ya'aqov, Israel. *Proceedings of the National Academy of Sciences of the United States of America, 99,* 2455–2460.

Goren-Inbar, N., Werker, E., & Feibel, C. S. (2002). *The Acheulian site of Gesher Benot Ya'aqov, Israel. The Wood Assemblage,* Vol. 1. Oxford: Oxbow Books.

Goren-Inbar, N., Alperson, N., Kislev, M. E., Simchoni, O., Melamed, Y., Ben-Nun, A., et al. (2004). Evidence of hominin control of fire at Gesher Benot Ya'aqov, Israel. *Science, 304,* 725–727.

Goren-Inbar, N., Grosman, L., & Sharon, G. (2011). The record, technology and significance of the Acheulian giant cores of Gesher Benot Ya'aqov, Israel. *Journal of Archaeological Science, 38,* 1901–1917.

Goren-Inbar, N., Sharon, G., & Alperson-Afil, N. (in preparation). *The Acheulian site of Gesher Benot Ya'aqov: The lithic assemblages, Vol. IV.* Dordrecht: Springer.

Goring-Morris, N. A., & Belfer-Cohen, A. (Eds.). (2003). *More than meets the eye: Studies on Upper Palaeolithic diversity in the Near East.* Oxford: Oxbow Books.

Gray, N.-M., Kainec, K., Madar, S., Tomko, L., & Wolfe, S. (2007). Sink or swim? Bone density as a mechanism for buoyancy control in early cetaceans. *The Anatomical Record, 290,* 638–653.

Grigson, C. (1982). Sex and age determination of some bones and teeth of domestic cattle: A review of the literature. In B. Wilson, C. Grigson, & S. Payne (Eds.), *Ageing and sexing animal bones from archaeological sites* (pp. 7–24). BAR International Series, Vol. 109. Oxford: Archaeopress.

Guérin, C., & Faure, M. (1988). Biostratigraphie comparée des grands mammifères du Pléistocène en Europe occidentale et au Moyen-Orient. *Paléorient, 14*(2), 50–56.

Guérin, C., Eisenmann, V., & Faure, M. (1993). Les Grands Mammifères du Gisement Pléistocène Moyen de Latamné (Valle

de l'Oronte, Syrie). In P. Sanlaville, J. Bensacon, L. Copeland, & S. Muhesen (Eds.), *Le Paléolithic de la Valée Moyenne de l'Oronte (Syrie)* (pp. 169–178). BAR International Series, Vol. 587. Oxford: Archaeopress.

Haas, G. (1970). *Metridiochoerus evronensis* n. sp. a new Middle Pleistocene Phacochoerid from Israel. *Israel Journal of Zoology, 19*(3), 179–181.

Hartman, G. (2004). Long term continuity of a freshwater turtle (*Mauremys caspica rivulata*) population in the Northern Jordan Valley and its paleoenvironmental implications. In N. Goren-Inbar & J. D. Speth (Eds.), *Human paleoecology in the Levantine Corridor* (pp. 61–74). Oxford: Oxbow Books.

Hawkes, K., O'Connell, J. F., & Blurton Jones, N. G. (2001). Hadza meat sharing. *Evolution and Human Behavior, 22,* 113–142.

Haynes, G. (1983). A guide to differentiating mammalian carnivore taxa responsible for gnaw damage to herbivore limb bones. *Paleobiology, 9*(2), 164–172.

Haynes, G. (1991). *Mammoths, mastodons, and elephants: Biology, behavior, and the fossil record.* Cambridge: Cambridge University Press.

Hernández Fernández, M., & Vrba, E. S. (2006). Plio-Pleistocene climatic change in the Turkana Basin (East Africa): Evidence from large mammal faunas. *Journal of Human Evolution, 50,* 595–626.

Hill, A. P. (1978). Taphonomical background to fossil man-problems in palaeocology. In W. W. Bishop (Ed.), *Geological society, London* (pp. 87–101). Special Publications 6. Edinburgh: Scottish Academic Press and University of Toronto Press.

Hill, A. P. (1980). Early post-mortem damage to the remains of some contemporary East African mammals. In A. K. Behrensmeyer & A. P. Hill (Eds.), *Fossils in the making* (pp. 131–155). Chicago: University of Chicago Press.

Hooijer, D. A. (1958). An Early Pleistocene mammalian fauna from Bethlehem. *Bulletin of the British Museum of Natural History, London, Geology, 3*(8), 265–292.

Hooijer, D. A. (1959). Fossil mammals from Jisr Banat Yaqub, south of Lake Hule, Israel. *Bulletin of the Research Council of Israel, G8,* 177–179.

Hooijer, D. A. (1960). A Stegodon from Israel. *Bulletin of the Research Council of Israel, G8,* 104–107.

Hooijer, D. A. (1961). Middle Pleistocene mammal from Latamne, Orontes Valley, Syria. *Annals of the Archaeology of Syrie, 11,* 117–132.

Hooijer, D. A. (1965). Additional notes on the Pleistocene mammalian fauna of the Orontes Valley. *Annales Archéologiques de Syrie, 15,* 101–104.

Horowitz, A. (1973). Development of the Hula Basin, Israel. *Israel Journal of Earth Sciences, 22,* 107–139.

Horowitz, A. (1979). *The Quaternary of Israel.* New York: Academic Press.

Horowitz, A. (2001). *The Jordan Rift Valley.* Lisse; Exton, PA: A. A. Balkema Publishers .

Horowitz, A. (2003). Elephant, horses, humans, and others; paleoenvironments of the Levantine land bridge. *Israel Journal of Earth-Sciences, 51*(3–4), 203–209.

Horwitz, L. K., & Monchot, H. (2002). Choice cuts: Hominid butchery activities at the Lower Paleolithic site of Holon, Israel. In H. Buitenhuis, A. M. Choyke, M. Mashkour, & A. H. Al-Shiyab (Eds.), *Archaeozoology of the Near East , Vol. V* (pp. 48–61). Groningen: ARC Publication.

Horwitz, L. K., & Monchot, H. (2007). Sus, Hippopotamus, Bos, and Gazella. In M. Chazan & L. K. Horwitz (Eds.), *The Lower Paleolithic Site of Holon, Israel* (pp. 91–109). American School of Prehistoric Research Bulletin 50, Peabody Museum of Archaeology and Ethnology. Cambridge, MA: Harvard University Press.

Horwitz, L. K., & Tchernov, E. (1989). The Late Acheulian fauna from Oumm Zinat. *Mitekufat Haeven, 22,* 7*–13*.

Isaac, G. L. (1967). Toward the interpretation of occupational debris: Some experiments and observations. *Kroeber Anthropological Society Papers, 37,* 31–57.

Isaac, G. L. (1983). Bones in connection: Competing explanations for the juxtaposition of Early Pleistocene artifacts and faunal remains. In J. Clutton-Brock & C. Grigson (Eds.), *Animals in archaeology* (pp. 3–19). BAR International Series, Vol. 163. Oxford: Archaeopress.

Kahlke, H. D. (1997). Die Cerviden-Reste aus dem Unterpleistozän von Untermassfeld. In R. D. Kahle (Ed.), *Das Pleistozän von Untermassfeld bei Meiningen (Thüringen)* (pp. 181–275). Monographien des Römisch-Germanischen Zentralmuseums, Vol. 40(1). Bonn/Mainz: Verlag des Römisch-Germanischen Zentralmuseums.

Kahlke, R.-D. (2007). Late Early Pleistocene European large mammals and the concept of an Epivillafranchian biochron. *Courier Forschungsinstitut Senckenberg, 259,* 265–278.

Kahlke, R.-D., & Gaudzinski, S. (2005). The blessing of a great flood: Differentiation of mortality patterns in the large mammal record of the Lower Pleistocene fluvial site of Untermassfeld (Germany) and its relevance for the interpretation of faunal assemblages from archaeological sites. *Journal of Archaeological Science, 32,* 1202–1222.

Kelly, R. (1995). *The foraging spectrum: Diversity in hunter-gatherer lifeways.* Washington, DC: Smithsonian Institution Press.

Kerbis-Peterhans, J. C., & Horwitz, L. K. (1992). A bone assemblage from a striped hyaena (*Hyaena hyaena*) den in the Negev desert, Israel. *Israel Journal of Zoology, 37,* 225–245.

Kersten, A. M. P. (1989). Age and sex composition of Epipalaeolithic fallow deer and wild goat from Ksar 'Akil. *Palaeohistoria, 29,* 119–131.

Kreutzer, L. A. (1992). Bison and deer bone mineral densities: Comparisons and implication for the interpretation of archaeological faunas. *Journal Archaeological Science, 19,* 271–294.

Kroll, von W. (1991). *The Strait-Tusked elephant of Crumstadt (near Darmstadt, Germany). A contribution to the osteology of Elephas (Palaeoloxodon) antiquus Falconer & Cautley (1847).* Ph.D. dissertation, Ludwig-Maximilians, University of München. (In German).

Kruuk, H. (1972). *The spotted hyaena.* Chicago: University of Chicago Press.

Kruuk, H. (1976). Feeding and social behaviour of the striped hyaena (*Hyaena vulgaris* Desmarest). *Ecological African Wildlife Journal, 14,* 91–111.

Kurten, B. (1965). The carnivora of the Palestine caves. *Acta Zoologische Fennica, 107,* 1–74.

Lam, Y. M. (1992). Variability in the behaviour of spotted hyaenas as taphonomic agents. *Journal of Archaeological Science, 19,* 389–406.

Lam, Y. M., & Pearson, O. M. (2005). Bone density studies and the interpretation of the faunal record. *Evolutionary Anthropology, 14,* 99–108.

Lam, Y. M., Chen, X., Marean, C. W., & Frey, C. J. (1998). Bone density and long bone representation in archaeological faunas: Comparing results from CT and Photon Densitometry. *Journal of Archaeological Science, 25,* 559–570.

Lam, Y. M., Xingbin, C., & Pearson, O. M. (1999). Intertaxonomic variability in patterns of bone density and the differential representation of bovid, cervid, and equid elements in the archaeological record. *American Antiquity, 64*(2), 343–362.

Laws, R. M. (1968). Dentition and ageing of the *Hippopotamus. African Journal of Ecology, 6*(1), 19–52.

Leakey, M. D. (1971). *Olduvai Gorge, excavations in beds I and II, 1960–1963.* Cambridge: Cambridge University Press.

Lee, R. B. (1979). *The !Kung San. Men, women and work in a foraging society.* Cambridge: Cambridge University Press.

Lemorini, C., Stiner, M. C., Gopher, A., Shimelmitz, R., & Barkai, R. (2006). Use-wear analysis of an Amudian laminar assemblage from the Acheuleo-Yabrudian of Qesem Cave, Israel. *Journal of Archaeological Science, 33,* 921–934.

Lister, A. M. (1996a). The morphological distinction between bones and teeth of fallow deer (*Dama dama*) and red deer (*Cervus elaphus*). *International Journal of Osteoarchaeology, 6,* 119–143.

Lister, A. M. (1996b). The stratigraphical interpretation of large mammal remains from the Cromer Forest-bed Formation. In C. Turner (Ed.), *The early Middle Pleistocene in Europe* (pp. 25–44). Rotterdam: A.A.Balkema.

Lister, A. M. (2004). Ecological interactions of Elephantids in Plesitocene Euraasia: Palaeoloxodon and Mammuthus. In N. Goren-Inbar & J. D. Speth (Eds.), *Human paleoecology in the Levantine Corridor* (pp. 53–60). Oxford: Oxbow Books.

Lister, A. M. (2007). Cervidae. In M. Chazan & L. K. Horwitz (Eds.), *The Lower Paleolithic site of Holon, Israel* (pp. 111–121). American School of Prehistoric Research Bulletin 50. Peabody Museum of Archaeology and Ethnology. Cambridge, MA: Harvard University Press.

Lupo, K. D. (2006). What explains the carcass field processing and transport decisions of contemporary hunter-gatherers? Measures of economic anatomy and zooarchaeological skeletal part representation. *Journal of Archaeological Method and Theory, 13*(1), 19–66.

Lupo, K. D. (2007). Evolutionary foraging models in zooarchaeological analysis: Recent applications and future challenges. *Journal of Archaeological Research, 15,* 143–189.

Lupo, K. D., & O'Connell, J. F. (2002). Cut and tooth mark distribution on large animal bones: Enthoarchaeological data from the Hadza and their implications for current ideas about early human carnivore. *Journal of Archaeological Science, 29,* 85–109.

Lyman, R. L. (1994). *Vertebrate taphonomy.* Cambridge: Cambridge University Press.

van der Made, J. (1998). Ungulates from Gran Dolina (Atapuerca, Burgos, Spain). *Quaternaire, 4,* 267–281.

van der Made, J. (1999). Ungulates from Atapuerca TD6. *Journal of Human Evolution, 37,* 389–413.

van der Made, J. (2001). Les Ongulés d'Atapuerca. Stratigraphie et biogéographie. *L'Anthropologie, 105,* 95–113.

Madsen, B., & Goren-Inbar, N. (2004). Acheulian giant core technology and beyond: An archaeological and experimental case study. *Eurasian Prehistory, 2*(1), 3–52.

Maglio, V. J. (1973). Origin and evolution of the Elephantidae. *Transactions of the American Philosophical Society, 63,* 1–149.

Marder, O., Halila, H., Gvirtzman, M., Rabinovich, R., Saragusti, I., & Porat, N. (1999). The Lower Palaeolithic site of Revadim. *Journal of the Israel Prehistoric Society, 28,* 21–53.

Marean, C. W., Spencer, L. M., Blumenschine, R. J., & Capaldo, S. D. (1992). Captive hyaena bone choice and destruction, the schlepp effect and Olduvai archaeofaunas. *Journal of Archaeological Science, 19,* 101–121.

Martínez-Navarro, B. (2004). Hippos, pigs, bovids, saber-toothed tigers, monkey, and hominids: Dispersals through the Levantine Corridor during Late Pliocene and Early Pleistocene times. In N. Goren-Inbar & J. D. Speth (Eds.), *Human paleoecology in the Levantine Corridor* (pp. 37–52). Oxford: Oxbow Books.

Martínez-Navarro, B. (2010). Early Pleistocene faunas of Eurasia and hominid dispersals. In J. G. Fleagle, J. J. Shea, F. E. Grine, A. L. Baden, & R. E. Leakey (Eds.), *Out of Africa I: Who? When? and Where?* (pp. 207–224). Vertebrate Paleobiology and Paleoanthropology Series. New York: Springer.

Martínez-Navarro, B., Rabinovich, R., & Goren-Inbar, N. (2000). Preliminary study of the fossil Bovidae assemblage from the late Lower Pleistocene archaeological site of Gesher Benot Ya'aqov (northern Israel). Abstracts of 2000 Meeting of the INQUA-SEQS "The Plio-Pleistocene Boundary and the Lower/Middle Pleistocene Transition: Type Areas and Sections": Bari.

Martínez-Navarro, B., Pérez-Claros, J. A., Palombo, M. R., Rook, L., & Palmqvist, P. (2007). The Olduvai buffalo Pelorovis and the origin of Bos. *Quaternary Research, 68*, 220–226.

Martínez-Navarro, B., Belmaker, M., & Bar-Yosef, O. (2009). The large carnivores from 'Ubeidiya (early Pleistocene, Israel): Biochronological and biogeographical implications. *Journal of Human Evolution, 56*, 514–524.

Martínez-Navarro, B., & Rabinovich, R. (2011). The fossil Bovidae (Artiodactyla, Mammalia) from Gesher Benot Ya'aqov, Israel: Out of Africa during the Early-Middle Pleistocene transition. *Journal of Human Evolution, 60*, 375–386.

Martínez-Navarro, B., Belmaker, M., & Bar-Yosef, O. (in preparation). Bovids from 'Ubeidiya.

Melamed, Y. (1997). *Reconstruction of the landscape and the vegetarian diet at Gesher Benot Ya'aqov archaeological site in the Lower Paleolithic Period*. M.Sc. thesis, Bar-Ilan University, Ramat Gan.

Melamed, Y. (2003). *Reconstruction of the Hula Valley vegetation and the Hominid vegetarian diet by the Lower Palaeolithic botanical remains from Gesher Benot Ya'aqov*. Ph.D. dissertation, Bar-Ilan University, Ramat Gan.

Mendelssohn, H., & Yom-Tov, Y. (1999). *Fauna Palestina. Mammalia of Israel*. Jerusalem: The Israel Academy of Sciences and Humanities.

Mienis, H. K., & Ashkenazi, S. (2011). Lentic Basommatophora molluscs and hygrophilous land snails as indicators of habitat and climate in the Early-Middle Pleistocene (0.78 Ma) at the site of Gesher Benot Ya'aqov (GBY), Israel. *Journal of Human Evolution, 60*, 328–340.

Miles, A. E., & Grigson, W. C. (Eds.). (1990). *Colyer's variations and diseases of the teeth of animals (revised edition)*. Cambridge: Cambridge University Press.

Moehlman, P. D. (Ed.). (2002). *Equids: Zebras, asses and horses: Status survey and conservation action plan* (pp. 2–10, 61–71). IUCN/SCC Equid Specialist Group, The World Conservation Union.

Moigne, A.-M., & Barsky, D. R. (1999). Large mammal assemblages from lower Palaeolithic sites in France: La Caune de L'Arago, Terra-Amata, Orgnac 3 and Cagny L'Epinette. In S. Gaudzinski & E. Turner (Eds.), *The role of early humans in the accumulation of European Lower and Middle Palaeolithic bone assemblages* (pp. 219–234). Monographien des Römisch-Germanischen Zentralmuseums, Vol. 42. Bonn/Mainz: Verlag des Römisch-Germanischen Zentralmuseums.

Monchot, H., & Horwitz, L. K. (2007). Taxon representation and age and sex distributions. In M. Chazan & L. K. Horwitz (Eds.), *The Lower Paleolithic site of Holon, Israel* (pp. 85–88). American School of Prehistoric Research Bulletin 50. Peabody Museum of Archaeology and Ethnology. Cambridge, MA: Harvard University Press.

Morin, E. (2004). *Late Pleistocene population interaction in Western Europe and modern human origins: New insights based on the faunal remains from Saint-Césaire, Southwestern France*. Ph.D. dissertation, University of Michigan, Ann Arbor.

Morlan, R. E. (1994). Bison bone fragmentation and survivorship: A comparative method. *Journal of Archaeological Science, 21*, 797–807.

Morton, A. G. T. (1995). *Archaeological site formation: Experiments in lake margin processes*. Ph.D. dissertation, Cambridge University, Cambridge.

Morton, A. G. T. (2004). *Archaeological site formation: Understanding lake margin contexts*. BAR International Series, Vol. 1211. Oxford: Archaeopress.

Moullé, P.-É. (1990). Les cervides de la grotte du Vallonnet (Roquebrune-Cap-Martin, Alpes-Maritimes). *Quaternaire, 3–4*, 193–196.

Moullé, P.-É. (1997–1998). Les grands mammifères de la grotte du Vallonnet (Raquebrune-Cap-Martin, Alpes-Maritimes). *Bulletin de Musée d'Anthropologie Préhistorique de Monaco, 39*, 29–36.

Munro, N. D., & Bar-Oz, G. (2005). Gazelle bone fat processing in the Levantine Epipalaeolithic. *Journal of Archaeological Science, 32*, 223–239.

Nielson, A. E. (1991). Trampling the archaeological record: An experimental study. *American Antiquity, 50*(3), 483–503.

O'Connell, J. F., Hawkes, K., & Blurton-Jones, N. G. (1988). Hadza hunting, butchering, and bone transport and their archaeological implications. *Journal of Anthropological Research, 44*(2), 113–161.

O'Connell, J. F., Hawkes, K., & Blurton-Jones, N. G. (1990). Reanalysis of large mammal body part transport among the Hadza. *Journal of Archaeological Science, 17*, 301–316.

O'Connell, J. F., Hawkes, K., Lupo, K. D., & Blurton Jones, N. G. (2002). Male strategies and Plio-Pleistocene archaeology. *Journal of Human Evolution, 43*, 831–872.

Oliver, S. J. (1989). Analogues and site context: Bone damages from Shield Trap Cave (24CB91), Carbon Country, Montana, U.S.A. In R. Bonnichsen & H. M. Sorg (Eds.), *Bone modification* (pp. 73–98). Orono, ME: Center for the Study of the first Americans.

Olsen, S. L., & Shipman, P. (1988). Surface modification on bone: Trampling versus butchery. *Journal of Archaeological Science, 15*, 535–553.

O'Regan, H. J., Bishop, L. C., Lamb, A., Elton, S., & Turner, A. (2005). Large mammal turnover in Africa and the Levant between 1.0 and 0.5 Ma. In M. J. Head & P. I. Gibbard (Eds.), *Early-Middle Pleistocene transitions: The land-ocean evidence* (pp. 231–249). *Geological Society*, Special Publications 247.

Outram, A. K. (2001). A new approach to identifying bone marrow and grease exploitation: Why the "indeterminate" fragments should not be ignored. *Journal of Archaeological Science, 28*, 401–410.

Outram, A. K., & Rowley-Conwy, P. (1998). Meat and marrow utility indices for horse (*Equus*). *Journal of Archaeological Science, 25*, 839–849.

Palombo, M. R. (2005). Biochronology of the Plio-Pleistocene mammalian faunas of the Italian peninsula: Knowledge, problems and perspectives. *Il Quaternario, 17*(2/2), 565–582.

Palombo, M. R., & Sardella, R. (2007). Biochronology and biochron boundaries: A real dilemma or a false problem? An example based on the Pleistocene large mammalian faunas from Italy. *Quaternary International, 160*, 30–42.

Palombo, M. R., Azana, B., & Alberdi, M. T. (2003). Italian mammal biochronology from Latest Miocene to Middle Pleistocene: Multivariate approach. *Geologica Romana, 36*(2000–2002), D 335–368.

Parfitt, S. A., & Roberts, M. B. (1999). Human modification of faunal remains. In M. B. Roberts & S. A. Parfit, *Boxgrove: A Middle Pleistocene hominid site at Eartham Quarry, Boxgrove, West Sussex* (pp. 395–415). London: English Heritage Archaeological Report 17.

Parfitt, S. A., Berendregt, R. W., Breda, M., Candy, I., Collins, M. J., Cope, G. R., et al. (2005). The earliest record of human activity in northern Europe. *Nature, 438*, 1008–1012.

Petraglia, M. D., & Potts, R. (1994). Water flow and the formation of Early Pleistocene sites in Olduvai Gorge, Tanzania. *Journal of Anthropological Archaeology, 13*, 228–254.

Pfeiffer, T. (1999). Die Stellung von Dama (Cervidae, Mammalia) im System pleisiometacarpaler Hirsche des Pleistozans. *Courier Forschungsinstitut Senckenberg, 211*.

Picard, L. (1952). The Pleistocene peat of Lake Hula. *Bulletin of the Research Council of Israel, G2*, 147–156.

Picard, L. (1963). The Quaternary in the Northern Jordan Valley. *Proceedings of the Israel Academy of Sciences and Humanities, 1*, 1–34.

Picard, L. (1965). The geological evolution of the Quaternary in the central-northern Jordan Graben. *American Geological Society Special Papers, 84*, 337–366.

Pickering, T. R. (2002). Reconsideration of criteria for differentiating faunal assemblages accumulated by hyenas and homininds. *International Journal of Osteoarchaeology, 12*, 127–141.

Pickering, T. R., & Egeland, C. P. (2006). Experimental patterns of hammerstone percussion damage on bones: Implications for inferences of carcass processing by humans. *Journal of Human Evolution, 33*, 459–469.

Pohlmeyer, K. (1985). *Zur Vergleichenden Anatomie von Damtier (Dama dama), Schaf (Ovis aries) und Ziege (Capra hircus). Ostelogie und Postnatale Osteogenese*. Berlin: Verlag Paul Parey.

Pokines, J. T. (1998). Experimental replication and use of Cantabrian Lower Magdalenian antler projectile points. *Journal of Archaeological Science, 25*, 875–886.

Potts, R. (1988). *Early hominid activities at Olduvai Gorge*. New York: Aldine de Gruyter.

Potts, R., & Shipman, P. (1981). Cut marks made by stone tools on bones from Olduvai Gorge, Tanzania. *Nature, 291*, 577–580.

Rabinovich, R. (1990). Taphonomic research on the faunal assemblage from the Quneitra site. In N. Goren-Inbar, *Quneitra: An open air Mousterian site in the Golan Heights* (pp. 189–219). Qedem Vol. 31, Jerusalem: Institute of Archaeology, Hebrew University.

Rabinovich, R. (1998). *Patterns of animal exploitation and subsistence in Israel during the Upper Paleolithic and Epi-Paleolithic (40,000–12,500 BP) based upon selected case studies*. Ph.D. dissertation, The Hebrew University of Jerusalem, Jerusalem.

Rabinovich, R. (2002). Man versus carnivores in the Middle-Upper Paleolithic of the Southern Levant. In H. Buitenhuis, A. M. Choyke, M. Mashkour & A. H. Al-Shiyab (Eds.), *Archaeozoology of the Near East, Vol. V* (pp. 22–39). Groningen: ARC Publication.

Rabinovich, R., & Biton, R. (2011). The Early–Middle Pleistocene faunal assemblages of Gesher Benot Ya'aqov—inter site variability. *Journal of Human Evolution, 60*, 357–374.

Rabinovich, R., & Horwitz, L. K. (1994). An experimental approach to the study of porcupine damage to bones. In M. Patou-Mathis (Ed.), *Outillage peu élaboré en os et bois de Cervidés IV: Taphonomie/bone modification 8* (pp. 97–118). Treignes: Éditions du Centre d'études et de documentation (*Artefacts 9*).

Rabinovich, R., & Lister, A., (in preparation). Cervid remains from Gesher Benot Ya'aqov, Israel.

Rabinovich, R., & Tchernov, E. (1995). Chronological, paleoecological and taphonomical aspects of the Middle Paleolithic site of Qafzeh, Israel. In: H. Buitenhuis & H. P. Uerpmann (Eds.), *Archaeozoology of the Near East II* (pp. 5–44). Leiden: Backhuys Publishers.

Rabinovich, R., Bar-Yosef, O., & Tchernov, E. (1997). "How many ways to skin a gazelle" – butchery patterns from an Upper Palaeolithic site, Hayonim cave, Israel. *Archaeozoologia, VIII/1, 2 (N 15/16)*, 11–52.

Rabinovich, R., Bar-Gal, G., & Marder, O. (2005). Taphonomy of elephants from the Lower Paleolithic site of Revadim Quarry (Israel). In L. D. Agenbroad & R. L. Symington (Eds.), *The World of Elephants – Proceedings of the 2nd International Congress* (pp. 665–667). Hot Springs: Mammoth Site Scientific Papers, 4.

Rabinovich, R., Gaudzinski-Windheuser, S., & Goren-Inbar, N. (2008a). Systematic butchering of fallow deer (*Dama*) at the early Middle Pleistocene Acheulian site of Gesher Benot Ya'aqov (Israel). *Journal of Human Evolution, 54*, 134–149.

Rabinovich, R., Yom-Tov, Y., Ashkenazi, S., Mienis, H., Zohari, I., & Biton, R. (2008b). Sub-program No. 2 – Zoology. Habitat and zoological biodiversity reconstruction of a 0.78 Ma old site along its 100 kyr sequence, in view of climate change. First scientific

progress interim report for the Israel Science Foundation, Center of Excellence Program, pp. 30–63.

Reed, K. E. (1998). Using large mammal communities to examine ecological and taxonomic structure and predict vegetation in extant and extinct assemblages. *Paleobiology, 24*(3), 384–408.

Roberts, M. B., & Parfitt, S. A. (1999). *Boxgrove: A Middle Pleistocene hominid site at Eartham Quarry, Boxgrove, West Sussex*. London: English Heritage Archaeological Report 17.

Rodríguez, J. (1999). Use of cenograms in mammalian palaeocology. A critical review. *Lethaia, 32*, 331–347.

Rodríguez, J. (2004). Stability in Pleistocene Mediterranean mammalian communities. *Palaeogeography, Palaeoclimatology, Palaeoecology, 207*, 1–22.

Rodríguez, J., Alberdi, M. T., Azanza, B., & Prado, J. L. (2004). Body size structure in north-western Mediterranean Plio-Pleistocene mammalian faunas. *Global Ecology and Biogeography, 13*, 163–176.

Roebroeks, W. (2001). Hominid behaviour and the first occupation of Europe. *Journal of Human Evolution, 41*, 437–461.

Roebroeks, W., & van Kolfschoten, T. (1994). The earliest occupation of Europe: A short chronology. *Antiquity, 68*, 489–503.

Rogers, A. R. (2000). On the value of soft bones in faunal analysis. *Journal of Archaeological Science, 27*, 635–639.

Sachs, L. (1984). *Angewandte Statistik. Anwendung statistischer Methoden*. Heidelberg: Springer.

Saltz, D. (2002). The dynamics of equid populations. In P. D. Moehlman (Ed.), *Equids: Zebras, asses and horses: Status survey and conservation action plan* (pp. 118–123). IUCN/SCC Equid Specialist Group, The World Conservation Union. Gland, Switzerland and Cambridge: IUCN.

Schick, K. D. (1986). *Stone age sites in the making, experiments in the formation and transformation of archaeological occurrences*. BAR International Series, Vol. 319. Oxford: Archaeopress.

Schick, K. D. (1987). Modeling the formation of early stone age artifact concentrations. *Journal of Human Evolution, 16*, 789–807.

Schick, K. D. (1992). Geoarchaeological analysis of an Acheulian site at Kalambo Falls, Zambia. *Geoarchaeology, 7*(1), 1–26.

Schick, K. D. (2001). An examination of Kalambo Falls Acheulian site B5 from a geoarchaeological perspective. In J. D. Clark (Ed.), *Kalambo Falls prehistoric site: The earlier cultures: Middle and Earlier Stone Age* (pp. 463–480). Cambridge: Cambridge University Press.

Schiffer, M. B. (1972). Archaeological context and systemic context. *American Antiquity, 37*, 156–165.

Schiffer, M. B. (1976). *Behavioral archaeology*. New York: Academic Press.

Schiffer, M. B. (1983). Towards the identification of formation processes. *American Antiquity, 48*, 675–706.

Schiffer, M. B. (1987). *Formation processes of the archaeological record*. Albuqerque: University of New Mexico Press.

Schulman, N. (1967). Remarks on the Quaternary in the Northern Jordan Valley. Israel. *Journal of Earth Sciences, 16*, 104–106.

Schulman, N. (1978). The Jordan Rift Valley. *10th International Congress of Sedimentology, 2*, 57–94.

Selvaggio, M. M. (1998). Evidence for a three stage sequence of hominid and carnivore involvement at FLK Zijanthropus, Olduvai Gorge, Tanzania. *Journal of Archaeological Science, 25*, 191–202.

Sharon, G. (2007). *Acheulian large flake industries: Technology, chronology, and significance*. BAR International Series, Vol. 1701. Oxford: Archaeopress.

Sharon, G., & Goren-Inbar, N. (1999). Soft percussor use at the Gesher Benot Ya'aqov Acheulian site? *Mitekufat Haeven, 28*, 55–79.

Sharon, G., Feibel, C. G., Belitzky, S., Marder, O., Khalaily, H., & Rabinovich, R. (2002). The drainage destruction at Gesher Benot Ya'aqov 1999: Archaeological and geological implications. Eretz Zaphon, *Atiqot 1*–19**.

Sharon, G., Alperson-Afil, N., & Goren-Inbar, N. (2011). Cultural conservatism against variability in the continual Acheulian sequence of Gesher Benot Ya'aqov, Israel. *Journal of Human Evolution, 60*, 387–397.

Shipman, P. (1981). *Life history of a fossil: An introduction to taphonomy and paleoecology*. Cambridge, MA: Harvard University Press.

Shipman, P. (1986). Scavenging or hunting in early hominids: Theoretical framework and tests. *American Anthropology, 88*, 27–43.

Shipman, P., & Rose, J. (1983). Evidence of butchery activities at Torralba and Ambrona: An evaluation using microscopic techniques. *Journal of Archaeological Science, 10*, 465–474.

Shipman, P., & Rose, J. (1984). Cutmark mimics on modern fossil bovid bones. *Current Anthropology, 25*, 116–177.

Shoshani, J., Goren-Inbar, N., & Rabinovich, R. (2004). A stylohyoideum of *Palaeoloxodon antiquus* from Gesher Benot Ya'aqov, Israel: Morphology and functional inferences. In G. Cavarretta, P. Gioia, M. Mussi, & M.R. Palombo (Eds.), *The World of Elephants – Proceedings of the 1st International Congress* (pp. 665–667). Rome: Consiglio Nazionale delle Ricerche.

Shoshani, J., Ferretti, M. P., Lister, A. M., Agenbroad, L. D., Saegusa, H., Mol, D., et al. (2007). Relationships within the Elephantinae using hyoid characters. *Quaternary International, 169–170*, 174–185.

Silver, I. A. (1969). The ageing of domestic animals. In D. Brothwell & E. Higgs (Eds.), *Science in archaeology* (2nd ed., pp. 283–302). London: Thames and Hudson.

Simmons, T. (2004). "A feather for each wind that blows": Utilizing avifauna in assessing changing patterns in paleoecology and subsistence at Jordan Valley archaeological sites. In N. Goren-Inbar & J. D. Speth (Eds.), *Human paleoecology in the Levantine Corridor* (pp. 191–206). Oxford: Oxbow Books.

Soligo, C., & Andrews, P. (2005). Taphonomic bias, taxonomic bias and historical non-equivalence of faunal structure in early hominin localities. *Journal of Human Evolution, 49*, 206–229.

Speth, J. D. (1987). Early hominid subsistence strategies in seasonal habitats. *Journal of Archaeological Science, 14*, 13–29.

Speth, J. D. (2004). Hunting pressure, subsistence intensification, and demographic change in the levantine late Middle Paleolithic. In N. Goren-Inbar & J. D. Speth (Eds.), *Human paleoecology in the Levantine Corridor* (pp. 149–166). Oxford: Oxbow Press.

Speth, J. D., & Clark, J. L. (2006). Hunting and overhunting in the Levantine Late Middle Palaeolithic. *Before Farming, 3*, 1–42.

Speth, J. D., & Tchernov, E. (1998). The role of hunting and scavenging. Neandertal procurement strategies, new evidence from Kebara Cave (Israel). In T. Akazawa, K. Aoki, & O. Bar-Yosef (Eds.), *Neandertals and modern humans in Western Asia* (pp. 223–239). London: Plenum Press.

Speth, J. D., & Tchernov, E. (2001). Neandertal hunting and meat-processing in the Near-East: Evidence from Kebara cave 2001. In C. B. Stanford & H. T. Bunn (Eds.), *Meat-eating and human evolution* (pp. 52–72). Oxford: Oxford University Press.

Spiro, B., Ashkenazi, S., Mienis, H. K., Melamed, Y., Feibel, C., Delgado, A., et al. (2009). Climate variability in the Upper Jordan Valley around 0.78 Ma, inferences from time-series stable isotopes of Viviparidae, supported by mollusk and plant palaeocology. *Palaeogeography, Palaeoclimatology, Palaeoecology, 282*, 32–44.

Spiro, B., Ashkenazi, S., Starinsky, A., & Katz, A. (2011). Strontium isotopes in *Melanopsis* sp. as indicators of variation in hydrology and climate in the Upper Jordan Valley during the Early-Middle Pleistocene, and wider implications. *Journal of Human Evolution, 60*, 407–416.

Stekelis, M. (1960). The Paleolithic deposits of Jisr Banat Ya'qub. *Bulletin of the Research Council of Israel, G9*, 61–87.

Stekelis, M., Picard, L., & Bate, D. M. A. (1937). Jisr Banat Ya'qub. *Quarterly of the Department of Antiquities of Palestine, 6*, 214–215.

Stekelis, M., Picard, L., & Bate, D. M. A. (1938). Jisr Banat Ya'qub. *Quarterly of the Department of Antiquities of Palestine, 7*, 45.

Stiner, M. C. (1994). *Honor among thieves: A zooarchaeological study of neandertal ecology*. Princeton, NJ: Princeton University Press.

Stiner, M. C. (2005). *The faunas of Hayonim Cave (Israel): A 200,000-year record of Paleolithic diet, demography, and society*. American School of Prehistoric Research Bulletins, Vol. 48, Peabody Museum of Archaeology and Ethnology. Cambridge, MA: Harvard University Press.

Stiner, M. C., Arsebuk, G., & Howell, F. C. (1996). Cave bears and Paleolithic artifacts in Yarimburgaz Cave, Turkey: Dissecting a palimpsest. *Geoarchaeology, 11*(4), 279–327.

Stiner, M. C., Achyuthan, H., Arsebuk, G., Howell, F. C., Josephson, S. C., Juell, K. E., et al. (1998). Reconstructing cave bear paleoecology from skeletons: A cross-disciplinary study of middle Pleistocene bears from Yarimburgaz Cave, Turkey. *Paleobiology, 24*(1), 74–98.

Stiner, M. C., Barkai, R., & Gopher, A. (2009). Cooperative hunting and meat sharing 400–200 kya at Qesem Cave, Israel. *Proceedings of the National Academy of Sciences of the United States of America, 106*(32), 13207–13212.

Stuart, A. J., & Lister, A. M. (2001). The mammalian faunas of Pakefield/Kessingland and Corton, Suffolk: Evidence for a new temperate episode in the British Early Middle Pleistocene. *Quaternary Science Reviews, 20*(16–17), 1677–1692.

Tappen, M. (1992). *Taphonomy of a Central African Savanna: Natural bone deposition in Parc National des Virunga, Zaire*. Ph.D. dissertation, Harvard University, Cambridge.

Tappen, M. (1994). Bone weathering in the tropical forest. *Journal of Archaeological Science, 21*, 667–673.

Tchernov, E. (1962). Paleolithic avifauna in Palestine. *Bulletin of the Research Council of Israel, 11*, 95–131.

Tchernov, E. (1973). On the Pleistocene Molluscs of the Jordan Valley. *Proceedings, Israel Academy of Sciences and Humanities, 11*, 1–46.

Tchernov, E. (1980). *The Pleistocene birds of 'Ubeidiya, Jordan Valley*. Jerusalem: Israel Academy of Sciences and Humanities.

Tchernov, E. (1981). The biostratigraphy of the Middle East. In P. Sanlaville & J. Cauvin (Eds.), *Préhistoire du Levant* (pp. 67–97). Paris: Colloques Internationaux du C.N.R.S. No. 598.

Tchernov, E. (1988). The biogeographical history of the southern Levant. In Y. Yom-Tov & E. Tchernov (Eds.), *The zoogeography of Israel* (pp. 159–250). Dordrecht: Dr. Junk Publishers.

Tchernov, E. (1992). Eurasian-African biotic exchange through the Levantine Corridor during the Neogene and Quaternary. In W. Koenigswald & L. Werdelin (Eds.), *Mammalian migration and dispersal events in the European Quaternary* (pp. 103–123). Vol. 153, Courier Forschungsinstitut Senckenberg, Stutgartt: E. Schweizerbart'sche Verlagsbuchhandlurg.

Tchernov, E., & Shoshani, H. (1996). Proboscidean remains in southern Levant. In H. Shoshani & P. Tassy (Eds.), *The Proboscidea. Evolution and palaeoecology of elephants and their relatives* (pp. 225–233). Oxford: Oxford University Press.

Tchernov, E., & Tsoukala, E. (1997). Middle Pleistocene (early Toringian) carnivore remains from northern Israel. *Quaternary Research, 48*, 122–136.

Tchernov, E., Horwitz, L. K., Ronen, A., & Lister, A. (1994). The faunal remains from Evron Quarry in relation to other Lower Paleolithic hominid sites in the southern Levant. *Quaternary Research, 42*, 328–339.

Thieme, H. (1996). Altpaläolithische Wurfspeere aus Schöningen, Niedersachsen-ein Vorbericht. *Archäologisches Korrespondenzblatt, 26*, 377–393.

Thieme, H. (1997). Lower Palaeolithic hunting spears from Schöningen, Germany. *Nature, 358*, 807–810.

Transboundary Species Project – Background Study Hippopotamus Appendix 2. http://www.nnf.org.na/RARESPECIES/InfoSys.

Trueman, C. N., & Martill, D. M. (2002). The long-term survival of bone: The role of bioerosion. *Archaeometry, 44*(3), 371–382.

Tsoukala, E. S. (1991). *Contribution to the study of the Pleistocene fauna of larger mammals (Carnivora, Perissodactyla, Artiodactyla) from petralona cave (Chalkidiki, N. Greece) Preliminary report* (pp. 331–336). Comptes rendus de l'Académie des sciences. Paris, Vol. 312, Series II.

Valla, R. F., Khalaily, H., Valladas, H., Kaltnecker, E., Bocquentin, F., Cabellos, T., et al. (2007). Les Fouilles de Ain Mallaha (Eynan) de 2003 à 2005: Quatrième Rapport Préliminaire. *Journal of the Israel Prehistoric Society, 37*, 135–379.

Vaufrey, R. (1951). Etude Paléontologique, I: Mammifères. In R. F. Neuville (Ed.), *Le Paléolithique et le Mésolithique du Désert de Judée* (pp. 198–217). Paris: Institute de Paléontology Humaine, Mémoire 24.

Vekua, A. K. (1962). *The Lower Pleistocene mammalian fauna of Akhalkalaki. Tbilisi.* (in Georgian)

Vekua, A. K. (1986). The Lower Pleistocene mammalian fauna of Akhalkalaki (Southern Georgia, USSR). *Palaeontographia Italica, 74*, 63–96.

Vekua, A. (1995). Die Wirbeltierfauna des Villafranchium von Dmanisi und ihre biostratigraphische Bedeutung. *Jahrbuch des Romisch-Germanischen Zentralmuseum Mainz, 42*, 77–180.

Verosub, K. L., Goren-Inbar, N., Feibel, C. S., & Saragusti, I. (1998). Location of the Matuyama/Brunhes boundary in the Gesher Benot Ya'aqov archaeological site. *Journal of Human Evolution, 34*, A22.

Voorhies, M. R. (1969). *Taphonomy and population dynamics of an early Pleistocene vertebrate fauna, Knox Country, Nebraska.* University of Wyoming, Contributions to Geology, Special Papers 1.

Voorhies, M. R. (1970). Sampling difficulties in reconstructing Late Tertiary mammalian communities. In E. L. Yochelson (Ed.), *Proceeding of the North American paleontological convention* (pp. 454–468). Lawrence: Allen Press.

Vrba, E. S. (1975). Some evidence of chronology and palaeoecology of Sterkfontein, Swartkrans, and Krom-draai from the fossil Bovidae. *Nature, 254*, 301–304.

Washburn, S., & Lancaster, C. S. (1968). The evolution of hunting. In R. B. Lee & I. DeVore (Eds.), *Man the hunter* (pp. 293–303). Chicago: Aldine.

Werker, E., & Goren-Inbar, N. (2001). Reconstruction of the woody vegetation at the Acheulian site of Gesher Benot Ya'aqov, Dead Sea Rift, Israel. In B. A. Purdy (Ed.), *Enduring records: The environmental and cultural heritage of wetlands* (pp. 293–303). Oxford: Oxbow Books.

Yeshurun, R., Bar-Oz, G., & Weinstein-Evron, M. (2007). Modern hunting behavior in the early Middle Paleolithic: Faunal remains from Misliya Cave, Mount Carmel, Israel. *Journal of Human Evolution, 53*, 656–677.

van Zeist, W., & Bottema, S. (2008). A palynological study of the Acheulian site of Gesher Benot Ya'aqov, Israel. *Vegetation History Archaeobotany, 18*, 105–121.

Zohar, I., & Biton, R. (2011). Land, lake, and fish: Investigations of fish remains from Gesher Benot Ya'aqov (Paleo-Lake Hula). *Journal of Human Evolution, 60*, 343–356.

Site Index

Subject Index